ERRITORIES.
THIRD EDITION
Refer to this map as G.S.G.S. 3961
HONG KONG, North Sheet
Third Edition
MIRS BAY
(TAIPANG WAN)
大鵬灣
Tolo Channel
Plover Cove
Port Shelter
Rocky Harbour
HIGH ISLAND
Junk Bay
Ninepin Group
Tung Lung or Lamtong Island
Po Toi Group
Po Toi Island
Peng Chau Hoi
INDEX TO ADJOINING SHEETS
INDEX TO SHEETS
Scale 1/80,000
TO GIVE A GRID REFERENCE ON THIS SHEET
REFERENCE
Scale 1/80,000 or ·792 Inch to 1 Mile.
Houses with gardens.
Built up area. Camps, Cemeteries, etc.
Communications and Main Topographical

Southern District
Officer Reports

Royal Asiatic Society Hong Kong Studies Series

Royal Asiatic Society Hong Kong Studies Series is designed to make widely available important contributions on the local history, culture and society of Hong Kong and the surrounding region. Generous support from the Sir Lindsay and Lady May Ride Memorial Fund makes it possible to publish a series of high-quality works that will be of lasting appeal and value to all, both scholars and informed general readers, who share a deeper interest in and enthusiasm for the area.

Other titles in RAS Hong Kong Studies Series:

Reluctant Heroes: Rickshaw Pullers in Hong Kong and Canton 1874–1954
Fung Chi Ming

For Gods, Ghosts and Ancestors: The Chinese Tradition of Paper Offerings
Janet Lee Scott

Hong Kong Internment 1942–1945: Life in the Japanese Civilian Camp at Stanley
Geoffrey Charles Emerson

The Six-Day War of 1899: Hong Kong in the Age of Imperialism
Patrick H. Hase

Watching Over Hong Kong: Private Policing 1841–1941
Sheilah E. Hamilton

The Dragon and the Crown
Stanley S.K. Kwan with Nicole Kwan

Public Success, Private Sorrow: The Life and Time of Charles Henry Brewitt-Taylor (1857–1938), China Customs Commissioner and Pioneer Translator
Isidore Cyril Cannon

East River Column: Hong Kong Guerrillas in the Second World War and After
Chan Sui-jeung

Resist to the End: Hong Kong, 1941–1945
Charles Barman, edited by Ray Barman

Southern District Officer Reports

Islands and Villages in Rural Hong Kong, 1910–60

Edited by John Strickland

香港大學出版社
HONG KONG UNIVERSITY PRESS

Hong Kong University Press
14/F Hing Wai Centre
7 Tin Wan Praya Road
Aberdeen
Hong Kong

ISBN 978-988-8028-38-2

British Library Cataloguing-in-Publication Data
A catalogue record for this book is available from the British Library.

Secure On-line Ordering

http://www.hkupress.org

The Press is grateful to Tim Ko and James Hayes for providing the photographs in this book.

Printed and bound by Liang Yu Printing Factory Ltd., Hong Kong, China.

Contents

Illustrations

Photos

Maps

Foreword

This book, the tenth in our series, is unique in that it is a compilation of largely unpublished material from a number of different sources. The title of the book is deliberately misleading. The south of today's Hong Kong, in the middle of which sits Hong Kong Island, can in no way be described as remote. However, in the days of Hong Kong as a British colony, the Southern District comprised most of the islands, apart from Hong Kong itself, as well as the area around present-day Tseung Kwan O and the south of the Sai Kung peninsula. This was a very large district to administer and much of it was indeed remote in the earlier half of the last century. The challenges of looking after such a large and disjointed area required remarkable men to fill the post of District Officer South.

John Strickland has combined masterfully the written reports of six such remarkable men and created the present volume. His achievement is to make the ordinary into something extraordinary, transforming reports that were potentially dry into a compendium that is eminently readable. For the most part, the raw material for this book was not intended to be published. Rather, the reports were meant to be a detailed summary of the present conditions that these men faced. Put together in one volume they now provide an extremely rare glimpse into a vanished past. Strickland's own love for and knowledge of the countryside of Hong Kong is evident in the way he has merged the elements of the various reports into a geographically logical sequence. Many of the villages here described have disappeared without trace, but what does remain is a very valuable description of what the majority of Hong Kong's land area used to be like.

The Royal Asiatic Society Hong Kong Branch and the Hong Kong University Press are very proud of what they have achieved so far with the Studies Series. More and more people, both here and abroad, are finding that Hong Kong and its unique history and culture provides a rich and fascinating field of study. An increasing number of schools are including the history of our city and its surroundings in their curricula, for which we should be able to take some credit. We will continue to bring to the public original works that will enhance this area even further.

The publications in the Studies series have been made possible initially by the very generous donation of seeding capital by the Trustees of the Clague Trust Fund, representing the estate of the late Sir Douglas Clague. This donation enabled us to establish a trust fund in the name of Sir Lindsay and Lady Ride, in memory of our first vice president and his wife. The Society itself added to this fund, as have a number of other generous donors.

The result is that we now have funding to bring to students of Hong Kong's history, culture and society a number of books that might otherwise not have seen the light of day. Furthermore, we continue to be delighted with the agreement established with Hong Kong University Press, which sets out the basis on which the Press will partner our efforts.

Robert Nield
President
Royal Asiatic Society, Hong Kong Branch
February 2009

1
Introduction
by John Strickland

This book comprises the reports written about the rural areas of Hong Kong in the 1910–20 and 1950–60 periods by district officers responsible for the Southern Administrative District, Eric Hamilton, Walter Schofield, S. H. Peplow, Paul Tsui, Austin Coates and James Hayes.

These men, all "cadet officers" (now called "administrative officers") were members of the small group that constituted the core of the governing system in the then British colony, regularly moving between different jobs so as to gain experience and to be able to see the big picture when making decisions.

These accounts differ in time and purpose. Hamilton's extracts come from letters to James Hayes written in 1958, Schofield's from an article requested by Hayes for publication in our *Journal*.[1] Paul Tsui's come from his unpublished autobiography,[2] and from a series of reports written around 1950, under circumstances which are unclear, but for which James Hayes has provided background information. The bulk of the village reports are otherwise provided by Austin Coates and James Hayes, and date from inspection visits made in the mid and late 1950s. Hamilton, Schofield, and in part Tsui, wrote in reminiscence of times past. Coates intended his *Summary Memorandum* (as he entitled it) for his successors in office. Hayes wrote for his own enlightenment, keeping his village notes in manuscript for the last fifty years. Peplow wrote when he was still in post.

At the time they were serving, none of the district officers were aware that their predecessors had conducted similar surveys and had made a record of their findings, although when James Hayes put his notes in order in 2008, he did have Austin Coates's reports in front of him. Only Austin Coates compiled a tidy document at the time. This is now lodged in the Public Records Office. Paul Tsui's village reports were found in an unmarked loose folder in an office drawer by one of James Hayes's colleagues in the 1980s and were passed to Hayes thereafter.

The reason for writing the reports was no doubt primarily as an *aide-mémoire* for the officers themselves. At the same time, one guesses that proper documentation of one's work was a standard discipline for administrative officers. Only Austin Coates explicitly stated that he was putting his notes in order for the benefit of his successors.

In the belief that the primary interest of readers will be the description of the villages and life in them, all the material has been reordered so as to put alongside each other the different author's commentary on the same regions, groups of villages and individual villages.

At the same time, these reports give a fascinating insight into how colonial civil servants viewed their jobs and discharged their responsibilities in the first half of the twentieth century.

Southern District was somewhat of a misnomer, comprising most of the islands of Hong Kong other than Victoria Island, i.e., Lantao, Cheung Chau, Lamma, Peng Chau, Tsing Yi, Leung Shuen Wan, Kau Sai Chau; plus the Hang Hau Peninsula (which we now think of as Tseung Kwan O), the Clear Water Bay Peninsula and the southern part of the Sai Kung Peninsula. Much of this area is of course to the west and northeast of Hong Kong, rather than to the south.

Presumably on account of the normal access route being by boat from Tai Po, the northern part of the Sai Kung Peninsula was part of the North Administrative District.

In the period covered by these reports, there were few roads in Southern District. Access was thus either by boat or on foot. With Hong Kong's mountainous terrain, the villagers must have been hardy people.

Eric Hamilton was a convivial soul who, besides his career in the Hong Kong civil service between 1911 and 1945, wrote to the newspapers on cricket and anything whimsical which took his fancy, under the pseudonym of "R. Abbit". He served as assistant district officer and district officer (Southern District) in 1917–23. He finished up as superintendent of imports and exports, and was interned for four years during the Japanese occupation. His long letters to James Hayes, written from Britain in 1958, now in the Hong Kong Public Records Office, are delightfully written and most evocative of the Southern District, its inhabitants, and his work in his day.

Walter Schofield (1888–1968) obtained an MA degree from Liverpool and Oxford Universities. He was cast in a more serious mould than Hamilton, though as the latter remarked in one of his letters, what he said and wrote always merited attention. Schofield was the same age as Hamilton and they both joined the colonial service in the same year. Schofield succeeded Hamilton as district officer (Southern District) in

1923 and served again in that position in 1930. After twenty-seven years in the Hong Kong civil service in various appointments, he retired to the UK in 1938, after spending the last eight years of his service in the magistracy. Throughout his career, he pursued (with official support) what was no doubt his first love, the geology and archaeology of Hong Kong, of which he was a pioneer.

S. H. Peplow's contribution is extracts from his book, *Hong Kong: About and Around*, published by the Commercial Press Ltd. Peplow was land bailiff (Southern District) in 1927 and district officer (Southern District) in 1930 when the book was first published.

Paul Tsui came from a humble background but managed to acquire a degree from the University of Hong Kong before the Second World War. During the war he served with the British Army Aid Group around what is now Huizhou in Guangdong Province. After the war he was the first Chinese to be appointed as a cadet officer in the British colonial administration. He retired in 1973 after thirty-eight years, having achieved the highest rank in the Hong Kong civil service. Because Paul Tsui's life was so colourful, this book contains in Chapter 9 a rather fuller biographical note than that provided for the other authors.

Austin Coates was the son of Sir Eric Coates, the well-known conductor and composer of light music. He only spent eight years in the Hong Kong administrative service. As can be seen from his reports, he already had a clear talent for writing and published his first book while still serving in Hong Kong. He spent a further five years in the colonial civil service in what is now Malaysia before he resigned to take up writing as a full-time occupation. His books on Southeast Asia are well-known and popular.

In his 32-year career with the Hong Kong government, half of it spent at one time or another in the New Territories district administration, James Hayes consistently demonstrated a passion for the rural areas of Hong Kong, the culture of its Chinese inhabitants, and the recent history of the colony. He served as an office-bearer of the Hong Kong Branch of the Royal Asiatic Society for many years, and was its president (1983–89) before he retired to Australia (although he regularly visits Hong Kong). He is also a prolific author of books and articles, and lectures on many facets of life in the Hong Kong–Canton region.

Chapter 9 of this book includes fuller biographies of the district officers.

VILLAGE IDENTIFICATION

Most of the villages described in this book can still be found on modern maps. The index in the *Hong Kong Directory 2008*, published by Universal Publications Ltd., lists many of them in its index together with a page number and a letter/number grid reference, albeit the names on the map are sometimes in Chinese only or use different spelling from this book.

The fifteen sheets of the 1:20,000 scale maps published by the Survey and Mapping Office of the Hong Kong SAR government are similarly comprehensive, but have no index.

The villages that are not found on these maps will often be located in this book within a group of other villages which are locatable on the maps, so the reader can guess the villages' approximate location.

Austin Coates identified his villages by map grid references which have now been made obsolete. These were the old military topographical maps, Series GSGS, No. L8811 at 1:25,000 scale, with contour intervals at 10 metres, and based on pre-war mapping. For restricted use up to 1962, they were not replaced by new maps until the 1970s.[3] A formula to relate the two systems is also elusive. The references are included in this book for the historical record.

In extremis Austin Coates's grid references can be plotted and used to work out the location of villages that cannot be identified from modern maps.

James Hayes identifies his villages by references to the pages of J. T. Wakefield's *Gazetteer* published in 1960.

MEASUREMENTS OF AREA

Our authors used a variety of names for the unit of area where 1 mu = 1 taochung = 0.067 hectare = 0.165 acre, i.e., 6 mu/taochung = 1 acre. In this book the spelling has been standardized as mu and taochung.

CHINESE NAMES

Other than measurements of area, Cantonese names and words are generally left as spelt by the author to illustrate the variety of romanizations in use at the time. Sidney Lau from the Chinese

Language Section of the Hong Kong Government Training Division, the compiler of *A Practical Cantonese-English Dictionary* published in 1977, had yet to set the standard for romanization of Cantonese.

MAPS

The challenge was to find maps to illustrate this book that showed the terrain of Hong Kong at the time the authors describe. Since the 1950s there have been enormous changes as a result of reclamation, reservoirs, levelling of hills and realignment of waterways.

The maps that have been used are taken from the Central Library Multimedia Information System (http://hkclweb.hkpl.gov.hk/hkclr2/internet/eng/html/frm-adv_srch.html) and have been scanned at very high resolution.

The 1902 maps come from "Colony of Hong Kong, New Territory, Kowloon Extension agreement dated 9th June 1898" published by the Southampton Ordnance Survey Office in the United Kingdom. With all the contours and most of the villages marked, the maps must have been a remarkable cartographic achievement given the lack of aerial photographic surveys at that time. It must be remembered that the New Territories were only ceded to the United Kingdom in 1897.

The 1902 maps were printed on eleven sheets, three across by four down, with the northwest corner (Deep Bay) omitted. Unfortunately sheet 7, covering Kowloon, northern Hong Kong Island, northeast Lantao and the western Sai Kung Peninsula, is missing from the Libraries collection.

The 1902 maps have been used in this volume to show the locations of villages referred to, with the 1951 map used to substitute for the missing sheet.

The 1957 maps come from "Hong Kong and The New Territories" published by Hong Kong University Press and Macmillan Limited, London in 1957. It was printed by W. & A. K. Johnston and G. W. Bacon Ltd. in Edinburgh. This map alone is a fascinating source of information about the territory in the 1950s. It was based on "Geographical Section, General Staff, no 3961. Published by the War Office 1936. 4th edition with Chinese characters added 1957". Land use is colour-coded on the map under categories: Woodland, Scrub, Rough Grass and Scrub under 12 inches, Badlands heavily eroded, Arable, Swamp, Houses with gardens, Built up area, Camps, Cemeteries, etc., Agriculturally unproductive. This classification was undertaken by T. R. Tregear (BSoc [Econ], PhD [Land]) of the Land Utilisation Survey.

A footnote to the 1957 map states: "The area of Chinese territory on this sheet is compiled from material of doubtful value, but is probably a fair general representation of the country", telling us something about relations across the border at the time.

In this volume the 1957 maps are used to give "the big picture" of the colony of Hong Kong, Lantao Island and the Sai Kung and Hang Hau Peninsulas as a whole. The maps of the Lesser Islands are also extracts from this source.

The 1951 map is from "Hong Kong and the New Territories" published by the Hong Kong Public Works Department in 1951. This simple black and white map shows the coastline, rivers, roads, ferry routes and some villages but no contours. This map has been used to substitute for the missing sheet 7 of the 1902 map.

Notes

1. Published in the *Journal of the Hong Kong Branch of the Royal Asiatic Society*, Vol. 17 (1977), pp. 144–156.
2. This can be found at http://www.galaxylink.com.hk/~john/paul/paul.html. The biographic notes on his life come from the same source.
3. See J. T. Cooper's article, "The Mapping of Hong Kong", *Journal of the Hong Kong Branch, Royal Asiatic Society*, Vol. 9 (1969), pp. 131–140.

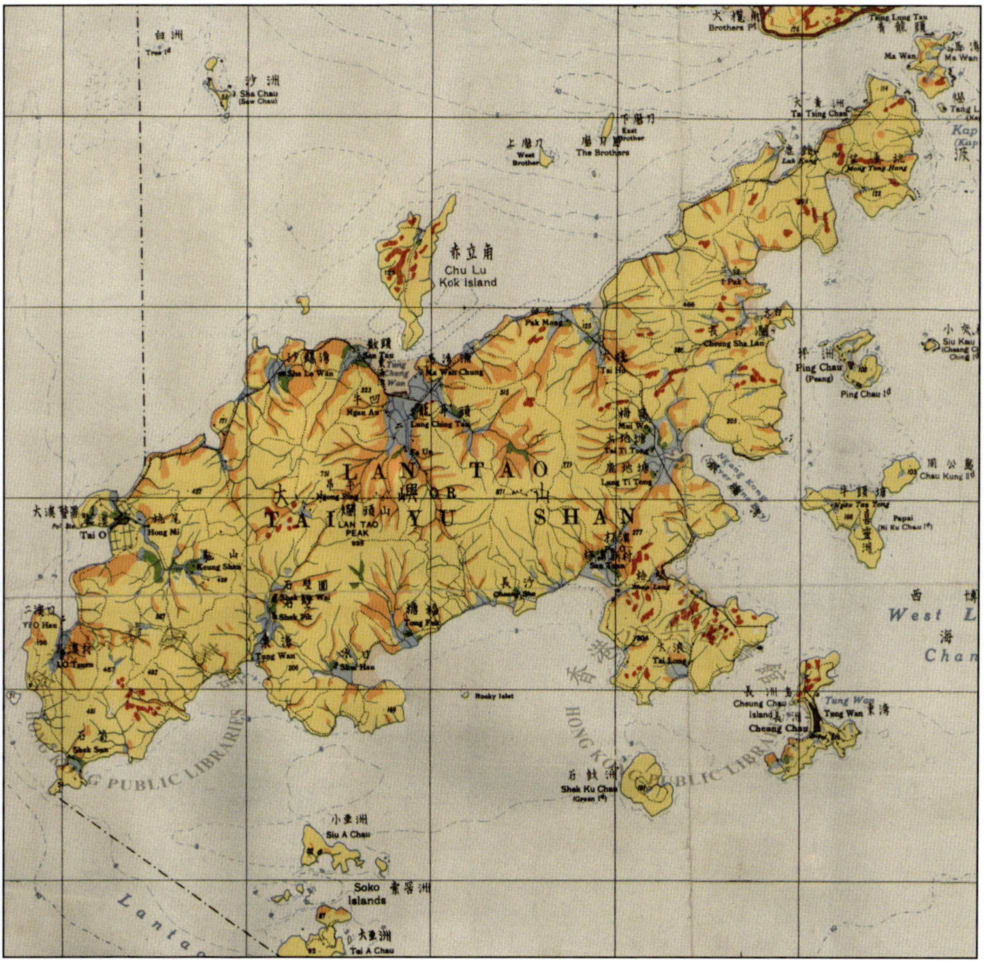

Lantao Island, 1957
[Scale 1: 80,000 or 0.792 inch = 1 mile]

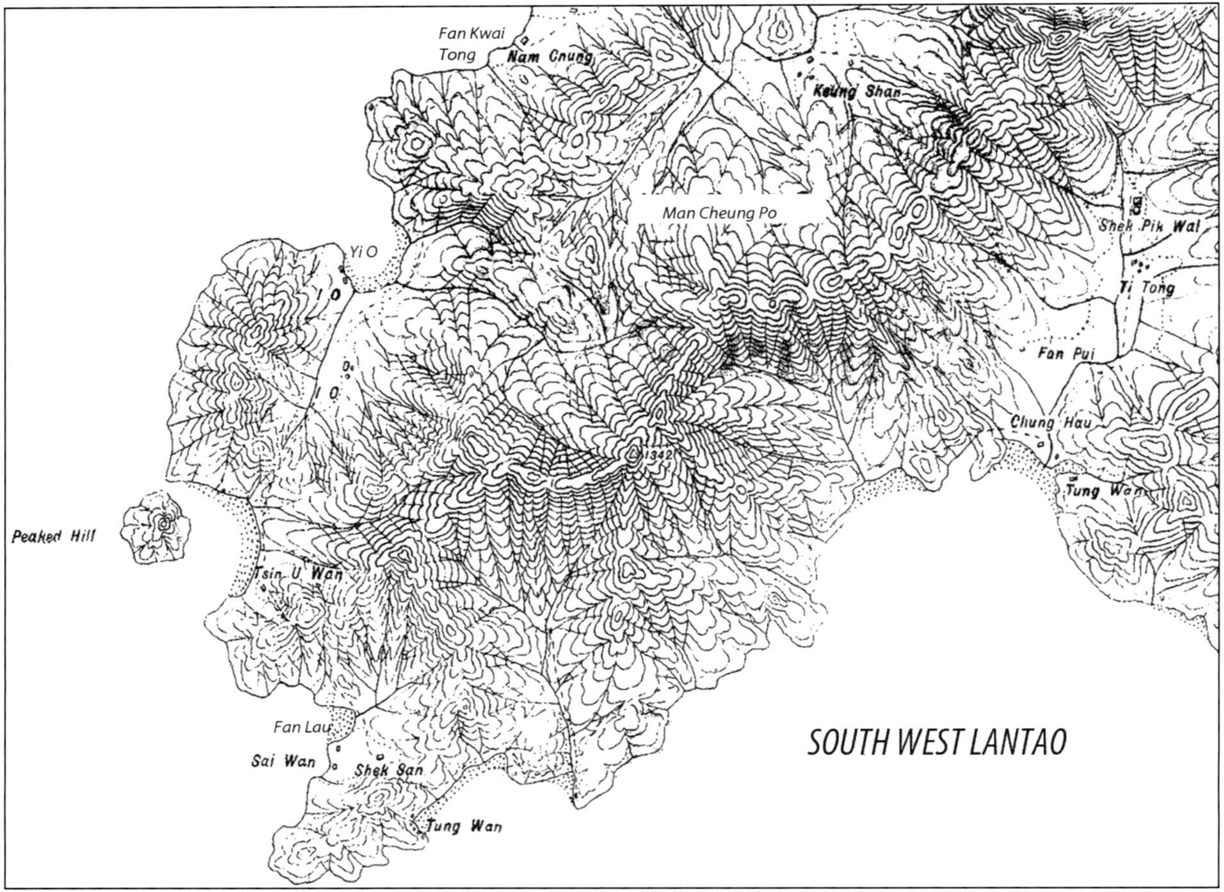

Southwest Lantao Island including the Shek Pik valley, 1902
[Scale 1: 31,680 or 2 inches = 1 mile]

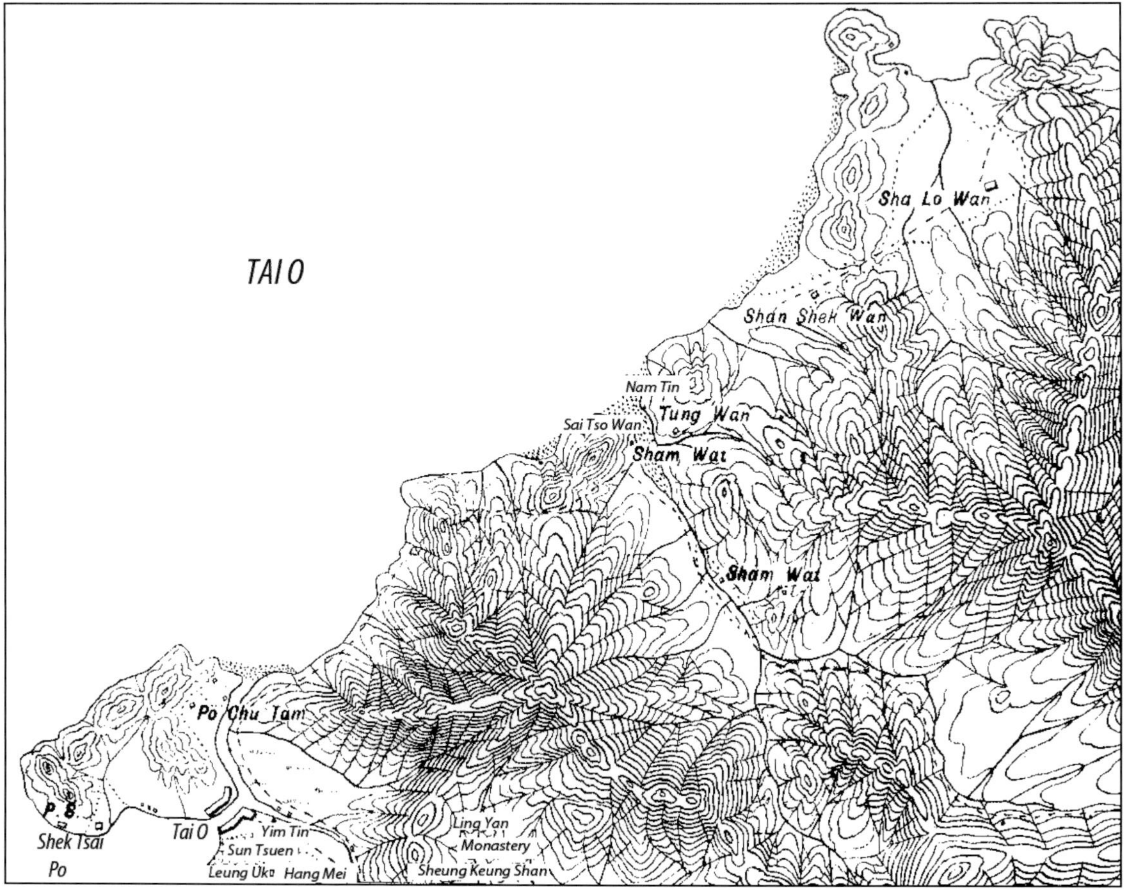

The north coast of Lantao Island from Tai O to Sha Lo Wan, 1902
[Scale 1: 31,680 or 2 inches = 1 mile]

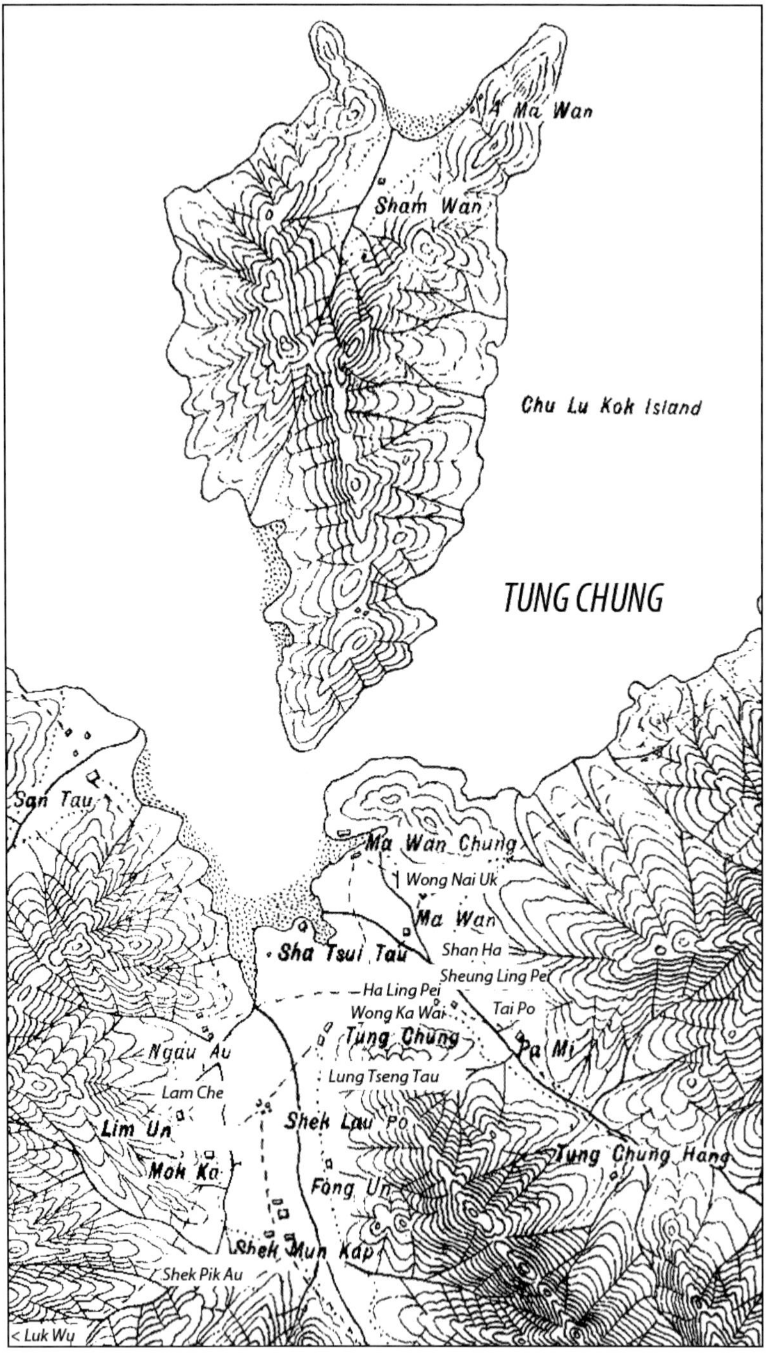

The north coast of Lantao Island: The Tung Chung valley, 1902
[Scale 1: 31,680 or 2 inches = 1 mile]

The north coast of Lantao Island: The Pak Mong valley, 1902
[Scale 1: 31,680 or 2 inches = 1 mile]

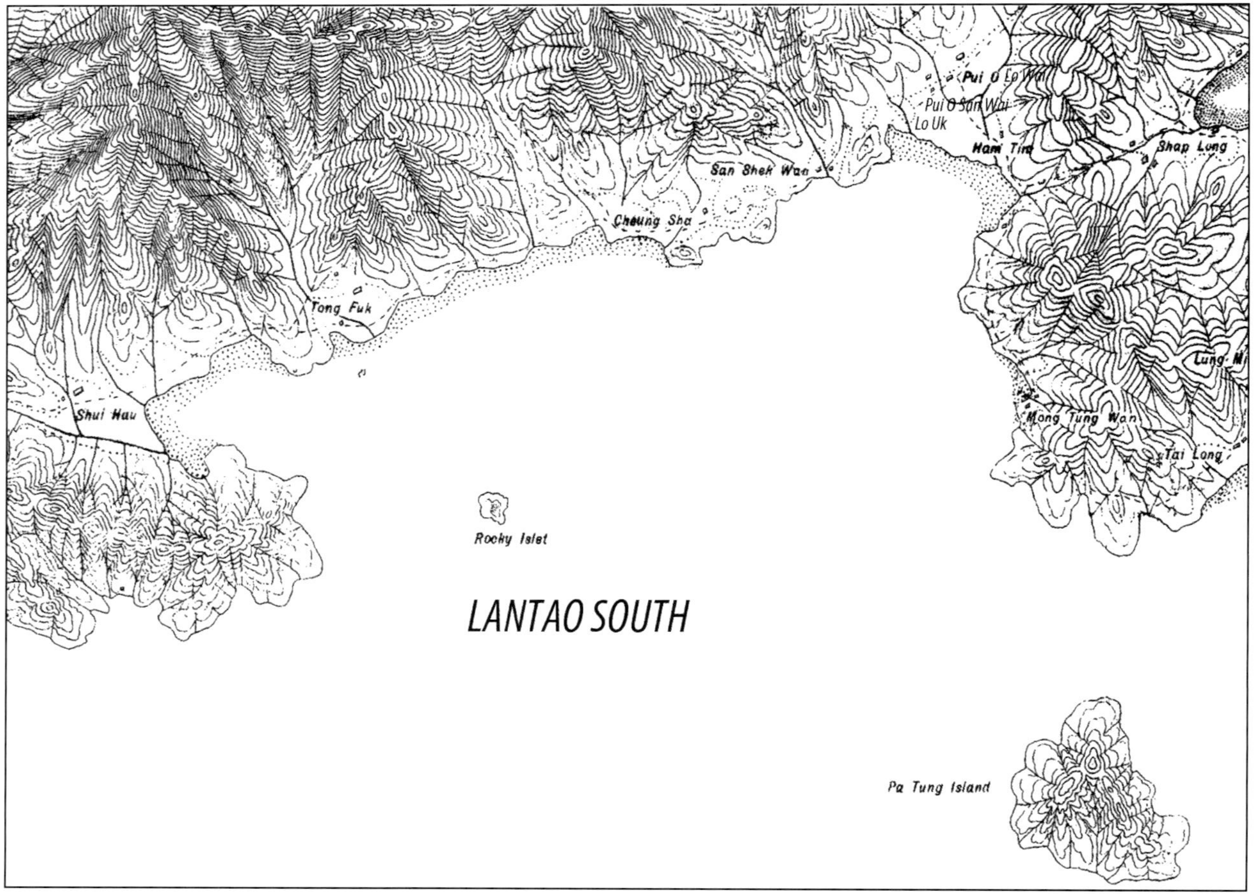

The south coast of Lantao Island: Tong Fuk to Pui O, 1902
[Scale 1: 31,680 or 2 inches = 1 mile]

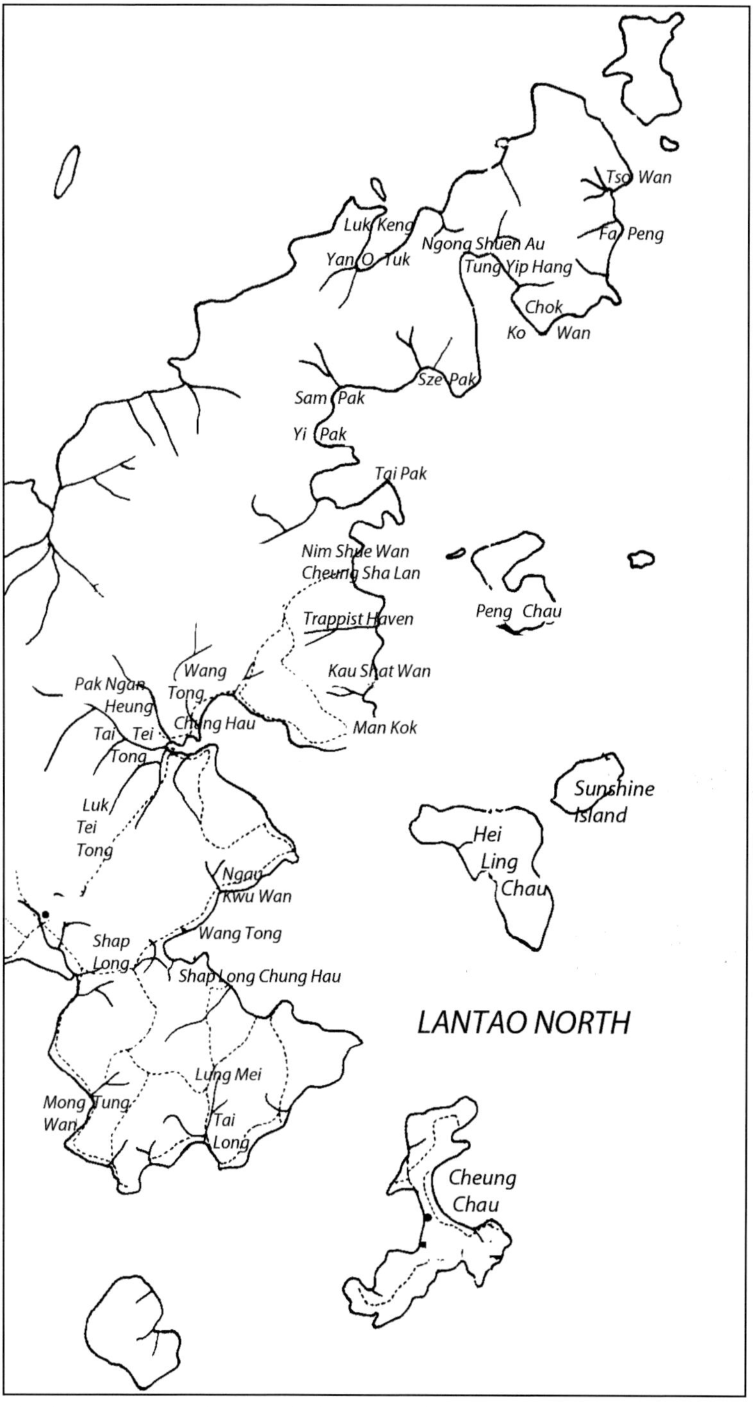

Northeast Lantao Island including the Chi Ma Wan Peninsula and Mui Wo, 1951
[Scale 1: 120,000 or 0.594 inch = 1 mile]

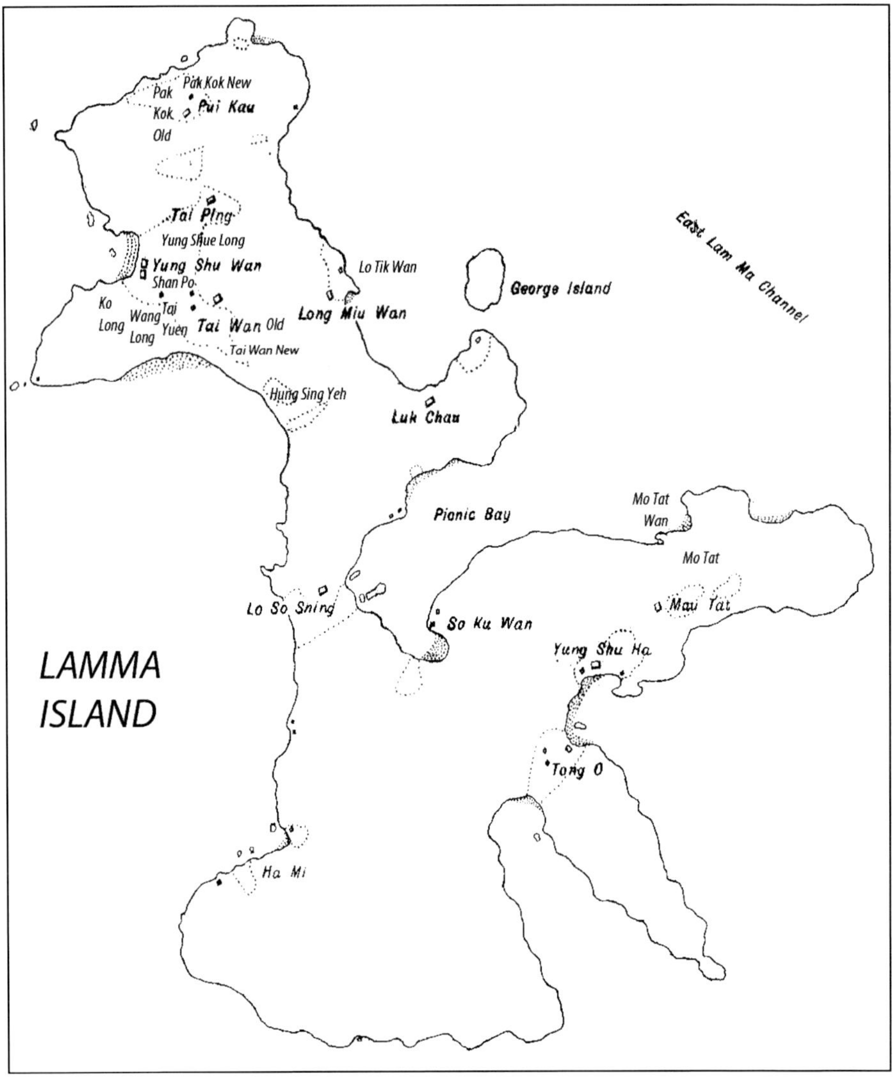

Lamma Island, 1902
[Scale 1: 31,680 or 2 inches = 1 mile]

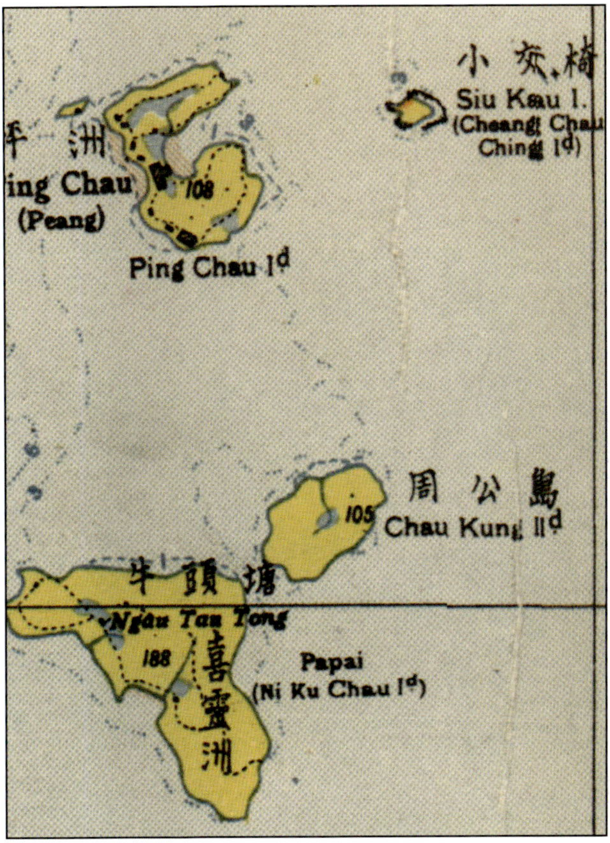

The Islands Peng Chau and Hei Ling Chau, 1957
[Scale 1: 80,000 or 0.792 inch = 1 mile]

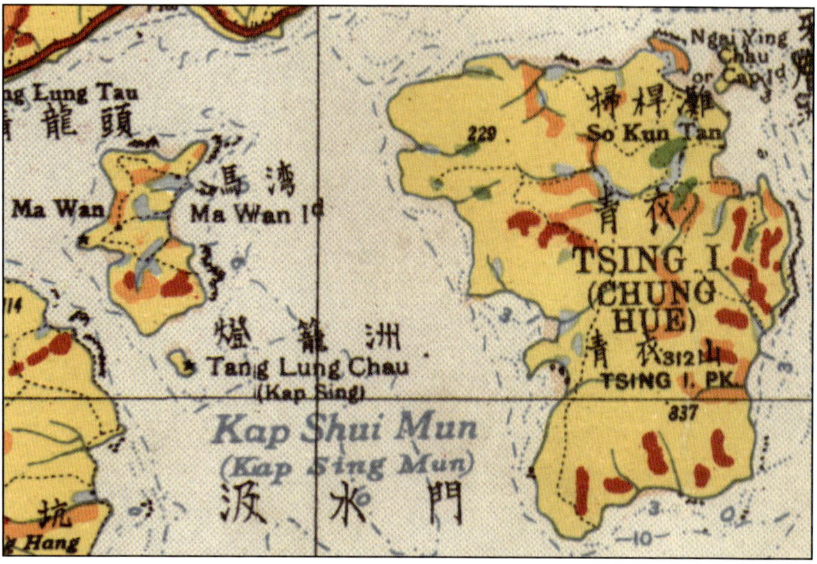

The Islands Tsing Yi and Ma Wan, 1957
[Scale 1: 80,000 or 0.792 inch = 1 mile]

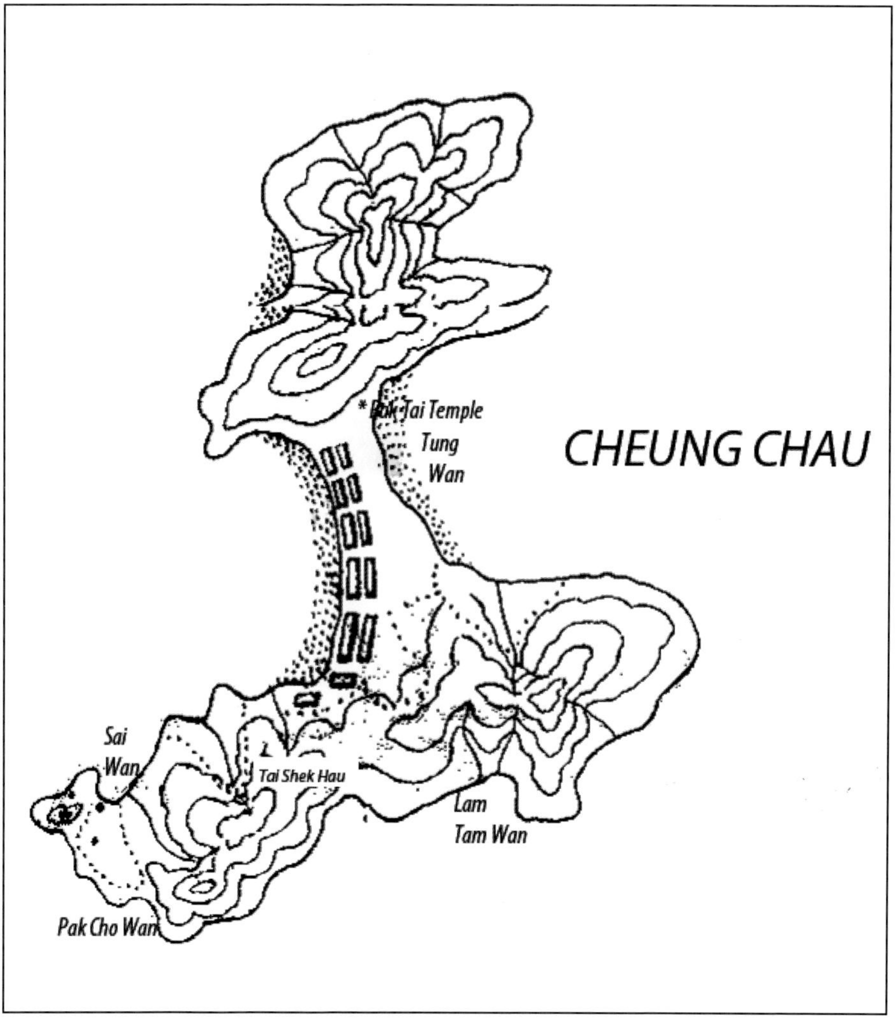

The Island Cheung Chau, 1902
[Scale 1: 31,680 or 2 inches = 1 mile]

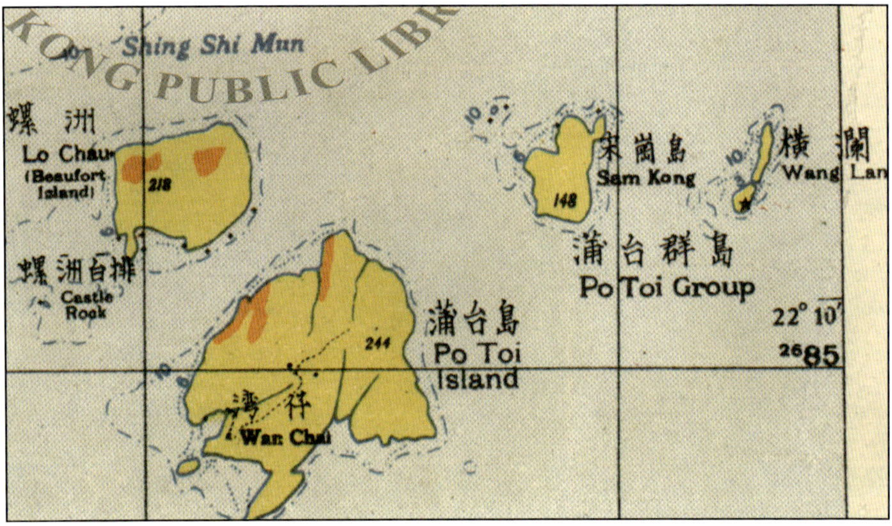

Po Toi Island, 1957
[Scale 1: 80,000 or 0.792 inch = 1 mile]

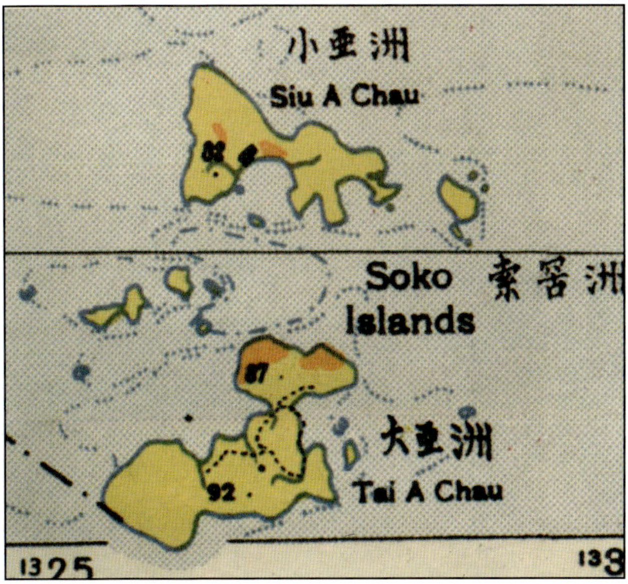

Soko Islands, 1957
[Scale 1: 80,000 or 0.792 inch = 1 mile]

The Sai Kung Peninsula, 1957
[Scale 1: 80,000 or 0.792 inch = 1 mile]

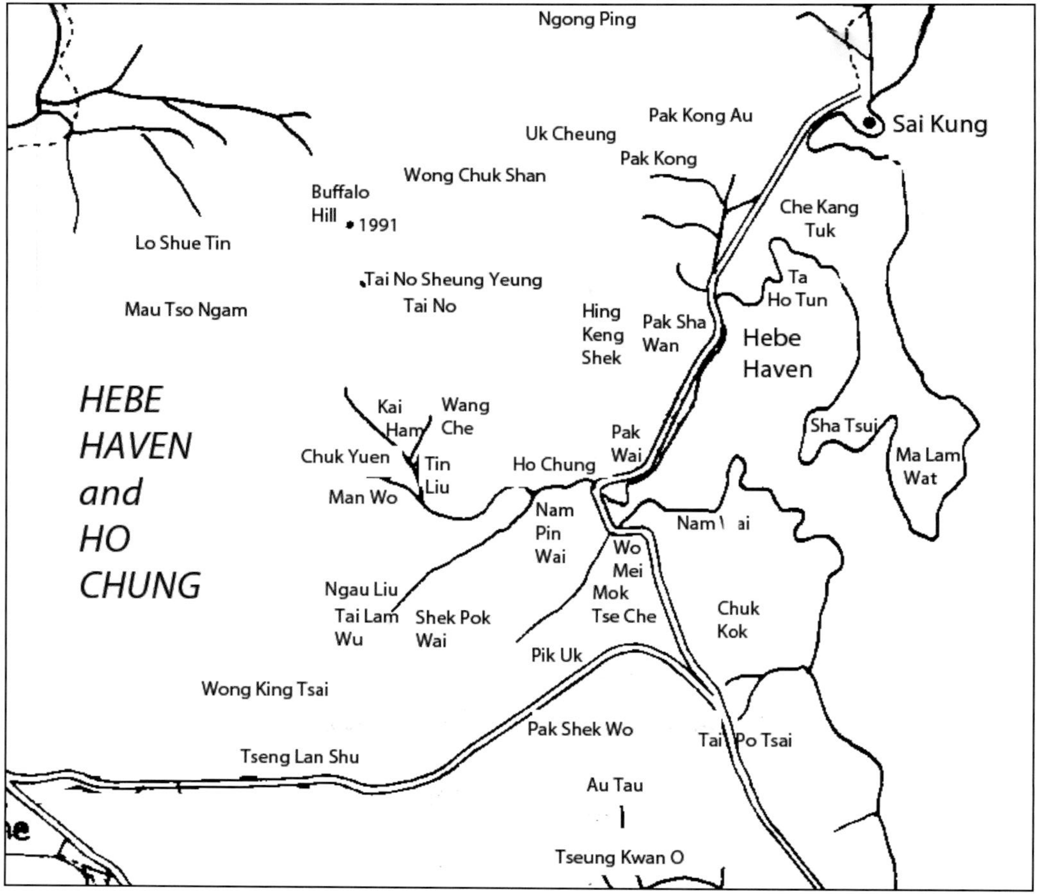

The Sai Kung Peninsula: The Ho Chung and Pak Sha Wan clusters of villages together with the hills above and the route to Kowloon, 1951
[Scale 1: 120,000 or 0.594 inch = 1 mile]

Sai Kung town and the nearby Pak Kong, Sha Kok Mei and Tai Wan village clusters, 1902
[Scale 1: 31,680 or 2 inches = 1 mile]

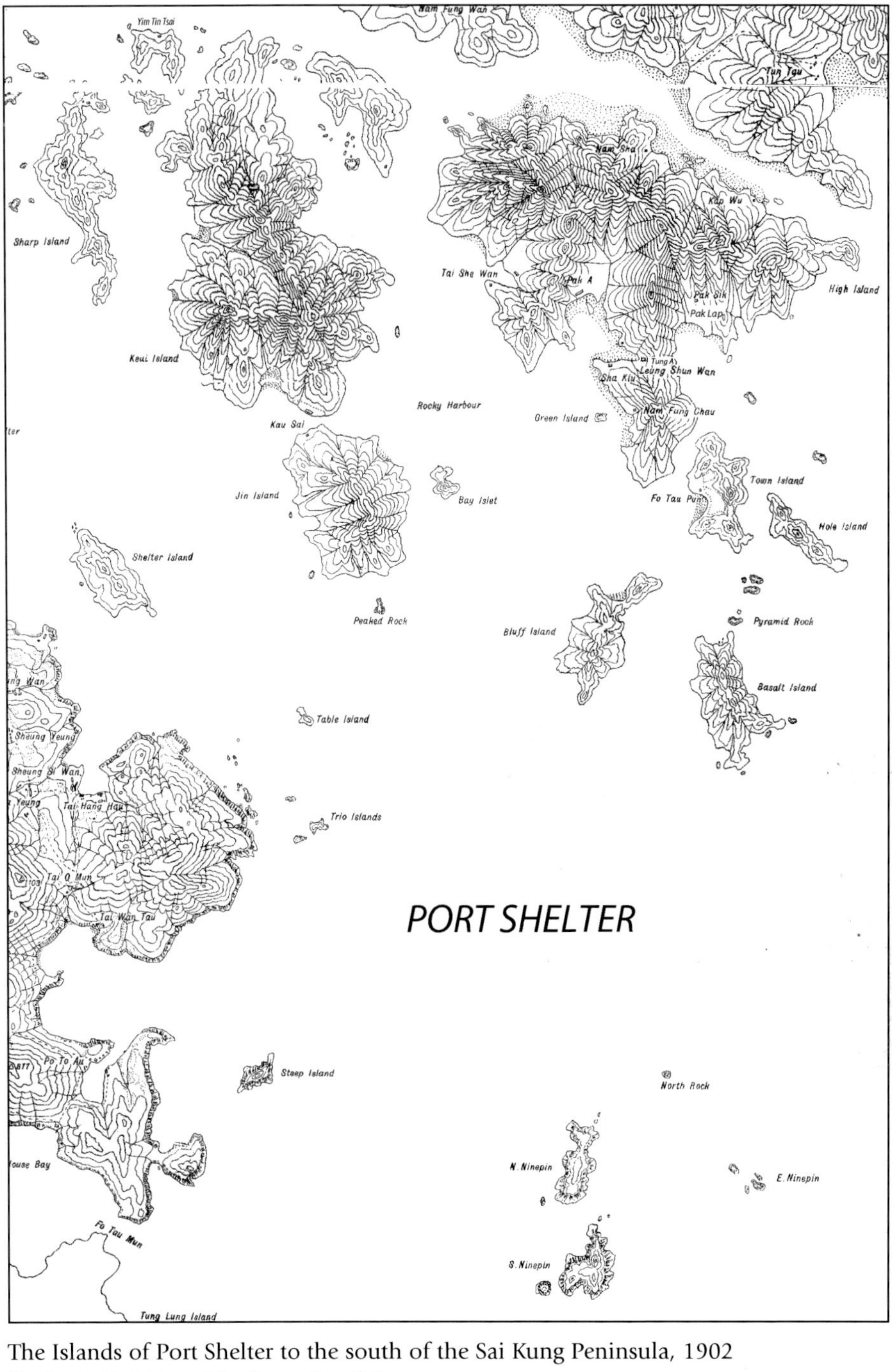

The Islands of Port Shelter to the south of the Sai Kung Peninsula, 1902
[Scale 1: 31,680 or 2 inches = 1 mile]

The Hang Hau (now known as Clear Water Bay) Peninsula together with Junk Bay, 1902
[Scale 1: 31,680 or 2 inches = 1 mile]

To the villagers in the Territory the District Officer is the Government. Whatever happens he is their own particular official, and what is more, their friend. He does his work with a minimum of legal formality, and a maximum of simplicity which has tended to inspire confidence, and to wean the people to our own matter of fact systems. Perhaps the PWD want to resume certain areas for making a reservoir or a new road. The DO is asked to confirm this, and to watch the people's interest in regard to compensation etc. He, it may be said, is the officer responsible for the welfare and happiness of the whole district. From the foregoing one might be tempted to think that these officers are men who know the peculiarities of the Chinese and how to deal with them. On the contrary, these posts are generally given to young Cadets after they have passed their necessary examinations in the Chinese language. They are given to them for two reasons. The first being to enable them to get a good idea of the habits and customs of the people, and the other, to get a thorough knowledge of the Territory in general.

Kaifong

In practically every village in the New Territory there are the Village Elders or "Kaifongs". These people are elected by the villagers yearly, and form what may be called for the want of a better name, a Town Council. They are under the direct supervision of the District Officer, and books are kept showing the expenditure of all public monies. They are generally men of mature age and known merit, whose duty it is to assist in the public welfare of the village. Their moral influence in the maintenance of the public peace, and their knowledge in the decision of questions concerning local customs, disputed successions, fung shui and such like, have often been of great assistance to the District Officer in charge.

Village Life

The majority of the people live in small villages. Nearly all are surrounded by small groves of trees which are carefully preserved, and, outside of these, by stretches of cultivation which present a very bare appearance, as they are usually devoid of any trees or hedges or grass land to give variety to the view. To see one village is to see the lot, they are all so much alike. One is often tempted to ask why the Government do not get a few standard block plans made, so as to give the villagers a variety of houses to choose from. Some of the houses are built of stone obtained from the hillside, and others simply of mud and sand moulded into bricks. The average size of a village house is 40 feet long and 12 feet wide, one storey, and they are built at the north end facing the south. A wooden partition divides the long room into two, one being used as a sleeping compartment. The doors are of wood held together by a cheap lock, and the floor is merely the earth beaten flat. If the occupants own any pigs or poultry, there may be a small shed at the back of the building for them, but more often than not they are kept in the back of the house.

The construction of village roads is of the simplest, the most pretentious being of granite slabs laid on raised pathways about three or four feet wide. Outside the villages the roads are simply paths about two feet wide, worn by the feet of the villagers. Great care should be taken when walking out in the country; always keep to these paths. In and around most of the villages there are wild cats and pigs, and to stop these animals from destroying the crops, the Chinese put down steel traps of the old so-called "man trap" style, and they are liable to inflict serious injury to anyone caught in them.

Good wells are found in all the villages.

The average villager does not set great store by cleanliness, or better housing. He finds himself unable to understand our aims and ideas, or on dismal conditions of unrest. He frankly dislikes our iconoclastic spirit, our want of imagination, and our blindness to all the forces of nature. He fears the inquisitions of the Police, or for that matter any other Officer in the Government service. Sometimes I have had to find a certain man in connection with Government business. Seeing four or five villagers together, I have asked for him by name. Immediately they want to know what he is wanted for. After about a quarter of an hour's talk explaining, one of them will step forward and say he is the man. It is useless asking why he did not say so before. Everything must be explained beforehand to them and, should it be a serious matter, one will simply be told that the man is not there; he has gone to the country or some such excuse, and in nine cases out of ten, the person so informing you is the one you want. The villager dreads the Sanitary Board, but he does recognize some solid advantages from British rule — chiefly in the security of life and property.

God of the Earth

Probably the most common shrine in China is that of the To Tei, or God of the Earth.

At every entrance to a village one will be seen. He is represented by a smooth stone, having no particular size or shape. He is a local deity, and is regarded as a kind of constable in the next world. He must be informed promptly on the death of an adult in the village. Sometimes children are given in adoption to this god. The procedure followed is to write the name of the child stating whether it is male or female, upon a piece of red paper, and place it upon the shrine together with wine, food and incense sticks, asking for the gods protection.

In addition to the shrines placed at each entrance to the village, smaller shrines are sometimes built in the walls of the houses, near the entrance, thus asking for protection for that particular building.

Domestic Life

The domestic life of the villager in the New Territory does not differ much from that of Chinese in other parts of China. Nor has it altered much during the few years of British occupation; if anything, it falls rather behind the general standard of freedom and enlightenment in the Canton province. For this south corner of the San On District, which is now the New Territories, was a remote and rugged country, far from the seat of Government and learning in Canton. Before the cession of Hong Kong, it was little touched by external influences, and even now the customs and habits of the people are probably little changed from what they were a hundred years ago.

It may be easy to administer, but its old established customs and institutions must not be lightly changed or affronted, and the necessary innovations have to be introduced with the greatest delicacy. In the New Territory as elsewhere the continuous descent in the male line is the foundation for many of their habits and customs. Respect for old age and experience is the second characteristic. The father of the family is the supreme head, and his sons come next in position and estimation. The mother of the family reigns supreme by virtue of her share in the continuance of the family over the feminine establishment, and often proves a tyrant to her daughter-in-law.

One way of defence which a Chinese wife has at her command is herself. If she has a fluent tongue, she can generally hold her own. If she is able to raise a storm about any trivial thing and keep it up, her position is secure. A Chinese woman in a temper wants a lot of beating.

It is the object of every living male to provide himself with an heir; if he dies before marriage or has no male issue, he must be provided with an heir by adoption.

Crops in the New Territories

The population of the Territory is for the most part agricultural, and rice forms the chief crop and staple article of food. It is grown wherever sufficient water can be obtained, and the fields are laid out and irrigation channels constructed with extraordinary care and skill to utilise to the full the available supply of water. The rice is usually sown in March and two crops obtained before November. There are three main qualities of rice, the best being grown on somewhat raised and therefore easily drained soil. It is sent to Hong Kong for export, as it fetches a higher price abroad than the Chinese will pay here. Foreign rice is bought for their own consumption as it is cheaper and rather preferred to the local product.

Sugar cane is grown to a small extent in the north of the Territory. The cane grows to a height of from eight to ten feet and is cut down when ripe, and the long stalks crushed in a mill, composed of two large, rough round mill stones revolving inwards, and usually worked by four oxen, two at each end of a large beam in the form of a yoke, the centre of which descends into the mill, and makes it revolve; the oxen working this like a capstan. The sugar falls into a vat beneath, and the squeezed cane is sold to make torches, and as fuel. Some years ago, an up-to-date sugar mill was supplied by Government to the farmers in order to encourage the industry, but it was rejected for the characteristic reason that it squeezed the cane too dry and thus spoilt it for further use.

Peanuts are grown in fields where there is insufficient water for rice and fetch a fair price. They are used chiefly for making oil.

Sweet potatoes are grown as a winter crop on the rice fields.

Pineapples are grown on sheltered hillsides in different parts of the Territory, and on Tsing I Island, but the greatest quantity is grown in the valley stretching up from Tsun Wan (Tsuen Wan) to Shing Mun. This industry was taken up vigorously in the first few years of our administration, and there seemed reason to believe that a good demand for canning the fruit would spring up, but this has not been realised. The area under cultivation in 1929 was about 224 acres. For registration purposes 10,000 plants are reckoned as one acre. This gives a total of 2,240,000 pineapple plants. Assuming only one pineapple to each plant, and sale for an average price of five cents each, we get a total of $112,000 per year.

Fruit-growing has not hitherto been successful. A business venture was tried at Castle Peak soon after the Territory was taken over. A large stock of fruit trees was imported from California, apples, pears, lemons, peaches, etc., but at no time during its career was it a success. The reason for this is perhaps not so much unsuitability of soil or climate as slack management and unskilled labour.

Lai-chi trees are planted in and around many of the villages, but the fruit yielded is of rather uncertain quality. Besides Lai-chis, there are a few oranges, limes, pomeloes, bananas and mangoes. But these have suffered in nearly all cases from want of attention, and no serious attempt seems to have been made by the natives to cultivate fruit with a view to supplying the Hong Kong market.

Animal Life

The commonest of the wild animals in the New Territory is the deer, which abounds in all the hilly districts. All are of the small species known as hog-deer, and do not possess horns. As they are very destructive to crops, and as their flesh is much esteemed and fetches a good price, they are hunted vigorously by the villagers, both with dogs and traps; they are not shot, as it is important to keep them alive to send into market.

Heavy damages are constantly being reported at various places on Lantao Island.

Game is not abundant; but snipe are fairly plentiful in certain districts in the autumn, and a good many quail with occasional partridge and wild duck can be obtained in the winter season.

Otters are to be found in some of the streams on Lantao and the mainland, notably in the Shing Mun River.

Wild Pig. In 1918 a report stated that considerable damage was done to the crops near Ping Kong by a herd of wild pig. An attempt was made to shoot some of them, but they succeeded in escaping from their lair just ahead of the guns. The herd was afterwards sighted near Wai Tau and probably made their way to the ravines of Tai Mo Shan. Both last year and this year the writer has been asked by the villagers of Sham Tseng, near the 13th milestone, to make up a party to shoot some of them on account of the damage done to the crops there. They have even offered to supply dogs and coolies free.

Tigers. There are well authenticated cases in which tigers have visited portions of the New Territory and even the Island of Hong Kong. Two tigers were knifed by Chinese in a cave in the hills near Sham Shui Po.

Cattle have been killed in large numbers, especially on Lantao Island, where some 60 or 70 were killed during 1911, apparently by some beast with claws and tracks similar to a tiger or panther. It was reported by one native living in a hut on Lantao that a tiger was seen by him dragging a chain, and it is not impossible that the beast in question might be a tiger escaped from a local menagerie. It would live mostly on deer, but occasionally pounce on isolated herds of cattle; nor is there any reason to doubt that it could swim over from one island to another. In one case in May 1911, a number of cattle had been killed in the south of Lantao, and the remainder of the herd, thirty in number, were sent over for safety to a small island half a mile away; but within two days 16 of them had been killed or badly wounded. An expedition went out there three days later, but by that time the beast had probably returned to the thick cover afforded by the Lantao hills. It was said to have been seen again early in 1912, both on Hong Kong Island and Lantao.

Cattle are employed everywhere for ploughing. The strongest and most valued are the water buffaloes, which are especially suited for hard work in the low-lying parts. They are of a variety of the wild buffalo, and come originally from India, whence they spread both eastwards over China and westwards as far as Greece and Italy. They have long horns, thick smooth brown hides and big feet, and are very tame in the hands of the native, but shy and sensitive to strange sights and smells, a fact that renders them somewhat alarming to the foreigner. The other variety of cattle is the short horned cow, usually of a brown or reddish colour, which is in common use throughout the district.

Bees are kept economically throughout the Territory. Their hives are made, as a rule, out of cylindrical rattan baskets which can often be seen suspended over the doorways, out of the way of vermin, ants and cockroaches. The bees swarm about April, when the Lichee and other trees are in blossom, and the honey is sold almost entirely to the local chemist, by whom it is utilised as medicine.

No Chinese village is complete without its compliment of pigs, poultry, and in low-lying parts, also ducks; all of which cost very little to keep. They do the scavenging and fetch good prices in the Hong Kong markets.

Foxes. 20-year-age foxes were plentiful in the hilly parts of the Territory, and they too were very destructive not only to the crops, but also to game and poultry. Although not so numerous now, there are still some to be found. In the *South China Morning Post* last September, a small paragraph stated: "There were several interesting incidents in local golf during the past week. AB killed a cobra at the ninth hole of the Fanling Old Course. BC played a ball on to the 15 green, which was promptly seized and removed by a large red fox."

Fishing

Of all sections of the Chinese, the most interesting, and the least known to foreigners is the fishing population: and of all, they have been least affected by foreign influence. Their life, methods, and their materials are now much the same as they were several hundreds of years ago. The bulk of the fishing is carried on by large junks, which trawl in pairs, with a long net some 250 feet in length. It is very wide in the middle and tapers to two narrow ends where it is attached to the two junks. Such junk's will go to sea for any time up to 10 days, and to a distance of 100 miles or more from land; they carry a large supply of salt in which each catch of fish is thoroughly salted and stowed away. The fish is dried on board or onshore, and sold to the shops in the nearest port, which act as agencies for the big salt fish firms in Hong Kong. The season for the fishing extends from about October to May; during the other months the winds are unreliable and typhoons are feared.

The chief home of large junks is Macau, where there is a convenient and extensive anchorage, good fishing grounds, and cheaper living than near Hong Kong. Their chief resorts in our Territory are Aberdeen, Shaukiwan, Cheung Chau and Tai O. The large bulk of the fish trade is in salt fish, which is exported from Hong Kong in all directions, but in the immediate neighbourhood of Hong Kong much of the fish can be delivered fresh. This is caught during the night and sent in by launches to reach the market before dawn Life on a junk is altogether one of gambling and speculation, and therefore dear to the heart of the Chinese. A junk may return with fish to the value of $2,000 or $3,000 or almost empty, and in either case all on board share in the fortunes of the junk. Living such a life, junkmen are naturally open-handed and free with their money, good customers for the shops which will give them credit for many years in bad times, in the hope of a good season coming at last. They are honest and dependable and give little trouble to the Police; their domestic life is harmonious and free from the bickerings which are such a feature of life onshore. On the other hand they are simple folk, and their education and their standards fall somewhat short of those of the land population. They are very superstitions, and the numerous shrines and temples along the coast bear witness to their piety. On this account they readily fall victims to the wiles of the fortune teller, and averters of evil spirits.

Below the large junks in size come a variety of small craft of schooner or cutter rig, which fish in the more protected waters of the Colony. As a rule they work singly with smaller trawl nets, but sometimes go out in pairs, as in the "Wong Fa" (yellow spawn fish) season in November and December. Lastly come the innumerable small craft which swarm in every sheltered cove, and emerge to fish either with small nets or with lines. The most notable of these are the Hoklo boats of whom many have settled in our waters. About April or May hundreds of these craft some 20 feet long, with high bows and sterns, and very shallow draught take advantage of a fair wind and descend upon these coasts. Three man each boat, of whom when fishing, one rows, and the other two work the net. The Hoklos are for the greater part connected with shops and fish dealers in Hong Kong, and surrounding ports; but many also come down as freelancers. When fishing is bad, the latter may become a menace to the more peaceful craft, whom they surprise and rob at sea on dark nights. They occupy mat-sheds at various convenient spots near good sandy beaches, where they can draw up their boats and dry and mend then nets. Fish are also caught in the large stake nets which are a familiar feature of the surrounding coasts; these are of two kinds, the offshore nets, worked from a sort of crow's nest, and the inshore nets fixed on four poles driven into the sea bed. Both nets are of similar construction, being large and square, with rather a small mesh to catch everything, and with a pocket in the centre, out of which the fish is taken from underneath. The net is fastened to four bamboo poles fixed into sockets, and is worked up and down by means of a windlass. The chief season for these nets is from June to September. It is a rather precarious living, and the nets are often worked at a loss.

The larger junks trawl for every kind of fish, except the porpoise, which is neither caught nor eaten, and the turtle, which the Chinese as a rule refuse to catch or eat, though they sell and eat their eggs, which are found in abundance on beaches around the coast during summer time. There are certain seasons and certain methods for catching various fish which deserve mention.

Whales, have occasionally visited the waters of Mirs Bay and in July 1905, one was struck by a passing steamer, and the carcase washed ashore at the top of Tolo Harbour, 10 miles from Tai Po; none appear to have been seen since.

Sharks come into the neighbourhood, one was reported to have been seen in June last year near Sham Tseng, Tsun Wan district, and they are fished for occasionally about June in the vicinity of Ling Ting Island, to the southwest of Hong Kong. A line about a mile long is laid down in the evening attached to floats, and fastened down by anchors at each end. To this are attached long hooks at intervals of every ten yards or so, baited with morsels of fish or meat. If a shark catches one of the hooks, he wears himself out with pulling at the line, which is taken up in the morning.

Wong Fa, are caught in large quantities during November and December at the mouth of the Canton River near Tai O, Lantao Island. They come in with the tide in immense shoals, and all the craft from the neighbourhood sally out to catch them. On 20 October 1927, 600 piculs, roughly 36 tons, were caught. The boats go out in pairs with large nets having sinkers at one end and floaters at the other. The men detect the presence of the fish by keeping their ears at the side of the junk, and throw in the nets when they hear them coming.

At Tai O scouts are sent out in small flat bottomed boats. Alter rowing out about two or three miles, they lay with their ear pressed to the bottom of the boat listening, and when the fish are heard they signal to the junks in the harbour, which immediately proceed to the places indicated.

Chu Yu, a kind of herring with a slight taste of mackerel is caught in January, February and March, by nets of the same kind as the above: but their coming is detected by their movements along the surface on dark nights.

Hak Chong Yu (black pomfret) is caught during the summer months by means of decoy fish, made of sandalwood and painted white, which are dragged through the water and followed by these fish, which are then easily caught; or boards painted white are laid out on the side of the boat, and the fish leap on to these and fall into the boat.

Acetylene lights are used at night in sheltered waters for catching fish. Curiously enough, in some cases as with Cheng Lun Yu (a sort of small herring) they are used in conjunction with the noise of wooden clappers on the side of the boat to frighten the fish into a net spread in front: in other cases they are used to decoy fish towards the boat, when they are speared or scooped up in nets.

Shrimps are caught by means of a close-woven trawl net during the summer months in the waters around Lantao Island. The shrimps are made into a paste, which is potted and sent to Hong Kong, whence it is exported in large quantities.

Shell fish are highly esteemed by the Chinese. They are caught by means of a long line let down with hundreds of empty shells attached. The fish occupy these and when the line is pulled up, they are removed, and the line let down again. In some parts they are also picked up by means of bamboo forks some 20ft long with iron prongs, with which the men feel along the bottom, and thus pick up the fish.

Crabs and lobsters are caught by means of a long narrow net let down to the sea bed, or in bamboo trap baskets, with bait inside.

There are large oyster-beds in the shallow waters of Deep Bay.

Most of the large junks employed in fishing around these shores come from Macau and Canton.

But there is good business in the building of junks and sampans at various places in the New Territory. The chief business is carried on along the shores at Sham Shui Po and Cheung Sha Wan in the northwest of the Peninsula, where the bulk of the boatbuilding of Hong Kong is carried on.

Nets are made entirely by the fishermen and their families, the string being twisted out of fibrous grass found on the hillsides.

INTRODUCTION (Walter Schofield)

My first introduction to the Southern District took the form of journeys by Water Police launches to various parts of it during the summer of 1919, when I lived for three months in the Water Police station quarters before my first leave. After it I sometimes repeated such voyages for purposes of geological research, on which I embarked with Government encouragement. A professional geological survey of the Colony was being planned in order to help in developing the resources of the Empire after the 1914–18 war, and to most people the Colony's geology was, quite understandably, a sealed book. The coasts and islands of the Southern District afforded many instructive sections, often showing the relations of different rock and mosses in a nearly undecayed state, which except in stream beds could hardly be seen anywhere else in the days before great motor roads cut the hills. This work enabled me to prepare a preliminary report on the Colony's sedimentary rocks and granite batholiths, which was presented in 1923 not long before the Canadian geologists began their labours.

In 1922, while I was working as second assistant to the Secretary for Chinese Affairs and deputy registrar of marriages, on the first floor of the Post Office building, Mr Wynne-Jones, the District Officer South, whose office was just above mine on the second floor, went to hospital with appendicitis and I was instructed at ten minutes' notice to go upstairs and do his job till he got better. As I had coveted the job for sometime, and had told my chief so (then the late ER Halifax), I was delighted.

In those days one of the District Officer's duties was to sit in his office as magistrate for the Southern District, excluding New Kowloon and the Lyemun area. This court usually functioned from 9 to 10 a.m. and might bring anything from a complicated murder to a petty assault case: the former, with its formalities, always ticklish for an inexperienced lay magistrate. The next job was to interview people sent for by the District Officer, deal with any disputes brought up by the parties or the Police, and hear any land cases fixed for that morning. On Monday, Tuesday and Thursday afternoons the longer cases could be heard: failing these, there were always land deeds and registers to sign, files to deal with, or minutes to write. At the end of the day the ledger, cash books and receipts would come in for checking.

In my time most of the cases that came to my office were from the nearer islands. New Kowloon, and the Tsun Wan district. Another class of case nearly always taken there was resumptions which I always considered the most distasteful and unpleasant task a District Officer can be expected to perform: for those resumptions in 1917 were usually paid for at a quarter cent a square foot, and those in 1926 at three and a half cents a foot. I never felt that money could in any way make up to a peasant for the loss of most or all of his land. Nearly always they wanted land in exchange, which it was rarely possible to find. I may remark here that when Mr Ruttonjee started the brewery at Sham Tseng about 1926 he secured the land for it partly by leasing a piece of foreshore from Government and reclaiming, and partly by leasing agricultural land from the villagers who were mostly surnamed Fu for a fixed term at a yearly rent, thus giving them a regular income and a right of re-entry on their land in default of payment, which seemed to me a very fair arrangement, though the raising of foreshore levels made a terrible mess of the fields.

My first spell at the District Officer South ended in about four weeks; but in March 1923 I left the Secretariat for Chinese Affairs for good and became "Lord of the Isles", and not a mere substitute. This gave me the chance to carry out researches without applying for Police launches, so I expect the appointment pleased the Water Police! It was the custom for the District Officer South to hire a big launch from a Chinese firm to take him, his bailiff, and his Chinese demarcator to Cheung Chau and Tai O on alternate Wednesdays if business there demanded his presence, or there were enquiries to make, or local applications for land to consider. For this he got a large travelling allowance, I think $1,200 a year, which I believe I nearly used up every year, though I don't remember asking for supplementary votes. Some District Officers seemed to pride themselves on saving as much as possible of this vote, but I always thought it a District Officer's duty not only to see as much of his district as he could, but to let its inhabitants see him.

The collection of revenue from the District, excluding lots held directly from the Government as Inland Lots granted through the Land Office, was always a most important duty of the District Officer. The Crown rent was collected both at the District Office and at various outstations, chiefly Tai O, Cheung Chau, Tsun Wan, and the small police quarters at Yung Shue Wan on Lamma. The outstation collections were done by a shroff accompanied by an Indian constable, and in the remotest places the Water Police gave their assistance. As a rule I believe the Water Police brought back the shroff and the money, though I think the ordinary ferry conveyed him to the scene of action. The collecting was done at the local police stations. It always began about the end of July, after the first rice crop, and went on at full blast till about October: Defaulters were dealt with early the next year. Licence fees for forestry, squatters fees, and pineapple plantation licence fees were usually paid before midsummer at the District Office.

In 1925, the year of the big Communist-inspired Nationalist general strike, the office shroff was transferred on promotion. His substitute was a young fellow fresh from the Treasury, who took advantage of the disturbance and the preoccupations of his superiors to embezzle part of the receipts, and finally absconded three months after the strike began. A former District Officer South remarked to me later that he had always been worried by the possibility of this kind of thing happening to him, and the almost total impossibility of keeping a tight check on shroffs when frequent absence from the office, sometimes all day, is part of the District Officer's duty. Luckily his security just covered his defalcations: and another shroff in the same racket was caught out by me and part of his loot recovered before I handed him over to the Police: thus I was able to show that on balance Government had in the end not lost a single cent. Both shroffs were arrested and sentenced later. I then spent a good deal of time, especially on voyages to the islands, drawing up rules for the financial guidance of my successors, but Mr Wynne-Jones, who took over from me in late 1926 thought them too cumbrous and discarded them.

INTRODUCTION TO PAUL TSUI'S VILLAGE REPORTS (James Hayes in March 2009)

The purpose of this brief introduction is to ascertain when and why the detailed socio-economic village reports contained in this section were produced. They are an important primary source, and some enquiry into how and why they came into being is warranted.[1]

I turned first to Paul Tsui's very detailed autobiography. Chapter 19, the last of the personal chapters, dealing with his time as District Officer, South, 1950–51, seemed the most likely place for information, but contained nothing. It provides an introduction to the District, but there is no account of his work there. The chapter stops abruptly, and is very likely unfinished.

I then found, from his Chapter 16 on the British Military Administration, Hong Kong, October 1945–September 1946, that Paul had worked in the New Territories during that period, and checked there for any likely clues.[2]

During that time Paul had been deployed to assist the civil affairs officer who advised the brigadier in charge of the mainland side of the harbour on matters affecting the Chinese population. This area had included the New Territories, and Paul had soon found himself working exclusively there, as part of the small team working for John Barrow, the very experienced pre-war D.O. North.

The chapter mentions a situation report written by Ken Barnett following a week-long visit to Lantao and other islands in the Southern District by police launch. Also, how Paul had produced a similar report on the Sai Kung side of the District, which, he says, had earned Barnett's approval, as being "more penetrating" than his own. This had led to various *ad hoc* assignments, all requiring investigative reports to be written, but as these were mainly to do with supply, trade and industry, it seems unlikely that they would have included the village reports.

Essentially, the village reports are concerned with land, livelihood, and village society, and combine economic data with other detail useful for senior officers requiring information on the New Territories (NT). Since it seems unlikely that such information would have been requested by John Barrow, the head of the post-war District Administration from 1946 until his retirement in 1952, with his long and close knowledge of the NT, it seems we must look elsewhere for the reasons that led to their production.

What follows is taken mainly from relevant archival material, Colonial Secretariat file 1/3184/50, now held by the Hong Kong Public Records Office.[3] The District Commissioner's printed annual departmental report for the financial year 1949–50 is also of some assistance, but only marginally.

In 1949, arising out of concerns expressed by the Development Officer, a Colonial Secretariat-chaired committee was tasked with reporting on land tenure in the NT, and whether rack-renting of immigrant vegetable farmers was taking place.[4] Needless to say, the District Commissioner was one of its members and closely involved. He had arranged for a number of Pilot Surveys on the ownership of land in the NT, which had confirmed that the average holdings were very small (about one acre). He had also established that press reports of large-scale rack-renting and eviction of NT farmers had been greatly exaggerated (DCNT's annual departmental report for the year 1949–50: paras. 20 and 30). Nonetheless, it had been conceded by all concerned that insufficient information was available on the farming villages.

No doubt with this in mind, during this period there had been correspondence with Professor Firth of the London School of Economics (LSE) in regard to a possible economic and social survey of the NT. Though not then serving in the District Administration, Paul had been consulted on the situation in view of his home village connections, his work in the NT during the period of military administration, and his attendance at Professor Firth's lectures whilst attending the "Devonshire Course" at Oxford University for colonial cadets in 1947–48. His name occurs frequently in the papers, and it is also clear that he had a great interest in the subject, even conducting some enquiries of the kind in his home area in his spare time. The LSE people were keen to involve him in any survey, but were told it was unlikely he could be spared for the work, especially as (in March 1950) he had just been appointed District Officer, South. On the basis of the papers I have seen, it would appear that, in the end, the NT-wide survey project was not proceeded with.

What can we deduce from the Southern District village surveys printed here? Some are originals, in Paul's own hand-writing. Others were in typescript. These latter were second or third copies, not the original typescripts. There is no covering memorandum or indication as to where they were sent, if indeed this had been the case. It seems most likely that the reports were written by Paul himself, during his one-year tenure as District Officer, South, and with or without the assistance of a few persons from the District Office.

What is still not yet clear, is whether they were done "off his own bat" or at the direction of his superiors. It is also unclear whether there were more reports which have either not survived, or not yet come to notice. A fuller scrutiny of the relevant Secretariat file, and a search for the originals and other copies of the reports in other surviving Secretariat or departmental files might shed more light on these points.

Whatever the case may be, there can be no doubt that the surviving reports are valuable documents, and probably unique of their kind. They were prepared by a capable person with a real "feel" for the task: someone who, on his own admission, had been tempted to take up an academic career in anthropology at one point during his attendance on the "Devonshire" course mentioned above (*Autobiography*, Chapter 17).

INTRODUCTION (Austin Coates)

This Memorandum has been written with the idea of providing District Officers with some general background information about the Southern District. When I first became a District Officer I felt the need of some kind of manual of information, and I think it is likely that succeeding officers in the same post may feel likewise.

Most of the material recorded was collected in the course of a tour undertaken during March and April 1955 when, accompanied by several members of the District Office staff, I visited more or less every village in the Southern District. Information regarding clans and their historical background, including generation periods, etc., is as given by the elders of the villages concerned. Wherever possible cross-checks were made, but I should warn anyone reading this Memorandum that it must not be taken as by any means accurate and final, it is not meant as an end-all statement on the District, but as a tentative beginning to assembling standard background information, to which I hope successive District Officers may feel inclined to add and supply corrections whenever they find inaccuracies.

Nearly the whole of the Memorandum was written out of office hours and in haste. It is therefore no masterpiece of literature. It is also as unsystematic as the jottings of a mediaeval monk. It consists of information about the subjects which interest me personally, and runs on as the different subjects occurred to me.

Due to illness, I was unable to complete it. The sections on the Saikung region, the Hang Hau Peninsula, Lamma Island and Lantau Island are more or less complete, but the Tsun Wan section is extremely sketchy and some of the Lesser Islands have been left out.

As well as providing background information, the Memorandum is intended to give, for record purposes, some idea of the state of the District in 1955, and much of it is therefore descriptive of things which today are commonplace, which in ten year's time may not be. Each section deals with a different region and is preceded by a general description of that region. At the beginning are some general observations and facts about the District which I hope may be of interest to Government officials outside the District Office, since as well as including various recommendations for the future, they may serve to give some idea of what the District Office does and has to think about.

I would like to record my appreciation of the services of the following members of the staff who accompanied me on the tour, and helped to assemble the material used here:

Mr Lo Cho Chi	Interpreter and Secretary
Mr Leung Ngai Kuen	Interpreter
Mr Fung Kwok Hing	Amanuensis
Mr Wong Wai Yuen	Amanuensis

INTRODUCTION (James Hayes)

My notes on these familiarization visits, now transcribed in full from the two small notebooks in which they have rested for the last fifty years, seem to require some accompanying commentary, to assist present-day readers.

First and foremost, unlike Austin Coates's Summary Memorandum, which, as stated in his Introduction, was intended to assist his successors in post, my notes were intended purely for my personal edification. I had not known of Coates's work, which was safely locked away in the District Commissioner's safe, and did not come across it until long after. Had it been available, I might not have persevered with my own note-taking.[5]

Secondly, whilst my notes cover much the same ground as the Summary Memorandum, there are major points of difference. Coates had been two years in post, and was able to provide long descriptive commentaries about the townships and sub-districts. As a newcomer to the District, I was, of course, quite unable to provide this kind of overall scrutiny and information. In any case, my purpose was essentially different and more limited in scope. On the other hand, I find that my notes on individual villages are sometimes fuller than in the Summary

Memorandum, as with the Pak Tam Chung and Tai Mong Tsai groups of villages in the Sai Kung region, and the Tung Chung villages on Lantao. Purely for the purpose of comparison, each has its uses, not least in Austin's case because his personality enlivens his account.[6]

As it was, acutely aware of my ignorance in practically everything connected with the New Territories, and with no other guidance than my predecessor's handover notes, my greatest need at the outset was to get to know the District and its people. This conviction drove me to visit practically all its 180 villages and hamlets within six months of becoming District Officer in mid-October 1957. And this, in turn, resulted in the notes which are reproduced in this volume.

There was no set programme of visits, nor could there be. In such a far-flung District with few roads, and many areas accessible only by sea, visiting was a very time-consuming business. I had my everyday work to do, and this was both varied and in abundance. There was, besides, and taking priority at all times, my duty to facilitate the civil engineering investigations in progress at remote but long settled Shek Pik on Southwest Lantao island, where, if found feasible, the government was fervently hoping to construct a large new reservoir, to ease Hong Kong's chronic water supply problem.

The notes were taken down from the village representatives and other men who met my small party when we arrived in each place.[7] The subjects of enquiry were of my own choosing. They focused on the kind of information I needed in order to understand how local people lived, and their needs, and hence, to do my job. They were also motivated by my keen interest in local history, about who they were and where they had come from, and their institutions, such as temples and shrines, ancestral halls, schools, and the like, since this information, too, was a means to better administration.

The figures quoted, on population breakdown, cultivated land, and the like, were provided on the spot. They were seldom verified thereafter. With the decision in mid 1958 to proceed with the reservoir forthwith, our work at Shek Pik had gained momentum, with the pressing need to recover all leased land from its owners and negotiate terms for village removals and resitings. Visits to villages after the initial round would be much fewer, and dictated largely by need. However, the population figures provided on our visits are close enough to those listed in Coates's record, and in J. T. Wakefield's invaluable *Gazetteer*, published officially in 1960.[8] However, they were bound to be similar, being taken from the same sources, and by the same means! What is more remarkable is that any information was to hand.[9]

The figures on land and livelihood are necessarily crude. I wanted them for general guidance, and there was no time to go into details, by lineage and family, of how much land was owned and how much was being rented from within the village, or from owners living elsewhere, etc., etc. Coates' figures are subject to the same caveats. We were not making the detailed surveys which would have given a more accurate picture.

Neither are our returns for lineages comprehensive or very satisfactory. The details were not always taken, and for lineages other than those to which the Village Representative (Village Representative) or other person volunteering the information belonged, their accuracy is questionable.[10] But it is striking how, in 1957–58, in some of the longest-settled villages, a few old lineages were represented by only one or two surviving families, whereas others of the same depth of settlement now had numerous descendants.

The notes show that, beyond question, all but a few villages predated the 1898 Lease, and that their resident lineages had lived there for generations: and indeed, in a great many places, for centuries. Also, that despite massive post-war immigration into Hong Kong, there was, in 1957–58, still only a handful of new settlements of agriculturalists in the Southern District. However, a closer scrutiny shows some internal movement within the District during the Lease, with people moving out of some villages to establish branch hamlets in the same area.[11] The entries show that the population was either Hakka or Punti (Cantonese speakers), with a preponderance of the former in its eastern parts.

Had we realized it at the time, this was the last period in which it would be possible to record the traditional rural economy of the region. Based on two-rice-crop subsistence farming, eked out by coastal fishing, various cash crops, and, for some families, remittances from absent menfolk, the process of transition had already begun. Over the next ten to fifteen years, development and modernization would force a total change of lifestyle, and especially for village women and girls. Rural society would never be the same again.

The notes provide an overall impression of what life had been like, and in most places still was. Besides the information recorded about crops, fields, and livestock, the answers given to my other questions provide more insights into the condition of the people in the villages.

One, prompted by the government's post-war drive to provide sufficient primary schools across the New Territories, brought much information on education for boys and girls, or the absence of it still. Factors emerged, including the traditional attitude that there was no need to educate girls, and there were deterrents like poverty, the need for extra hands on the family farm, or the hesitation to send young children to school along rugged hill paths. There was yet a lack of schools in some places, and insufficient provision of places in others. Overall, by 1957–58, it is painfully clear that the situation in the District was still far from satisfactory. Even in areas like Clear Water Bay, close to Kowloon and connected to it by public transport, the notes show that it was downright shocking.

The responses to my enquiries as to whether the villagers wanted any of the materials supplied by the District Administration and the Kadoorie Agricultural Aid Association to assist them with local public works projects, were more positive.[12] .The intention here was to encourage them to make improvements to irrigation dams and channels, to replace unsafe or inadequate potable water supplies from streams or wells with clean water piped in from the adjoining hills, and to help them improve existing footpaths to all-weather condition. Almost everywhere, village heads had projects in mind. It was clear that existing conditions left much to be desired, and that our assistance was both desired and appreciated by persons now willing and eager to help themselves.[13]

Our discussions brought information on other salient aspects of rural life. In many outlying places, people were still entirely reliant upon local ferries for travel and marketing produce and livestock, with attendant dangers.[14] And in mainland areas away from the existing roads, and on the islands, inter-village travel was mainly on foot, by old footpaths across hills and streams. As for municipal services in the townships, too much was still being left to their own managerial organizations (the Kaifongs) with some outside assistance.[15] In regard to medical and welfare services in the District, the government's provision was equally inadequate.

The notes of my initial talks with the Peng Chau Rural Committee show how energetic its old Kaifong of local shopkeepers had been, indeed still was, in its new guise, in providing local services, including an electricity supply, and how keenly it was pressing for the government and the public utilities to take them over. They illustrate very well the transitional nature of the time.

Like the expansion of the District Administration put in hand in 1959–60, the much needed programme of improvements in new government buildings and services introduced at the close of the decade was long overdue.[16] Both had been urged by Ken Barnett, my first District Commissioner, but not heeded in the corridors of power until his successor, Ronald Holmes, confirmed that they were urgently required to cope with the steadily worsening situation brought on by immigration and development in the mainland New Territories.[17]

The notes refer to depopulation in some places.[18] Removals from older sites are recorded, ascribed to deaths caused by malaria or other endemic disease.[19] These moves were always associated with adverse fung shui, in which, good or bad, I had soon learned there was a universal belief. There is also mention of wartime losses, due to burnings carried out by Japanese troops in reprisal for suspected involvement in guerilla activities. And the visual evidence, of ruined village sites and abandoned fields, confirmed all these reports.

I found that, in the 1950s, and for long before, men had often worked outside the village. Besides some working in Hong Kong or Kowloon, not a few were still employed at sea, or had gone on contract to Nauru and North Borneo post-war. I met seamen and overseas workers from an earlier generation who were now retired, and heard of others who had gone to work in Singapore or Malaya and other parts of Southeast Asia (the Nanyang) pre-war, and even of a few families which had settled in Borneo, but not all of these had returned.

As for the people themselves, what did they ask for? Requests for post-registrations of birth, to enable younger men to get a passport to go to United Kingdom for employment in Chinese restaurants, were common in some areas. Employment for young men was often raised, especially in minor government jobs. Widows in need, and likely to qualify for assistance from the Kadoorie Brothers, were brought to my attention. Medical referrals or follow-ups were asked for. There was the assumption that all these requests were in order, and that it was part of my duty to assist and facilitate. This was also the official view.[20]

The village notes are not all of a uniform depth or quality. I did not always ask all the questions in my usual range of interests. So many of the visits involved walking to, and between, settlements, helped out by boat travel where necessary or available, that we might sometimes have had to push on. Then, too, I might have forgotten, or been lethargic, or else the Village Representative might have been absent and my listeners apathetic! But, taken overall, the notes have something to communicate: not least because, in so many ways, they are a faithful mirror of the past, as much as of the moment.

Now that the greater part has been transcribed, I have been struck by one curious fact. Although the villages were wholly rural, and the District Officer was the land authority for the District, it is remarkable how few mentions there are of anything relating to land, or reference to any complaint or dispute. The general lack of development aside, a principal reason for this seeming anomaly is, I think, that life in the villages was, and for so long had been, very self-contained. All knew precisely who owned what and where, and public opinion was still a potent force, constituting a "damper" on unwarranted encroachment on other people's entitlements.

The notes do contain something I had almost forgotten. Along with the recorded facts, the earlier notebook contains "asides" which show how quickly I came to like Chinese country-folk. Some of my jottings refer to the Village Representatives, others to elderly villagers. They include the following comments: The Village Representative of O Tau, "a very nice man": the Village Representative of Nam A, "a good, intelligent chap, who came all round with us"; the 74-year-old Village Representative of Ngau Kwu Long, Lantao, "a nice old boy": and the like. The Village Representative of Sha Tsui, Sai Kung, was set down as "a sensible, courteous Village Representative". I had noted appreciatively that three persons from Shan Liu, including the Village Representative, had come to Nam A to guide me to their own village, and that the folk at Au Tau, near Tseng Lan Shu, on Clear Water Bay Road from Kowloon, were "nice people". The language used is banal indeed, but it indicates that they had made a very favourable impression.[21] Invariably welcoming, but not overtly deferential, the village people of the day had an inbred natural courtesy and ease of manner. Essentially, they were comfortable with themselves and who they were, and it showed.

Unfortunately, I am not a photographer. Nonetheless, I have some photographs of people and places from that time which help to illustrate what I saw during my visits.

However evocative these images may be, they cannot communicate the wonderful experience of walking through a countryside of great natural beauty, enhanced by human activity and effort. Both in the fields and on the hillsides, there were persons of both sexes at work, and the evidence of their forbears' industry and efforts was all around. Wherever we looked, there was this connection between the people and their past, which made the experience so special. And though not always so very far from a landing place or a road, many places possessed a remoteness in time and spirit; whilst in others, a reported sad and distant past seemed still to resonate.

In time, I would learn much more about the District and its people: especially in those places to which my work brought me more frequently, as at Shek Pik, and on Lantao generally.[22] But the knowledge and insights acquired during these initial visits would serve me well.

More, they had opened the door to a lifelong fascination with China and the Chinese. It is this which, looking back, gives these notes a special interest for me. And, perhaps, for those readers bitten by the same infectous bug!

Notes

1. To the best of my recollection they were found by one of my friends and colleagues in the Land Executive grade who, knowing of my interest in historical papers, forwarded them to me after finding them loose in a drawer in the District Office in which he was serving at the time (probably in 1981–82).
2. Strictly speaking, civil administration was re-established in May 1946.
3. Being unable to consult this file in person, I am grateful for the help given by Professor Chris Arriess of the University of Calgary and Mr Bernard Hui of the Public Records Office, in making copies of some of the relevant papers available for study.
4. Also expressed, had been the Development Officer's wider concern that the ongoing Communist land reform in China would highlight the unreformed situation in the New Territories, requiring the Hong Kong authorities to review the situation there as a matter of expediency and some urgency.
5. See Chapter 5, note 10.
6. He was well-liked and remembered in the District, because of his obvious interest in his work, but perhaps also because his theatrical manner struck a chord with the village people. They loved opera shows and, when the need arose, could themselves become suitably dramatic!
7. There had always been village headmen, but in 1948 the District Administration introduced more formality into the system, by providing for appointments as "Village Representatives". See note 3 above.
8. The full title is *A Gazetteer of Place Names in Hong Kong, Kowloon, and the New Territories*. It was published by the Government Printer, and the foreword is dated December 1960. There was a reprinting about 1969. Mr Wakefield's name does not appear, but I know he was the compiler, when serving as Chief Assistant Secretary for Chinese Affairs in the 1950s. Prior to this, he had also been District Officer South.
9. This has set me wondering. Were they available simply because the village representatives needed figures to hand, to satisfy "busybodies" like successive district officers, or were there other reasons? The last Hong Kong census had been in 1931, and the first post-war census was not taken until 1961. The clue may lie in Mr Barnett's mention that an "unofficial census

of the population of the New Territories was taken in March 1955..." (see page 3 of the printed *Annual Departmental Report of the District Commissioner New Territories 1955–56*). No information is provided on means, but hopefully this can be found in the archives. It is curious that Austin Coates' tour of villages was made at precisely this time (his preface states March and April 1955) but there is no hint of the count being part of the reasons for his visits). Together with statistics on landownership and tenancy in the New Territories supplied by the Director of Agriculture, Forestry and Fisheries, Mr Barnett's population figures were included in a table which he circulated among the District Officers. A copy is included in the appendix to these notes.

10. Some of the replies in regard to length of settlement clearly refer to the period since the first ancestor came into the province, and not to the current village. This is not uncommon, but it misled Austin into making some projections into local history on this basis. I believe this was one of the reasons why the District Commissioner kept the Summary Memorandum in his safe!
11. I suspect that there was more movement than I was able to uncover here. This would have caused problems for their descendants when, approaching the end of the Lease, and in accordance with the provisions of the Sino-British Joint Agreement of 1984, the District Administration was drawing up its list of pre-1898 villages, so that indigenous inhabitants could benefit from the provisions inserted in their favour.
12. The Kadoorie Brothers also gave loans, funded projects like village orchards, and helped widows with grants, and gifts of livestock, channelled through the Agricultural and Forestry Department and the District Administration. Initially, these philanthropists had not thought that the government was doing enough in these, to them, important fields — especially with a Communist China over the way.
13. We had a recently appointed Assistant Inspector of Works, seconded from the Public Works Department, an able and enthusiastic young man, ideally suited to give technical advice where required.
14. On my initial visit to the Sai Kung Rural Committee, it was mentioned that several persons from Sai Wan and Tai Long had been drowned in transit to the township.
15. The Kaifongs were the traditional local management bodies comprising local shopkeepers, to be found in Cheung Chau, Peng Chau, Tai O, and Sai Kung.
16. The size of the buildings housing, say, a police post, a post office, a medical clinic, a public health office, and the like were such as to make them the largest and most imposing structures in the place. On Peng Chau, where they were sited on a new reclamation that included a proper ferry pier, the new building was known derisively as "Lam Shue-chun Buildings", after the then rural committee chairman.

17. A full account is given in his printed *Annual Departmental Report* for that financial year, and those immediately following.
18. Lower population figures are confirmed for some places — especially on Lantao island, by comparisons with the detailed returns of the 1911 Hong Kong Census, as illustrated in some of my footnotes.
19. There is little or no mention of this depopulation in pre-war official reports, but that malaria was a scourge is made plain by the strict measures taken to prevent sickness in the large workforces assembled for major public works in the 1930s, as for the Shing Mun reservoir and the Gin Drinkers Bay defence line. Malaria was still prevalent in some areas well into the post-war period.
20. Confirmed by the paternalistic tone of the statements in regard to welfare and employment contained in the District Commissioner's printed annual departmental reports for the first decade of the post-war years.
21. Not that I was undiscriminating. There were some uncharitable judgments in the notebooks which I have not transcribed into these published notes.
22. I should mention here that I was D.O. South until I went on overseas leave in November 1960. Upon my return in August 1961, I became D.O. Islands, because the Sai Kung portion of the District had been excised following an intended, but later aborted, proposal to build a new reservoir in Hebe Haven. Thus, my close connection with that area ceased at that time.

3
Lantao Island

LANTAO ISLAND (Eric Hamilton)

There were biggish patches of cultivation around Tung Chung and Tai O. My own view in those days was that the chief importance of Lantao Island vis-à-vis Hong Kong was: (a) the fishing industry based at Tai O (Wong Fa are delicious!) and (b) the salt pans at Tai O. These were rather mismanaged and I think we ought to have devoted more attention to them.

LANTAO ISLAND (S. H. Peplow)

Lantau. Lan — broken, Tau — head. So-called from the shape of the Lantau Peak, which has a cleft or as the Chinese say, is broken. It is sometimes called Tai Yu Shan, Big Island Mountain.

LANTAO ISLAND (Paul Tsui, written after he retired)

I love Lantao Island. As a small boy when still in school, I took part in a launch picnic which took us right round the island. It took the whole day to get round and a part of the voyage sailing to the south of the island was marked by pretty rough seas which caused me to be seasick. As an undergraduate at the University of Hong Kong, Professor Hsu Ti Shan of the School of Chinese Studies used to take us up to the monastery at Ngong Ping for long weekends to observe how the Buddhist monks lived and prayed. During the British Military Administration the Royal Navy brought me to visit the various parts of the island, once by Mine Sweeper, several times by Harbour Defence motor launches and twice by their sea planes called Sea Otters. The Marine Police took me there by their largest Police Launches (ocean-going tugs). On top of that I took the ferry. One of the routes took some three hours to Tai O via Ma Wan and Castle Peak. The other took just about an hour to Silvermine Bay for Mui Wo.

When I returned to serve as District Officer South I had my own departmental launch, an Australian-built light wooden tug; and when I entertained visitors (e.g., Mr Harold Ingram, at one time the Acting Governor for Aden who was commissioned to write a book on Hong Kong and its people for the Corona Library) I could use the Governor's barge or the Marine Department's de luxe lighthouse tender. I have also been offered the use of private pleasure craft owned by my friends. Once landed on an island, I would have no alternative but to walk. By established custom make-shift sedan chairs (a bamboo chair attached between two bamboo poles) would normally only be used by people who were too sick to walk.

By far the most memorable of all the trips I have made to Lantao was one in the company of a Mr Hum, the Port Works Engineer of the Public Works Department, whom I took out to take soundings of the sea bed fronting Tung Chung, with a view to ascertaining whether or not it was worth building a jetty for the convenience of the inhabitants living in the Tung Chung valley. We travelled on board my department launch, and the journey from Tsim Sha Tsui took about two hours. On the way, Mr Hum told me how an engineer would set about tackling the problem that we were about to tackle. He explained in great detail how technological soundings could be taken, but emphasised the importance of taking note of the movements of tides as well as water flowing from the Pearl River, the geological formation of the sea bed and beach. He had to assess the potential usage of the project, noting the number of people as well as the type of craft likely to use the facilities when built. Then he would work out mathematically how big the pier should be, how big the pillars of support should be, how much steel to reinforce the concrete and what sort of a mix the concrete should be to withstand the rusting, etc. After the pier is built, who would maintain it? And how much it might cost to employ one or more workers to maintain it? The lecture or rather the "tutorial" was so fascinating that it ended up with my saying how clever and marvellous he must have been to have worked his way up to qualify as an engineer. To my surprise, he responded by saying, "Mr Tsui, an engineer only deals with subject matter which is quantifiable, in that you can calculate accurately even the volume of a grain of fine sand. What is much more difficult is the task of an administrator like you, who has to deal with human beings, human nature, social values and social relations, all of which are unquantifiable. Your task as a political administrator is more complex and difficult than the tasks of an engineer.

Every part of Lantao is different and had its own characteristics. Mui Wo at Silvermine Bay, though having a few villages with good paddy fields surrounding them had developed into a retreat for the relatively wealthy. The most respected elder residing in the valley was Yuen Wah Chiu, who was more popularly known as Kau Yeh. He was reputed to be a retired "buccaneer" from the county of Tung Kwun but was highly respected, and for years he had enjoyed a tremendous reputation throughout the entire Pearl River Delta. Not far from his house was the house of Mr Yeung Tsun Tat, a returned emigrant from New Zealand who had been the managing director of the Hongkong and Yaumati Ferry Company Ltd. and himself the owner and proprietor of United Delivery, a removal firm which operated the largest fleet of lorries in Hong Kong. Nearby, a bungalow belonged to another Yeung family which was a leading importer, wholesaler and retailer of rice in Hong Kong. There was also a retired architect living in one of the bungalows at the far end of the bay. Tucked in at the back of the Mui Wo valley was a disused mining cave, from which silver was said to have been mined, and from which Silvermine Bay derived its name.

On top at the eastern peak of Lantao (Sunset Peak) used to be the summer resort for Methodist missionaries, where a number of cottages were built to house the missionaries when they came to spend their holidays.

Across the ridge to the northwest of Mui Wo was the village of Ngau Ku Long, believed to have been used by a band of guerrillas during the Japanese occupation and where the Japanese army committed a nasty atrocity shortly before the end of the Second World War.

Further beyond Ngau Ku Long was the Tung Chung valley where in the midst of some 10 villages are the remnants of an old fort with cannons mounted on a granite battlement. Behind the battlement was a small path leading up to the peak of Ngong Ping and its Buddhist monastery.

Further west from Tung Chung is the very large village of Sha Lo Wan, before Tai O.

Tai O is a substantial fishing centre near the western tip of the island, halfway between Hong Kong and Macao. A police station there forms an unmistakable landmark. Apart from two rows of shops forming the market town, Tai O had many huts (in fact disused boats turned into static dwellings) built on stilts over the beaches, some parts of which were always under water, giving visitors an impression of Venice of the East. Parts of the beach had long been enclosed to form pans for evaporating sea water into salt. The plentiful supply of salt helped to make famous the salt fish from Tai O.

About two miles to the south of Tai O at the western tip of the island is Fan Lau, where the Pearl River meets the Pacific Ocean, marked by the line formed by the different colours of the water.

Turning northeast from Fan Lau you reach Shek Pik, once an inhabited village with paddy fields surrounding it, badly infested with mosquitoes bringing malaria to its inhabitants. This village has since the late fifties been flooded by water of the Shek Pik reservoir.

Further northeast from Shek Pik are Cheung Sha and Tong Fuk, before Pui O. On crossing the ridge from Pui O you come back to Mui Wo and Silvermine Bay.

To climb to the top at the western peak of Ngong Ping with the Buddhist monastery, I used to take the longer route via Tai O on a three-hour walk, passing by Keung Shan and Chui Fung Au. Off the beaten track there was the Chi Ma Wan Peninsula to the southeast, with Tai Pak and Yi Pak to the northeast.

Since my days as District Officer, much modern development has taken place, particularly on the southern side of the island. You now have buses and taxis for hire to take you from one end of the island to the other. The Shek Pik reservoir and its catchwaters feature the southern slopes. A motor road along the south coast from Silvermine Bay takes you across to Tai O, with a side road branching off at the saddle between Shek Pik and Tai O to the monastery. A number of Correctional Service institutions are now along the southern coast of the island, Chi Ma Wan, Tong Fuk, Ma Po Ping, Cheung Sha and Shek Pik. Most of them were developed from the buildings left by the engineers who built the reservoirs and the catchwaters. Along the southern driveway you can see rows of beautifully finished bungalows designed for weekend leisure. The Buddhist monastery at Ngong Ping is bidding to become a cultural centre for Buddhism, a pride of Buddhist culture outside China and Thailand.

The Trappist monastery at Tai Shui Hang between Silvermine Bay and Discovery Bay came into existence in the mid-fifties.

The private housing schemes at Pui O and Discovery Bay came into existence in the early eighties.

LANTAO ISLAND (Austin Coates)

Lantao presents a number of contrasts with other parts of the New Territories. Its appearance is more rugged and majestic than, for instance, Lamma Island or the Sai Kung region. The larger part of the island consists of steep hills, and most of the population lives around the coast in a number of small valleys, many of which were once under the sea. Whereas in the Sai Kung region each such valley contains a group of villages whose inhabitants and way of life are somewhat similar, on Lantao no valley is like its neighbour. Each community has a different background and traditions. Whereas it is not difficult for the people of Sai Kung or Lamma, or even for the long-settled families of the Tsun Wan region, to think of themselves as a single community, this is far less easy for the people of Lantao, which is less homogeneous than other parts of the Southern District.

The total population is approximately 20,850. About 2,000 of these are fishermen who either anchor permanently at, or visit seasonally the large market town of Tai O, at the western end of the island, which is the chief centre of population, having 12,030 land-based inhabitants. Silvermine Bay, at the east end of the island, is the next largest centre, having 2,800 inhabitants, and the Tung Chung group of villages comes next with 981. All these three places are connected with Hong Kong by daily ferry services.

The population is predominantly Cantonese, but Hakka are to be found in most of the valleys. At Pui O they are the principal community. In the hills in the western sector of the island are the largest group of Buddhist settlements in the Colony, this part of Lantao being one of the principal beauty spots of the New Territories. There is a Chinese Trappist monastery at the east end of the island, opposite Ping Chau, a coterie of European bungalows very close to the top of the eastern peak, known as Sunset Peak, a large Anglo-Chinese forestry project near Tai Pak, a home for the disabled and destitute near Shap Long, and a wolfram mine near Sha Lo Wan. The ruins of Chinese forts, of some historical interest, can be seen at Tung Chung and Fanlao (Shek Sun).

Tai O is the market for its own adjacent villages, the Buddhist settlements of Luk Wu and Ngong Ping, and the villages of Yi O, Fan Lau, Fan Pui, Sha Lo Wan, San Tao, and to some extent Tung Chung and Shek Pik. A peculiarity of the rest of the island is that its main markets are on other nearby islands. Shek Pik, Shui Hau, Tong Fuk, Cheung Sha, Pui O, Tai Long and Shap Long all trade with Cheung Chau. Silvermine Bay, while an important transit place for sending goods to Hong Kong or Cheung Chau, is a poor shopping centre, and here again Cheung Chau acts as the chief market. Pak Mong also uses Cheung Chau's services. The small coastal villages and even smaller hill villages on the east coast between Mang Kok Tsui and Sei Pak all deal with Ping Chau, while around the northeast coasts Ma Wan is the principal transit place, being connected with Hong Kong by ferry. This last area is distinctive for its stake-net fishing. Ma Wan people own nets near their island, and on the north coast of Lantao most of the stake-net operators are from Tai Lam Chung, on the mainland (Yuen Long District).

Being situated at the mouth of the Pearl River, Lantao Island is very much easier to trace in history than Hong Kong is. Under Tai O, Tung Chung, Fan Pui and Shek Pik, in what follows, will be found some speculations regarding the early history of Lantao. There is archaeological evidence to support that in prehistoric times traders were using the Shek Pik valley, and it is not beyond the bounds of possibility that some of the people living today in these older villages are descendents of the pre-Chinese Maan tribes. Settlements of some sort evidently existed at Tung Chung and Tai O in, at the latest, Southern Sung times, and it is clear that at the fall of that dynasty, in 1279, a number of refugees, some of them members of the Imperial Court, settled at Po Chu Tam, one of the villages forming the present township of Tai O, and at Tung Chung, where there is at least one family still in existence which is almost certainly descended from those refugees. Shortly after this, the families at present occupying the Shek Pik valley arrived from Ma Tau Wei, on the Kowloon peninsula, the inference being that they were hiding from the newly established Ming dynasty, or else escaping from the petty tyrannies that obtained in Kwangtung between 1279 and 1368.

From 1513 onwards, when European records of the Pearl River start, there are fairly regular references to Lantao Island, and it is from these early records that we have its name, which in Cantonese is presumed to mean Cleft Head, and would be an appropriate name due to the shape of the Lantao peaks. This name is no longer used; in fact it is unremembered. The inhabitants consider that it is a foreign name, or else apply it solely to Sunset Peak, which they have heard European visitors calling Lantao. However, in the Portuguese, and in the early XVII century Dutch and English records, there is no doubt about its name being Lantao. This may, of course, have been the name which Nantao people used for it, and not necessarily the name used by the islanders themselves. Another problem is that, while peaks and villages and bays may have names, islands as large as Lantao seldom do. My personal view is that the Portuguese and other Europeans got the name from Nantao people, whose name for the peaks which are a prominent feature of their daily view was Lantao (Cleft Head), a very understandable name. It seems unlikely that the name ever referred to the island, although the Europeans, with their different ideas of geography, made it so. It also seems unlikely that the inhabitants of Lantao ever called their mountains Lantao, because when you live there you are not very conscious of the cleavage referred to; it is only from a distance that the head is obviously cleft, and, supporting the suggestion made about Nantao, only from the north or the south.

The present Chinese name for the island is Tai Yu Shan, or Great Island Mountain, once again a reference to the peaks which are its most prominent feature. But Tai Yu Shan certainly is used in the general geographical sense known to Europeans. When referring to the high peak one therefore uses its special name, Fung Wong Shan. There is a definite incongruity about the name Lantao Island, since it is evidently a name chosen from without, rather as incongruous as it would be were the United Kingdom called White Cliff Land; at a glance we would know where the name came from.

Macao was founded in 1557, and though there are only occasional references to actual villages on Lantao Island (and of course many Portuguese records are still unexamined) it is safe to assume that Tai O slowly began to grow from this time onwards. In answer to the question whether it was a pirate place, I would say that the neighbouring island, Ladroes (Thieves), as the Portuguese called them, were so infested with pirates that Tai O could not have carried on any settled existence without being closely associated with some groups of pirates to protect them from others.

From 1601 to 1627 the Dutch made attempts to seize Macao. The final naval battle which decided the issue was fought in the waters adjacent to Lantao in 1627, when Thomas Vieira routed a Dutch fleet.

It is most unfortunate that the earliest clans resident at Silvermine Bay have lost their clan records, because there is an interesting historical coincidence concerning them which cannot in this event be examined. One of these families, the To family of Pak Ngan Hung, have soldier ancestors, and allowing a 25-year (rather short) generation period they came to the village about 1585, where, so they say, they were allowed to settle tax free on demobilisation. In 1564 a fairly large number of Chinese soldiers who had received no pay for a long time, deserted and turned pirate. They proved themselves the worst menace the coast had seen for years; it was a particularly bad time for Chinese and Japanese piracy. With nine powerful ships they terrorised the entire Pearl River area, and blocked all sea transport to and from Canton. The Kwangtung authorities organised as strenuous a resistance to them as they were capable of, but that did not amount to much; and the credit goes to the organising official, Yu Tai Yau, who asked the Portuguese at Macao to give assistance. The Portuguese sank at least one of the pirate junks at the first encounter, and so thoroughly disturbed the others by the power of their armament, and the dexterity with which they used it, that the rest fled. Where did they go to? They could hardly have gone back to the Chinese army after all this. And from what is known of the varying lengths of family generation periods it would be possible to identify at least some of them as among the founders of Pak Ngan Heung. Only twenty years separate the two dates, and allowing a 27-year generation period, which is in some respects a better average for Cantonese families, the dates tally almost exactly. It is important to note furthermore that the To family does not say they were officers. When a family has officers in its ancestry, whether they were civil or military, it always knows it. The To elder simply says they were soldiers; and it is believed that the other early families have the same derivation. All told, I am tempted to link the founding of Silvermine Bay with the Portuguese rout of Chinese pirate-deserters in 1564.

Sha Lo Wan appears to be the next oldest village, and the possible foundation of 1670 suggests that the original inhabitants may have been in some way connected with the famous coastal withdrawal of 1663–69. The place enjoys an unenviable reputation as a hiding place of bandits in the past.

As elsewhere in the District, Lantao was affected by the increased scale of migration in the later part of the eighteenth century. Tung Chung was particularly affected, and migration into the island has continued on a small scale right into this century. Most of the Buddhist settlements date from the fall of the Chinese Empire (1912). The recent influx of refugees into the Colony has only affected Silvermine Bay, and one or two very small settlements on the coast, consisting of no more than a handful of people, with the one exception of Nim Shu Wan, opposite Ping Chau, where immigration has been on a larger scale.

Lantao is a good deal less remote than would at first appear; it is more accessible, for example, than the eastern parts of the Sai Kung region, and the Government's post-war interest in the New Territories made an early start here, new village schools being the first sign of this interest.

The network of rural schools required on the island is now almost complete, and provided the present standard of education can be maintained, and provided medical facilities can be slightly improved, the most important benefit which the Government can bring is the construction of some roads. The road most urgently needed is that already suggested as a Colonial Development and Welfare scheme, along the south coast. The villagers on this coast depend, for connection with a market, on a motor sampan (occasionally interfered with by the Police for overloading) and rowed sampans, operating between Lantao and Cheung Chau. Due to the rollers which wash this coast, it is at all times dangerous and sometimes impossible to reach certain villages, in particular Tong Fuk and Tai Long, except in the calmest weather. What is really required is a road from the ferry pier at Silvermine Bay, over the hills and down into the Pui O valley, thence along the coast to Cheung Sha, Tong Fuk, Shui Hau, and into the Shek Pik valley, climbing in zigzag over into the Keung Shan valley, and thence down to Tai O. The road so far suggested to the Secretariat is the first section of this, from Silvermine Bay to Shek Pik. Another road is required from Silvermine Bay to Pak Mong and Tung Chung, and a third from Silvermine Bay northwards, connecting with the Trappist monastery, the Lantao Development Company's forest, and the village of Tai Pak. All these roads are practical possibilities, and with persistence it should be possible to build them all within the next twelve years or so. Ferry services between Hong Kong and Lantao would tend to be concentrated on Silvermine Bay; buses and lorries would operate on Lantao; and the development of at least the eastern side of the island as a dormitory for Hong Kong with people commuting as they already do from Cheung Chau, may confidently be expected.

LANTAO SOUTHWEST

I O *(Yi O San Tsuen)* (202915) (Austin Coates)

Population 74, including 14 of recent arrival. Cantonese. Surnames: Kung, Laam, Cheng, Leung, Pun, Kwok, Tang, Ip, Tai. This was formerly a very much larger village than it is today. At the end of the last century, the villagers say, the population was about 600, and there is abundant evidence of former houses, steps, paths, and cultivation on a wider scale than at present. About 1900 there was a plague here, in which many died and hundreds fled elsewhere, never to return.

The village suffered a good deal under the Japanese, 4 men being executed for assisting anti-Japanese guerrillas. The place is slowly getting on its feet again, and I note considerable improvement in the six years I have known it. In 1945 nearly the whole population moved from the old village site, near the wreckage of the former town, down to the present site overlooking Tai O bay. The move was made for better fung shui. The oldest families are Kung and Cheng; a man of Cheng of the 10th generation is 57, giving a foundation date c. 1700, which accords with what I have mentioned about Tai O, growth dating from c. 1685.

64 taochungs of rice land, and a great many fields have fallen into neglect; a few pigs, poultry and vegetables. Kadoorie Agricultural Aid Association assistance is needed here. 3 small fishing boats, operated by the Chengs, and 2 ferry sampans.

A road is needed between Tai O and I O, running on if possible to Fanlao. This is being included in the district road programme. The village is very anxious to get a road, and has mentioned it to me on several occasions. A small bridge is required over the outflow from the waterfall, coming down northeast of the village from Mang Cheung Po.

Yi O (*Gazetteer* p. 72) (James Hayes)

The remaining native population of Cantonese farmers is now about 60 persons, but had been much higher, they thought about 300, pre-war. The depopulation was due to endemic malaria, and deaths from malnutrition during the Japanese wartime occupation. They had moved down from the old village higher up the valley in the early post-war years.

We visited the Kung family's partly ruined ancestral hall in the old village. The altar carried ancestor tablets, placed in successive rows down to the 6th and 7th generation. All were of ancestors born in the Ch'ing dynasty, with none after.

The present population of 13 families includes 7 men and 7 women who are too old to work, and 12 children of school age. It is not clear from my note how many children were attending school, or where, but interestingly the attendees included two girls of 10 and 7 years old. There are three widows, aged in their 40s–50s, and all have paddy fields left by their husbands.

There is clearly far more land available than there are hands to work it. They exchange their own rice for cheaper quality imported rice. They said that moneyed people would pay $60–70 a picul for locally grown rice, whereas they bought imported rice at $30. They grow vegetables on an eighth of their cultivated fields, mainly watermelon in the spring, which they sell to Hong Kong.

In their new location, eleven of the 13 families rear pigs in flimsy pigsties, but only one family has a sow at present. The families own 20 water buffaloes, formerly used in the fields. They are sold for meat when they get old.

My visit was made with Mr Gus Borgeest, in connection with a possible rural resettlement scheme. He has been involved in several such projects, the best known being on Sunshine Island (the local name is Chau Kung) near Peng Chau. He has the blessing of Mr Barnett, the District Commissioner.

There are already four newcomer families living across from the new village. Relations between the two groups were reported as being good, though not close, and with no common business or exchanges.

Mang Cheung Po *(Man Cheung Po)* (220917) (Austin Coates)

A small settlement high up in the hills, 40 minutes hard walking from Leung Uk. There are a few farmers; the principal dwelling is a Buddhist nunnery partly rebuilt in 1953. It was severely damaged by the Japanese in an engagement with guerrillas who were hiding there, and subsequently fell into decay. Population about 20.

Fan Lau (Walter Schofield)

Fortifications of a similar kind can be found on the other islands of the Southern District. Just above the village of Shek Sun at the west end of Lantau stands a Dutch fort built about 1610, rectangular in plan. A few cannon balls and other relics have been found in it, but it is very overgrown and needs clearing if any research is to be done there, or sightseers enabled to visit it. The old fort and cannon protecting the small yamen were repaired when E. W. Hamilton was District Officer, I think between 1927 and 1929: I remember that one room in the yamen was inscribed *shu shat* (library).

Fan Lau (202882) (Austin Coates)

Population 103, including one at the Seria oilfields and another on Nauru. 24 families. Surnames: Laam, Ho, Chan, who are Cantonese and were the earliest arrivals. Their first place of residence was the village of Shek Sun (204882), situated slightly east of Fan Lau. The founder family is Chan, but their records are lost. The Ho came soon after them, and a man of Ho aged 76 is of the 10th generation, giving a date of arrival c. 1680. The present village of Fan Lau, on the west coast of the small peninsula, was founded by Hakka immigrants around the time of the lease of the New Territories. The Hakka surnames are Leung and Ng. I am inclined to think that the Hakka are slowly ousting the Cantonese from control of this place.

On a small hill just south of the two villages is a disused fort which according to the San-On Topography, was built in 1817. This was two years after the first real, but unsustained, effort of the Kwangtung government to suppress the opium trade, and the fort was probably part of a general design to control foreign and pirate shipping in these waters. The date is subsequent to the disorders caused by the pirate Cheng Po Tsai, and to Anglo-American hostilities 1812–15. It has been fashionable in recent years to describe this as a Portuguese or Dutch fort, but there is no evidence whatever to support this, and from investigation I would say that either claim is nonsense. Local opinion and tradition support this. All the claims to foreign origin have been made by foreigners. The fort was used until 1899. It was then used as a village, and traces of domestic occupation still survive. In 1951 the villagers, at Mr Morrison's request, kindly burnt out the vegetation within the fort so that we could make a proper examination of the interior.

There is a small temple on the southwest coast, but no festival is held there. Fan Lau participates in Tai O's annual theatre show.

Apart from growing rice and sweet potatoes, the villagers own 4 stake-nets and 9 boats. There is a small business in breaming *(burning grass to remove marine deposits from the hulls of boats)*, and pigs are sold to Tai O.

The school is situated in the ancestral hall, and the village has many times requested the construction of a new school. As the hall will soon need expensive repairs, it will probably be cheaper for the Government to agree to build a school, and yet another approach to the Education Department is being made.

Fan Lau (Shek Sun) (*Gazetteer* p. 72) (James Hayes)

Though I have a clear remembrance of my visit to the village, its Tin Hau temple (an altar table bore the date 1784) and the Ch'ing dynasty fort on a hill above it, my notes on this village seem to have long gone missing.

Gazetteer has: "The Fan Lau or Shek Sun peninsula is a small inhabited promontory to the far SW of Lantao island. Population of the peninsula: 110, with the surnames Ho and Ng predominant (Hakka and Cantonese)."

Fan Kwai Tong (Foreign Devil's Pool) (212933) (Austin Coates)

A small village included with Tai O, on the south side of the bay. It was the earlier home of the Leungs of Leung Uk. The village representative of Yi O informed me that it got its present name from the fact that it was the camp of the British military forces that took formal occupation of this part of the New Territories in 1899.

SHEK PIK AND VALLEY

Shek Pik (Eric Hamilton)

I recall an amusing memory of my youth. Paddy May — a proper Prussian Governor — sent for me to see him when I was Assistant District Officer South. He began, "Hamilton, I am worried about the afforestation round Shek Pik Wei." By the grace of God I had been there the week before. The south knob of Lantao below the Tai O line was absolutely placid and quiet at the time and there was no call for an Assistant District Officer to butt in. Our theory was to be like the Almighty: revered but not seen except under urgent circumstances.

In my time there were only tracks, a footpath was a rarity, and approaches to villages were most intricate with the idea of baffling pirate raids. The route from the landing beach on south Lantao to Shek Pik simply vanished in one spot.

Shek Pik Wai (S. H. Peplow)

Shek Pik Wai. Shek — stone, Pik — a wall, defensive works, Wai — to encircle. A village surrounded by stone defensive works, a walled city.

Shek Pik (Walter Schofield)

On the south coast, near Shek Pik there is a very ancient rock carving on a cliff that was found quite recently.

Shek Pik (258920) (Austin Coates)

Population 179. Cantonese. Surnames: Fung, Cheung, Wong, Chan, Tsui, Chi. These people speak Cantonese with an extraordinary accent, unlike any other village in Southern District and very hard to understand. It is difficult to find out the age of this village. The Fung family, we know from Fan Pui, came to Shek Pik at about the same time as the Fungs of Fan Pui; they were close relatives, loosely described as brothers. The Cheung family, however, claim to have founded Shek Pik. None of the families have any clan records, although an examination of the relationship between the Fungs of the two villages may indirectly reveal more historical facts. It seems possible that at Shek Pik we are really in touch with "original earth" of the pre-Chinese period.

Shek Pik and Fan Pui are among the very few villages in the District which grow enough rice to keep themselves alive without outside connections, should it ever be necessary for them to do so. They have about 300 taochungs for 179 mouths. In addition, they have 7 stake-nets, each with a boat, and 11 sampans (one man). The valley has splendid water supplies, and in the driest season has abundance. Shek Pik owns one of the finest herds of cows in the New Territories and is a prosperous place, judged by proper rural standards. That is to say, by comparison with the flashy people of Mang Kung Uk the people of Shek Pik are miserable, but, unlike Mang Kung Uk, Shek Pik's economy is balanced and almost unaffected by changes in the outside work. The village sells pigs to Tai O, but gets its cheap rice and most of its supplies from Cheung Chau, from which there is a fairly regular village ferry service run by Shui Hau people for the joint benefit of Shui Hau, Shek Pik and Tong Fuk.

I formerly thought that Shek Pik would be a good place for a social survey, and mention will be found in the Secretariat files on this. I now withdraw that view. Shek Pik is not a representative New Territories village. There are 2 men working as coolies in Hong Kong, and another in Tai O. There are about 40 able-bodied men who are available for outside work, but only at strictly limited times of year, due to farming and fishing seasons.

The Village Representative is deeply religious, again in a primitive way, and is of influence on the rest of the village. The immediate cause of this interest in religion, which manifests itself chiefly in the care of the temple and the performance of all proper rites, is an outbreak of plague or some other disease c. 1928, when about 70 people either died or fled from Shek Pik. A search through old reports, etc., reveals nothing on this. The Village Representative told me how, being then about 24 years old and so weak with illness, that he was scarcely able to move, he dragged himself into the Temple in the old village and lay there, unable to move any further, praying for the safety of himself, his children (he had two sons at that time and now has about nine), and the village.

When many families were spared he regarded it as an act of Heaven for which he and the village should show everlasting gratitude. Although there are probably other and much older reasons for the religious spirit of the village, this goes far to explain the phenomenon of never passing through the valley without hearing temple drums and bells, and never enquiring for the Village Representative unexpectedly without being told he has gone to the temple, or is looking after the temple affairs.

After the plague, the village moved southwards to its present site; the old village was allowed to fall into decay, and only the temple in it, which is reputed to be 300 years old (and is probably a good deal older than that), is maintained. It was repaired this year at the village's sole expense. It is important that the Government should bear in mind, in its dealing with Shek Pik, this religious element. The area within the walls of the old village is used for grazing cattle. The new village is not walled.

There is a Government-subsidised school with its own building, erected in 1950. This year, with the cooperation of the Sappers and Miners, an excellent bridge has been built over the big stream that runs across the main track from Shek Pik to Tai O. Fan Pui is connected almost entirely with Tai O. Although the Fan Pui people are invited to weddings, etc., at Shek Pik, it is clear that in daily life the two villages agree to differ; the exact reasons for this need further enquiry. For reference to prehistoric sites, see under Fan Pui.

Shek Pik (*Gazetteer* p. 73) (James Hayes)

Population 200. Cantonese. Village Representative is Tsui Mun-hei.

This village, together with the smaller village of Fan Pui and the hamlet of Hang Tsai, are likely to be affected if the current engineering investigations into the feasibility of building a large reservoir in the Shek Pik Valley are successful.

The entries copied here from the notebooks are dated in early 1958, and are connected with the reservoir soon to be approved and built there. They are included because of their relevance for our work there at this critical early period. Any reference to "us" means the small team engaged on these duties, including my land bailiff, John Abbas, several equally trusty demarcators, and myself.

Tsui Kan-shing with Tsui Kan-hing, "a good example of how expensive registering a succession to property can be".[1]

There are over ten outstanding disputes over the ownership of land. These will have to be determined by holding land courts. Ask Mr Lo to make the necessary arrangements.

Must get onto the Department of Agriculture, Forestry and Fisheries to get them to update their survey of all agricultural properties and livestock, and especially the valuations, in case they are needed in a hurry.

One of the men of the Fung lineage of Fan Pui is working for Messrs Dragages, or rather for their contractor. The other day, he refused to remove stones from a path. Upon enquiry, this turned out to be because there are fung shui sensitivities in this particular spot. The contractor had not kept the Village Representative informed of where and what was to be done. Speak to Mr Dempsey of Dragages, asking that this be done in future for any work in areas round the village.

The extension of the South Lantao Road to Shek Pik would take another 18 months to complete, and there was no ferry service to and from Shek Pik. The villagers saw that the contractor's boat was coming and going regularly between the village and Hong Kong, and some of them ("they were quite friendly") had asked if I could request places on the boat for a few persons. They suggested once a week for up to 10–12 persons, with tickets to be given to the Village Representative for distribution. As their proposals were quite sensible, I promised to discuss the possibility with Mr Dempsey. They also asked me about getting work with the contractor. Without telling them, the note shows that I had wondered whether the contractor could, or should, contact the newly formed "council of elders" for names if he was willing to take on any village men.

There are some other "thinking aloud" jottings, including one on the various ways of recovering private land from the villagers. This was against the likelihood that the still continuing engineering investigations would prove successful, and require us to recover all the land holdings in the valley, as would prove soon to be the case. Looking back, I am not impressed![2]

There is a note on a discussion with the Village Representative on 30 January 1958, about their resettlement arrangements if they did have to move out. Among the topics raised with us were possible resite areas (Tsuen Wan was mentioned, as earlier attempts to find agricultural land on Lantao for a resite village and fields had not been promising), compensation for land and houses ("not enough"), re-housing (they wanted to visit Tsuen Wan to see the type of apartment blocks being built in the town), size of replacement flats, title deeds to be given upon completion of the new buildings, a subsistence payment to assist their livelihood following their removal ("three years' crop compensation" was requested), help with finding employment, a burial ground for the resited villagers, payment for a propitiation ritual before they leave the valley, and removal expenses for moving family and lineage graves. There was also a request for the apartments allocated to the different lineages to be grouped together.

On another page, I had jotted down the Village Representative's further proposals for a removal fee for each household of $500 each for the first and second members, with $200 each for additional persons; for exemption from payment of electricity and water charges, as well as rates ("we don't have to pay them here"); together with various points that had occurred to them in regard to the design and height of the new accommodation ("a storeroom for every two houses", and for some unknown reason, the buildings to be "no higher than four storeys").

Looking back, it is interesting to see how, at this early stage, and before the Government's decision to proceed with construction of the reservoir, some villagers were already putting forward ideas as to what would be required if they were asked to move out. Clearly, too, they had been in touch with the ex-Tai Lam villagers, removed to Tsuen Wan only two years before, from the site of the new Tai Lam Chung reservoir. Whilst indicating that, in principle, we would look favourably on some these points, I had asked the Village Representative to send in a letter, to help formalize their proposals for further consideration and discussion. A lot of water would flow under the bridge before the arrangements could be finalized, but this was surely an encouraging start. They were at least beginning to think about the possibility of removal and resiting, and discussing it among themselves.

There is also a list of villagers. Since it includes the two Village Representatives and men from each of the lineages it is most likely to be the "committee of elders" which the villagers established to vet the resettlement arrangements being made on behalf of the village. I never met its members formally, but its existence was made known to us, and it would be a useful forum for the villagers to resolve any differences of opinion inside the village. In practice, we worked through the Village Representatives, save when (as happened from time to time) they were out of favour and had temporarily lost their influence, in which case (and depending on urgency) we had either to hold a meeting with the whole village or bide our time.

Kong Pui (261919) (Austin Coates)

A subsidiary hamlet of Shek Pik. Cantonese. Surnames: Chi, Ho (2 families).

Hang Tsai (261916) (Austin Coates)

A subsidiary hamlet of Shek Pik, included in that population. Cantonese. Surname: Tsui (3 families).

Hang Tsai (James Hayes)

Hang Tsai's population is included with Shek Pik. Its three families of Tsui are of the same lineage as the Tsui lineage of the big village.[3]

Fan Pui (250914) (Austin Coates)

Population 56. Cantonese. Surname: Fung. This is one of the oldest families on Lantao Island, and one of the very few which still possesses a clan record. The Fungs are from Toishan district, Kwangtung. From there they moved to Tsun Wan, thence to Cheung Sha Wan, and thence to Ma Tau Wei. The present head of the village is 45, of the 21st generation at Fan Pui. He insists that a 30-year generation should be allowed, which gives a foundation date c. 1310. By the 27-year period, the date would be 1370. The Village Representative is so insistent about the 30-year period, however, that it may be that the assessment by this method is correct. The Fung family also peopled Shek Pik, and spread to Shui Hau and to Sun Tsuen in the Tai O area.

Fan Pui is a poor but hard-working village; it receives no remittances and is inclined to scoff at the people of Shek Pik as being people who do not really work for their living. In 1949 the village performed excellent work in conveying to the Police information about bandit gangs operating in the area; the information led to the arrest of a dangerous and troublesome group, and the Commissioner of Police considered a reward to the village to the extent of $2,000. Fan Pui wanted to erect a watch-tower. Negotiations between the village, the Police and this Administration were allowed to peter out in 1950; the Police were losing interest, and in this Office the matter appears to have been one which was not brought up, as it should have been, when Mr Tsui succeeded Mr Wakefield as District Officer. A recent move by the Police to limit the numbers of rifles and ammunition held by the village guards on Lantao led to a complaint from Fan Pui to me, and from this the reward, which the village is still expecting, was unearthed. It is now being dealt with.

The fields owned by the village are excellent, the irrigation good. With farming rice and fishing (there are 5 sampans) the people are doing about as much as they can. 2 wells need repairs.

Near the part of the shore where one lands to reach Shek Pik and Fan Pui is a small temple, of which the bell is dated Ch'ien Lung 30th year (1764). On the beach in front of Fan Pui is the prehistoric trading site archaeologically examined by Father Finn and Stephen Balfour before the war. On various rocks at the seaward end of the valley are carvings of bold conventional patterns, similar in strength to the best primitive designs in other parts of the world. There is in addition a tradition, which I first learnt about in Tai O, that foreigners used to come into this bay to obtain water and fresh food. I have not yet traced any confirmation of this from a Western source; it is possible that the English, and probably also the Dutch, watered at villages in this area while attempting to trade with Canton by bypassing Portuguese Macao, but in those times the northern side of Lantao Island was more frequented than the south. It may be that the tradition is more recent: when Lintin became the main opium anchorage (1821), and when British men-of-war began regular patrols at the mouth of the Pearl River, from 1796 onwards.

Fan Pui (*Gazetteer* p. 73) (James Hayes)

Population 59. Cantonese. Village Representative is Fung Ping-yau.

Tei Tong Tsai (*Ti Tong*) (287951) (Austin Coates)

Population 26, a Buddhist settlement situated about 1,000 feet above sea level on the main track between Tung Chung and Ngong Ping. The principal foundation is Faat Lam, started in 1929. Mr Ngan Sai Leung (the Silent Yen), a Hong Kong businessman and one of the more intellectual Buddhists in the community, has a small retreat here.

TAI O AND VALLEY

Tai O (Eric Hamilton)

I was drawn into the case of the murder of Sergeant Glendenning in Tai O police station in 1917. You will find a scathing letter in the *South China Morning Post* from F. B. L. Bowley ticking off the Government with its Carsonian Governor and its dilettante and poetical Assistant District Officer South (me!). Actually the Indian policeman had been sent to goal for larceny in the police station. God knows why he was sent back under arrest to collect his belongings, but after sleeping the night there, he got up, took a loaded rifle from the arms rack, shot Glendenning dead as he sat at his desk writing his morning report, went upstairs and raped his wife, set the police station alight, returned to bed, pulled the mosquito curtain and shot himself. By God's mercy a police launch came in just as the station started to burn and they rescued the wife.

Do you remember the Decameron story of the gardener with the nunnery who pretended to be deaf and dumb? I had a case that was very much akin to this at Tai O. He had been pinching their blankets and pawning them for money to gamble with in the multitudinous gambling shops of Tai O.

Tai O (S. H. Peplow)

Tai O. Tai — big, O — a cove. The big bay village. Probably one of the quaintest villages in the New Territory. It possesses a fairly large harbour, and during the Wong Fa (yellow spawn fish) season, lasting from the end of October to the middle of December, is one of the most important fishing centres in the Colony. The place presents a wonderful sight during this time. Hundreds of large junks from Hong Kong, Canton and Macau are packed closely together filling the harbour.

At break of day the listeners set forth in their small flat bottom boats and are eagerly watched by the fishermen.

There is really only one street in the village, the majority of the inhabitants living in mat-sheds built upon bamboo poles in the shallow part of the harbour. The support poles of the sheds are about eight feet high, and each shed has a short ladder reaching down to the water. The sheds are not erected haphazardly, but are laid out in rows forming streets or creeks, each street or passage way bearing its own name such as Main, 1st, 2nd or 3rd creek.

Small sampans are used by the mat-shed dwellers to go to and from the shore for their market produce. In addition, sampans fitted up as small shops selling dried vegetables, salt fish, matches, etc., are used by hawkers, and these row around thus saving the inhabitants from making unnecessary journeys ashore.

At a rough guess the population of Tai O would be about 4,000 and at least three-fourths of these live in the sheds, the remainder occupying the houses ashore.

Stone Figure at Tai O (S. H. Peplow)

At the base of the small hill upon which Tai O Police Station has been erected, there is a curious rock formation.

Viewed from the sea, there appears to be simply a collection of rocks jumbled together indiscriminately, but as one rounds the point, the stones form themselves into the shape of a human being, the feet rest on the seashore, and the body appears to be resting against the hillside.

The story is as follows. Years ago, long before British occupation, the island of Lantao was ruled by a petty chief or magistrate, named Kwan who according to the Chinese was an evil living and licentious man. He kept the villagers, especially the womenfolk, in such a state of terror that they were afraid to walk about after dusk. After about a year of utter misery, during which the women and girls were openly carried off and ill treated by this monster, the villagers decided to kill him. A meeting was called and arms of all descriptions were handed out to those present. That night when Kwan made his usual visit, the villagers with ferocious yells rushed upon him. He, seeing that the odds were all against him, turned and ran along the path and so gained the hill upon which the present Police Station stands. Here, further escape was impossible owing to the sea. So facing around he determined to sell his life dearly, but was totally outnumbered and beaten to a state of insensibility. The villagers then locked him up and dashed him down the hillside. By the time his body reached the base, the whole of his clothing had been torn off, and he lay naked in the position he is today.

After he had been thrown down the hillside the priests were called; these prayed to the gods that the body be turned into stone, and it was so.

After this, Tai O returned to its former tranquillity. A short time after the above events had taken place, it was noticed that the boys and girls living in a small village called Tong Ka Tsun, near Macau, were not deporting themselves properly, in accordance with Chinese customs. A meeting of the Elders was called, and it was decided that some evil influence was at work. The priests were consulted with a view to finding out the cause. The whole island was searched, but without result; so it was decided to explore the surrounding islands. A boat was obtained and the priests set off. After many searches they came to Tai O, and the stone figure of Kwan was discovered lying naked against the hillside facing Tong Ka Tsun. Here undoubtedly was the cause of all the trouble. It was there and then decided to destroy the figure, but unfortunately the tools in their possession were inadequate. So after a consultation they decided to mutilate the stone body by removing the male organ. This was done, and the party returned home, and informed the Elders of what had taken place. A great feast was held, the priests rewarded, and the gods thanked for their kindness and sagacity, and the village settled down to live, as the story book says, "happily ever after".

Tai O Piracy (S. H. Peplow)

On 25 March 1925, a gang of about 60 pirates raided the village of Tai O, Lantao Island. They murdered a woman, kidnapped two men, and robbed most of the houses in the village. The total loot was estimated at $28,000.

Some of the robbers were posted as piquets to guard the path leading to the Police Station. The attack was a complete surprise, and owing to the posting of piquets the villagers were unable to raise the alarm, or inform the police who were in entire ignorance of the raid until after the departure of the robbers. The Police Station is about a mile away from the village situated on a hill and can only be approached from the village by one path. The woman was shot when trying to give the alarm. It is suspected the pirates were ex-soldiers or bandits from Heung Shan or Shun Tak district.

Tai O (Walter Schofield)

Talking of New Territory police station siting, the Tai O station was originally to have been built close to the village, but the local elders put up representations against it and the presence of mosquitoes in the village may have provided an argument for its present siting beyond Shek Tsai Po. Silting of the harbour may also have influenced the Government. But I have heard that what influenced the villagers was the existence of gambling houses which yielded them a good profit, and they knew that with the police among them the hope of their gains would be gone. In 1925 they had their reward. A boatload of 60 pirates from the Delta landed at Po Chu Tam, marched along the creek-side road and plundered the village, murdering a woman and kidnapping two men. They got away without interference. Government promptly "locked the stable door" by stationing an armed Indian police guard, later replaced by village scouts, in a mat-shed close to the mouth of Po Chu Tam creek for several months, about 50 yards from the site of an old Chinese stone-built guard station dating from the era of Japanese piracy in South China. Apparently the Police knew nothing of the raid till all was over. I think all that happened was that the sergeant in charge was transferred to another station.

When I first took charge of the District Office, the "black gold" rush had been over for three years. The bottom having dropped out of the tungsten market with the coming of peace, but the lime-burning and sand-digging boom was in full swing because of the road-making and building then going on in Hong Kong and Kowloon. (These were times of anarchy in China.)

Thus I had to deal with one or two applications for land for lime kilns. These kilns were thickest on Ping Chau; but Nei Kwu Chau and Tsing Yi also had kilns, and another was put up at Hang Hau. This distribution is due partly to nearness to the market in Hong Kong, partly to the presence of coral in the shallow and then comparatively clean waters of the western approaches.

During my periods in office I made an attempt to get the Chinese communities and villages owning forest lots to look after them and to plant trees. Free seed was distributed and planting instructions given, and a forest guard appointed to supervise and watch results. The difficulties of forest conservation in such scattered and isolated areas were certainly formidable: one was that the boat people could land almost anywhere and steal trees; another, that the grass-cutters who annually collect fuel in autumn are quite likely to cut and take young seedlings; to say nothing of disease and caterpillar infestation, often very serious. One bad case was at Tai O where an entire hillside was laid bare at one swoop by its licensee instead of being cut in stages, and I told him to get it replanted. I don't remember the sequel, as I was transferred not long after. The denuded hillside faced west, and lay across the Po Chu Tam creek from Tai O market. Another great difficulty was to find forest guards who would do their job, a former A. District Officer North once minuted, "where forest guards abound, there do abuses much more abound!"

Another relic of old coast defences, close to Tai O, is the old Chinese guard station already referred to, outside Po Chu Tam creek, and quite ruined.

Tai O (Austin Coates)

Population 12,030, together with a floating population of approximately 2,000, the third largest town in Southern District. The town lies along the southern fringe of a hill which may at one time have been an island. The shallow bay in front of it was evidently reclaimed at a fairly early date, and now consists largely of salt pans. These pans are the largest in the colony and are at present owned by a single Hong

Kong businessman, who completed his monopoly by purchasing the last by public auction in 1954. Each pan is held on a 21-year lease, at the expiry of which the pan is auctioned. The town spreads across an important inlet, the main outflow of the large stream descending from Keung Shan and Ling Yan monastery.

Various people operate sampan ferry services between different parts of the inlet and town; the principal crossing place, in the centre of the town, is leased out by tender by the Tai O Residents' Association. The town is made up of three distinct sections: Tai O market, which occupies the central area and lies on both sides of the inlet; Shek Tsai Po (207944), a coastal village west of the market and facing the bay; and Po Chu Tam (215951), a smaller but possibly older village northeast of the market and having access to the coast on the north side. There is a ferry pier, constructed with Colonial Development and Welfare funds in 1952, situated at the head of the bay, about 20 minutes' walk from the town, and the Hongkong and Yaumati Ferry Company operates a daily service to Hong Kong (via Tung Chung, Castle Peak and Ma Wan), with an extra trip on Sundays and holidays and a special early morning service carrying fish to the Government Wholesale Fish Market at Aberdeen.

There is a Police Station situated on the hillside immediately overlooking the pier, the top of the same hill being occupied by a small observation station of the Royal Navy. There is a resident Government doctor, with nursing staff, working in rented premises in the middle of the town, and the Marine and Marketing Departments both have offices at Shek Tsai Po. There is a resident Health Inspector, with scavenging staff, and a small Fire Brigade unit equipped for land and water. There is a postal agent, but he does not handle parcels or registered mail, to obtain which a Tai O recipient is obliged to make a journey into Hong Kong. By leaving one's house in the centre of the town at 6.30 a.m. one can reach the General Post Office in Hong Kong at 10.40 and the District Office, Kowloon, at 11 a.m. (4½ hours).

Tai O is in a simple way fairly prosperous; it is noticeably cleaner than Cheung Chau and the much smaller Sai Kung. The main market is owned by the Government; the Residents' Association provide street lighting, and contribute to the costs of scavenging and the fire service. An electric light company was set up in 1954; this is the second venture of its kind, and looks like doing better than its pre-war predecessor. Radio Hong Kong is difficult to hear in Tai O, which in general relies on Macao for entertainment, and to a certain extent, Canton.

There are 7 primary schools, most of them run by various churches. There is a well attended Catholic church and Christian services of various kinds are held in other buildings. The Tai O Hotel, next to the market, is clean and well kept; the cost of an average room is $3 per night. There is a boarding house on the south side of the inlet. There are 2 restaurants and several tea houses.

Compared with Tsun Wan and Cheung Chau, Tai O has relatively few industries; it is chiefly important as a market town and rural shopping centre. There are boat building and repairing yards at She Tsai Po and a very large and profitable breaming beach controlled by a single family; shrimp paste is manufactured, and salt fish prepared by individual families. Formerly the salt pans were run at considerable profit by smuggling salt into China for sale at rates below the Government's monopoly price, but since the war the pans have been only partially exploited; they are at present run for the benefit of a few refugee clansmen of the owner, and the main reason why the latter completed his monopoly of the pans is said to be for luck; he survived the war in comparative comfort due to salt, and believes that as long as he owns the pans his luck will last.

To ascertain the age of Tai O would involve long research; from what has so far been done, it would appear the settlement started at Po Chu Tam at the end of the Sung dynasty; there is a stone tablet, the wording of which suggests this, in the Hau Wong Temple, from which it would appear also that the original settlers were refugees, Buddhists, and educated people. All told, it points to remnants of the Sung court at the downfall of the boy emperor Tai Ping (1279). It is said that after the final Sung-Mongol battle large numbers of the court fled to Lantao Island, where many of them committed suicide. From the evidence at Po Chu Tam it would appear that not all died in this way. It may, however, be that even this was not the first foundation at Tai O.

Inspector Smith of Tai O Police Station reported to me that the keeper of the Tin Hau Temple at Sun Tsuen (see below) claimed that his temple was older than the Hau Wong Temple at Po Chu Tam, which would point once again to the fishermen being the oldest inhabitants, with a probable land population of which no record survives, due chiefly, it may be, to illiteracy. Local tradition also says that Tung Chung is older than Tai O, a fact borne out by the age of families still existing at Tung Chung (see below). It appears that Tai O was a fishing village certainly from Southern Sung times. At the downfall of that dynasty it was peopled by a small number of refugees, probably persons of quality, who settled at Po Chu Tam. It remained a small place. It is not recorded that either Jorge Alvares (1513) or Perestrello (1514) visited it; their Chinese pilots took them straight to Nantao. Its prosperity probably dates from shortly after 1685, when K'ang Hsi opened the ports of China to non-tribute-paying foreigners. The bells in the two temples, the oldest dated objects, date from K'ang Hsi 32nd year (1692) at Sun Tsuen, and the same reign 38th year (1698) at Po Chu Tam.

Tai O's most urgent needs today are a post office and repairs to the salt pan sea-walls which, if left unrepaired, will allow flooding of half the town should there be a westerly typhoon. The Government proposes to build a health centre shortly, and the Association's requests include the building of one central school, with at least three middle school classes, more ferry services, a shelter on the pier, and two public lavatories. Funds are already being collected for the middle school, and appropriate action is being taken on the other matters. A bathhouse is also desirable.

One unusual feature of the population is the large numbers of fishing people who live in wooden houses built on stilts over the inlet. These are included in the land-based population figure, and this section of the people lives by fishing and collecting grass for sale to the boat breamers. Overseas remittances account for a fair amount of the town's prosperity.

Po Chu Tam (S. H. Peplow)

Po Chu Tam. Po — valuable, Chu — a pearl, Tam — a pool or lake. The story is that many years ago a valuable pearl was lost, and in time changed into a lake (this is a common supposition regarding the pearl). According to the Chinese, the moon shining at night time on to the water, turns the lake into the shape of a massive pearl.

Shek Tsai Po (S. H. Peplow)

Shek Tsai Po. Shek Tsai — small stones, Po — a mart. A village where the foreshore is composed of small stones, no sand.

Yim Tim *(Yim Tin)* (S. H. Peplow)

Yim Tim. Yim — salt, Tim — fields. A village where salt is obtained from sea water by evaporation, such as the village near Tai O.

Hang Mei (224943) (Austin Coates)

Incorrectly described on maps as Hong Mi, the source of Tai O's present water supply. A small village included with Tai O. There is a monastery here of the combined faith of Confucius, Lao Tse and Buddha. In the hills behind Hang Mei is the town's main cemetery.

Leung Uk (217937) (Austin Coates)

Population 96. Surnames: Leung, Chow, Man, Sin. 3 men overseas, one in New York, one in San Francisco, one at Seria oilfields. Vegetables, pigs, grass-cutting for sale to Shek Tsai Po for breaming. It would be desirable to run a small road round past this village from Tai O to Yi O, and if Tai O's water supplies can be improved from the waterfall near Yi O, a road could be combined with this work.

Sun Tsuen (220939) (Austin Coates)

Population 28. Surnames: Fung and Kwok. The Fungs are related to those of Fan Pui; and thus one of the oldest families on Lantao Island. The village is called Leung Uk Sun Tsuen, but it seems to me that it is in fact much older than Leung Uk. The Tin Hau Temple may be the oldest foundation in the region (see under Tai O) and is situated in a very fine position facing midway between the two arms of the bay. It was evidently once on the seashore, before the salt pans were reclaimed. Some newcomers, all Hakka.

Luk Chau (*Gazetteer* p. 100) (James Hayes)

(The editor has not been able to locate this village on the map.)

80 persons, all of the Yiu lineage. The Village Representative is Yiu Yau-kuen.

Have only 7–8 taochung of rice fields. Don't grow vegetables, only sweet potatoes.

They have 23 sows, and currently 210 pigs altogether. Pigs sell at $160 per picul at present.

All pigs are sold at Sai Wan (Kennedy Town). They are taken first to Aberdeen by sampan (no motor junk is available) and then to Kennedy Town by lorry.

It transpired that none of their pigs are vaccinated, and I will pass a request to Agriculture and Fisheries Department on the subject.

They kept some buffaloes pre-war, but these (they claim) were taken by Japanese soldiers, and they have never kept any thereafter.

There is no village forestry lot, but there are a few private orchards.

There are 20 children of school age, of whom 14 go to school at Sok Kwu Wan. However, it is a difficult journey on foot, which takes 40 minutes, and the younger children are not sent.

Two men are at sea.

Pre-war, one of the Yius was a big contractor in Yaumati, Kowloon. The firm was called "Yiu Fu Kei". He and a brother managed the business. He has since died.

Keung Shan (S. H. Peplow)

Keung Shan. Keung — ginger, Shan — hill. A village where ginger is grown on the surrounding hill sides.

Keung Shan (235933) (Austin Coates)

Population 106. Surnames: Wan (10 families), Cheung, Leung, Luk, Cheng, Mak. The Wans are old settlers and Hakka. The rest are recent arrivals, Cantonese. 166 taochungs of well-watered hill rice land. Considerable trouble has been spent in the past on these terraces, which are unusually high and made of stone. 1 villager abroad (Borneo), and another working as a restaurant foki in Tai O. The Wans have foolishly sold most of their fields to Ling Yan Monastery, Chuk Yuen Nunnery, and other Buddhist institutions. They are now working as tenant farmers on what was once their ancestral land. The pigs need improving, and Kadoorie Agricultural Aid Association help might be given. The rent paid to Chuk Yuen is 8 taos of paddy per taochung. A dilapidated village. A bridge is required over a stream; assistance was promised in 1948, but at the last moment the village was unable to provide labour. The request has been made again.

Sheung Keung Shan (239935) (Austin Coates)

Population included with Keung Shan. Hakka. Surnames: Mo. The head of the two families occupying this village, which is comparatively high and well concealed from all directions, is deputy Village Representative of Keung Shan. The Mo family is of historical interest. They are very simple countrymen, barely literate. Their clan record has been lost; but the present head of the family, who is 39, saw the record before the war, and remembers the outline of it, which is that the Mo family moved from Wai Chow and settled at Shataukok 23 generations ago. They only stayed at Shataukok for one generation, after which they moved here. They are thus 22 generations at Keung Shan and are almost certainly the oldest Hakka family in the Southern District. Their date of arrival may therefore be calculated with more care than in other cases. From various evidence I am inclined to favour a generation period of 27, which puts their arrival at Shataukok c. 1346, and at Keung Shan c. 1373. If a normal 25-year period is used, the arrival at Keung Shan is c. 1390, and if the 30-year period, which many Chinese families say is the correct one, is used the Keung Shan date should be c. 1315. Their family tradition is that they were refugees from persecution, and this might explain why they chose as a place of settlement such a high and well-concealed spot. They originally earned their living making charcoal, selling it to villages on the coast. From the inscription in the temple at Tung Chung I am inclined to think that the charcoal was sold to Tung Chung and Santao (formerly Sai Chung), and that Keung Shan was at one time (at the beginning of the Ch'ing dynasty) a bigger place than it is now. It may also once have been the general name for the whole area, including Luk Wu and Ngong Ping.

Keung Shan (*Gazetteer* p. 73) (James Hayes)

There are two villages, Upper and Lower, inhabited by lineages of Mo and Wan respectively, with a combined population of 103 persons in 24 families. My note says, "Upper one-third, Lower two-thirds". Probably Hakka.

The names of the two village representatives are Mo Tin-loi and Wan Shang.

There are 160 taochung of paddy fields and 20 of vegetable land. There are 30 brown cattle, 15 sows, but not many porkers at present.

All the men are reported to be living at home, with none abroad.

There is no school at Keung Shan, but 12 children are attending classes (presumably in Tai O) out of over 20 of school age.

They had applied to Kadoorie Agricultural Aid Association through the Agriculture and Forestry Department for water tanks and a filter, and asked me to chase up Agriculture and Fisheries Department on their behalf. They would like our Local Public Works materials to improve a footpath and footbridge, the work to be carried out later in the year. The Upper Village wanted cement to repair a well. They also requested some post-registrations of birth.

Besides the villagers, there are a number of non-indigenous farmers who came into the area pre- and post-war. There are also the residents of the many, mostly small and pre-war, Buddhist or Taoist religious houses in the Keung Shan–Luk Wu area. I visited some of these during my first visit to the area, but took no details.

Gazetteer gives a total population of around 200.

Ling Yan Monastery (232937) (Austin Coates)

Founded in 1926, a popular place of pilgrimage 25 minutes' walk from Tai O and only a slight climb. It has come to be surrounded by other lesser monasteries and Buddhist retreats, the total population of which is 75.

LANTAO CENTRAL PLATEAU (Eric Hamilton)

There was a point about Lantao which was always at the back of my mind. The Lantao plateau. I blame myself very much that I never investigated this personally. It was, I confess, sheer idleness, though admittedly I should have had to take weekends off to go into camp to do it.

Missionaries had one or two hill stations up there and one or two nunneries crept up that way, one supported regularly by voluntary contributions from Hong Kong Chinese prostitutes who, when age dictated retirement, used to retire there for the evening of their days.

There was always a feeling that we should develop this plateau as a holiday hill station. Apart from the ministers the only people who went up there were people who wanted to shoot deer — and there were deer there — the same little blighters who ate my lettuce crop above Severn Road.

You had to land at Tung Chung and walk up the hill with all the stores carried by coolies.

Ngong Ping (Walter Schofield)

(The editor has not been able to locate this village on the map.)

One of the interesting communities on Lantau was the group of Buddhist temples and *chai tong* or fasting halls on the well-known high plateau between Tung Chung and Tai O figuring as "Ngong Ping" on the maps. It lay at about 800 ft. above sea level and its members maintained a good pathway from Tai O across a stream and up the hill to their settlement and ran their buildings, somewhat in the manner of vegetarian youth hostels. They occasionally harboured strange characters, as might be expected in unsettled and revolutionary times. One such, I believe, was a big-scale opium smuggler and den-keeper who had operated in London and was nicknamed "Brilliant Cheung". I think he got banished from the Colony. The track from Tai O to Tung Chung was a favourite walk for many people: I unfortunately never did it.

Ngong Ping (267949) (Austin Coates)

A Buddhist settlement situated on a well-watered plateau 1,600 feet above sea level, approximately 2 hours' walk from either Tai O or Tung Chung, lying in the shade of the gaunt 3,000-foot Lantao Peak, second highest mountain in the Colony. At the time of the lease of the New Territories, this plateau was uninhabited. In 1912 the first Buddhist monastery was erected, Po Ming monastery, and was followed in 1918 by the today much larger Po Lin monastery, which has grown to be one of the richest foundations in the Colony, due to its quite justified but unexpected appeal to tourists, among them many Europeans. As a haven of Buddhist learning it cannot be compared with Chuk Lam monastery, Tsun Wan, but as a place of holiday pilgrimage it is more rewarding, to those who can face the long uphill walk.

Since the foundation of Po Lin Monastery, many other Buddhist retreats have been started, until Ngong Ping has become numerically the largest Buddhist centre in the Colony. Since the war other devotees of different character have arrived, and properties are held by the University (still undeveloped after 8 years). Mr Brook Bernacchi, barrister and Member of the Urban Council, owns an 11-acre farm on which he keeps cattle, goats, a wide range of poultry, and a horse, and grows tea, in addition to many vegetables. Two other European families maintain smaller house-and-garden lots. Population 157. Po Lin Monastery owns a number of rice fields which its denizens cultivate.

Ngong Ping (James Hayes)

Ngong Ping is not a village, but is well known for its Po Lin Buddhist Monastery, established about 1911, and the many small religious houses built on this pleasant high plateau on the mountainous spine of Lantao.

LANTAO NORTH COAST: TAI O TO TUNG CHUNG

Sam Wat (*Sham Wat*) (243962) (Austin Coates)

Population 9 (3 families), a Hakka hamlet situated in a bay on the north coast of the island about 40 minutes' walk from Tai O.

Sham Wat (*Gazetteer* p. 72) (James Hayes)

Population 34, in seven families. Families of Wan, Chau, and Leung live in the old settlement in the upper valley, and have only been here for between four to ten years. The other families live behind the beach. None of the old families who used to live here remain, and their abandoned rice fields are now largely used for grazing cattle.[4]

There are seven children of school age, of whom one is sent to the Wing Chor Catholic School in Tai O.

There are 40 taochung of rice fields there, but these, I think, are private land, much of it uncultivated. They also raise pigs with, currently, two sows and nine porkers, as well as many chickens. They appear to be rice farmers, since they say they have seven buffaloes (this could well be brown cattle). They exchange what rice they grow at Tai O, where they buy all their other necessities.

Tai O is one hour away by rowing sampan, and 45 minutes on foot. It takes half an hour to walk to Shan Shek Wan, and less by sampan.

As *Gazetteer* relates (p. 72), these people were persuaded to unite with the equally new community at Shan Shek Wan to form the Sham Shek Village. I am sure the initiative came from the latter place.

Sai Tso Wan (238966) (Austin Coates)

Population 16, Hakka hamlet on the west side of the Sam Wat bay. Surname: Ng. Date of settlement c. 1900. The family came from Shum Chun. They have a splendid forestry lot, but need assistance in improving water supplies. 2 stake-nets, each with a small sampan.

Nam Tin (244967) (Austin Coates)

Population 6 (2 huts), a Hakka hamlet on the northeast side of Sam Wat bay. There is a deceptive sandbank at the mouth of the bay, and it is wiser to anchor outside when visiting.

San Shek Wan (252978) (Austin Coates)

Population 12. Surname: Wong (Hakka), Tam (Cantonese). It is evident that this village formerly got water from a source high up in the hills north of Ngong Ping; the dam and open channel are marked on the 1:20,000 maps, but an investigation shows that due to a bad leak in the channel no water passes more than 20 feet from the dam. This is being brought to the attention of the Irrigation Engineer. The source never dries up.

San Shek Wan (*Gazetteer* p. 72) (James Hayes)

Population 47, in 12 households, rather than families as there are only eight women with 13 children. They are of the most diverse origin (see below), as the men had come to the area originally as workers in the wolfram mines in 1951. When this finished two years ago, they turned their attention to full-time farming of vegetables here, on 30 taochung of vegetable land.

They sell their produce at the wholesale vegetable market at Sai Wan on Hong Kong Island, using their own motor junk bought specially for that purpose. The charge for taking a big basket to market is $1.50.

They are also pig-breeders. They buy their rice and other necessities at Tai O.

They comprise Cantonese, Hakka, Hoklo, and one Mandarin speaker, and originate, by single households, from the following counties: The Cantonese are from Wan Fau (I'm not sure where this is in the province), Ko Yiu, San Wui, Chung Shan, and Kwong Ning. The Hakkas are from Ng Wah, Tsing Yuen, and Wai Chow. The Hoklos are from Kit Yeung (two households) and Hoi Fung. The sole Mandarin speaker is from Yunnan Province. They are a remarkably go-ahead lot who owe their cohesion to their unofficial Village Representative, whose name I forget, but whose energy and leadership were exceptional.

Sha Lo Wan (S. H. Peplow)

Sha Lo Wan. Sha — sand, Lo — a spiral univalve shell, Wan — a bay. A village near a bay where these shells are plentiful.

Sha Lo Wan (Walter Schofield)

One crime that often came before my court in the office was stealing sand for building. Sand collecting was regulated by a system of permits, allowing junk masters to collect sand at selected beaches, each junk having its own collecting beach. Sand shortage was serious from 1924 to 1926, when concrete was coming into fashion for building, and between the demands of builders, and the interests of New Territory cultivators of land behind the sandbanks, there was acute conflict, which sometimes grew into a shooting match. One such conflict took place at Sha Lo Wan in northwest Lantau; this village was very jealous of the fine sandbank protecting its fields, and had licensed gun owners; so the junk crews, who had no permit for that beach, were driven off without their sand. One of my duties was to discover and report beaches that could be dug without injury to cultivated land. Some of these have since then been completely worked out, notably on Sha Chau, as I found in 1938 during archaeological researches. Eventually the Public Works Department started a scheme for dredging and working sand from the sea bottom off Tai Lam Chung about 1929, which enabled the builders to get what they wanted. The beaches at Tai Long in Lantau and Tai Wan in Lamma were specially reserved for the waterworks filter beds because of the cleanness and high quality of the sand there.

Sha Lo Wan (260984) (Austin Coates)

Population 270. Cantonese. Surnames: Chan, Lee, Cheung, Man, Lau, Cheng, Kwan, Mok. A man of Lee aged 45 is of the 12th generation, giving a date of settlement c. 1670. This is a wealthy village, relying a good deal on remittances. 3 men are in the USA running restaurants, 6 in the Seria oilfields, 1 at Nauru, and several working in Hong Kong. The Lee family came from Tung Kun district, via Sai Heung, 200 taochungs, pigs, stake-net fishing, 10 small sampans and 2 larger ones for transport.

Wolframite was discovered in the hills above Sha Lo Wan, and between there and San Tao and San Shek Wan, in 1950, and when the Korean War sent the price of wolframite up, there was a rush of illegal miners, whose numbers had by 1952 risen to 3,000. In 1952 the mining area was brought under control by the Government, the price of wolframite fell, and by 1953 there was only one company, Structural Products Ltd, still operating. It seemed likely to this company that the only large amount of wolframite easily accessible lay within the Dragon's Pulse, but when the miners started working near it the villagers claimed that yet another piece of hillside was within their fung shui area and so should be included in the prohibited zone. At the same time, abundant evidence is held by the Tai O Police that had the mining company been willing to pay sufficiently to the village for the right to mine in the zone there would have been little objection, and might have been none. The District Commissioner (Mr Teesdale) however ruled that the fung shui zone should be enlarged on maps held by the company, which has in any case now gone out of business. A new company has been taken over the area; it is a subsidiary of Wheelock Marden and Co, and is keeping well away from the prohibited zone, having more dealings with San Tao than with Sha Lo Wan. It is evident that the new company expects to find only wolframite; only mysterious hints have come so far, but I infer from these that they are looking for gold. The company is working in a more scientific manner than its forerunner. Machinery is being installed, and the actual mining is not expected to start till the end of this year or the beginning of 1956.

A Police Post was set up in the village's school building during this time, and there were frequent disputes caused by miners trying to work in the Dragon's Pulse, the eminence rising from the forest behind the village.

Sha Lo Wan (*Gazetteer* p. 72) (James Hayes)

Population 290. Cantonese. Those above and including 16 years old, include 120 males and 80 females. There are around 40 boys and rather fewer girls.

There are said to be 50 men away at sea, all as firemen in the engine rooms of ocean-going ships. They say there are more people nowadays than previously.

They own 350 taochung of rice fields, some of them in Shan Shek Wan adjoining those they have owned for a long time. They do not grow enough rice, despite exchange, and have to buy from the Tai O shops to provide a supply for four months of the year. They buy all their vegetables from the immigrant vegetable farmers at Shan Shek Wan.

They have 110 brown cattle, "all very healthy" they say, but only two sows and 20–30 pigs.

Their main problem is with poor communications. They do most of their rice exchange and other business with shops in Tai O, but have to travel there by sampan or local junk. If there was a proper ferry pier, the Tai O ferries could call at the village.

There is a Pa Kong temple near the shore, with a repair tablet dated 1852. The bell is dated in Ch'ien Lung 39th year (1774). Curiously enough, it is stated thereon as having been presented to a Hung Shing

Temple, perhaps in error. Neither is this a Tin Hau temple, as stated in the *Gazetteer*.

TUNG CHUNG AND VALLEY

Tung Chung (Eric Hamilton)

The police station at Tung Chung was in an old Chinese fort with a lovely walled in compound.

I heard my cases under a huge tree there and always had to drink a large tumbler of goat's milk provided by the Indian sergeant in charge. He would have been awfully hurt had I refused. It might be alright with a half a pint of rum or whiskey, but I had not the heart to do it.

Tung Chung (Austin Coates)

The latest population figure for all the villages in this area, including the Hakka hamlets, refugees, and Buddhist settlements, is 981, of whom 70 are working abroad, and 15 are strangers of recent arrival. The traces of history stretch further back in Tung Chung than in other parts of Lantao Island. If, however, we accept the possibility that in certain areas the present inhabitants are pre-Chinese "original earth", we must confine any statement on the settlement of Tung Chung to this: at Tung Chung there are definite traces of Chinese settlement of an earlier date than that discoverable in any other part of Lantao Island. It may be that Chinese settlement at Po Chu Tam (see under Tai O) was of the same date or even slightly earlier; but of this settlement no definite trace can be found, the only evidence being a veiled statement in the temple there. By contrast, at Tung Chung there is the Ho family of Sheung Ling Pei, of whom a man of 63 is of the 24th generation period, giving a date of arrival c. 1292, which is just 13 years after the fall of the Southern Sung. This is, I consider sufficient evidence to say definitely that, the family arrived in or just after 1279, and that Chinese settlement in Tung Chung owes its origin directly to the fall of the Sung. The Ho family tradition is furthermore that the family were officials "long before Tung Chung fort was built", suggesting that his ancestor may have been one of the members of the Sung Court escaped to Lantao Island after the last defeat by the Mongols near Tsun Wan. The head of the Ho family, whom I interviewed, is a retired seaman, slightly deaf, and speaks reasonably good English. Again, it seems unlikely that the Ho ancestor would have brought his family with him on the catastrophic retreat from the mainland, and it may be that the fact that there is a Ho family living today is due to the presence of non-Chinese "original earth" at Tung Chung in 1279, and possibly for centuries before that date.

The principal place of settlement in the Tung Chung area is the place occupied by the three adjacent villages of Sheung Ling Pei, Ha Ling Pei, and Lung Ching Tao, which together form what is loosely called the village of Tung Chung. The Chinese fort built on the lowest stretch of hill between the two Ling Pei villages was constructed, according to Stephen Balfour (who appears to have obtained the evidence from the San-On Topograph, but does not quote to his sources) in 1817. It was built at the same time as the fort above Fanlao and Shek Sun, and the fort which formerly existed at Po Chu Tam, and would in this case be connected likewise with the first Chinese attempt (1815) to suppress the opium trade, and disorders caused by pirates and by hostilities of the Napoleonic War. The six guns on the battlement bear dates in the Ch'ia Chiing and Tao Kuang regions (with the exception of one, which was evidently miscast and bears no legible inscription). This accords with the date of construction being 1817. Local tradition asserts that the fort was definitely a military establishment and at no time enclosed any part of the civilian buildings. The parts of it that remain today are used as a school, quarters for teachers and an agricultural officer, the office of the Tung Chung Rural Committee, and a garden. Various Europeans have generated a myth that the fort was a reconstruction of an earlier Dutch or Portuguese fort, but there is no supporting evidence whatever, which in a place as old as Tung Chung there definitely would be if the story was true.

The English, and probably also the Dutch, certainly used Tung Chung as a watering place on numerous occasions from the seventeenth to nineteenth centuries, but they never settled here. Tung Chung presents the Administration with problems faintly discernable in other villages but nowhere more striking than here. The chief of these is the shocking idleness of the people. For the last 50 years or so, Tung Chung people have been going abroad to earn their living; a number of them are owners or part-owners of restaurants in New York, San Francisco and London, and many others are seamen. Remittances have thus solved all village problems. But whereas in Lamma Island, where a similar situation obtains, a great deal of pride and interest in the parent village remains, here there is little or none. In Lamma it is easy to get local public works undertaken jointly, in the usual fashion, with Government material and village labour. In Tung Chung it is impossible. So far as is known, Tung Chung has never undertaken a local public work. One particular work is most necessary; the reconstruction of the bridge connecting Ma Wan and Ma Wan Chung, which provides the main thoroughfare from the ferry stop to all the valley's villages, and to Ngong Ping. This bridge, built when Mr Kennedy-Skipton was District Officer, is now falling to pieces; but determined pressure by a succession of post-war District Officers has failed to arouse the slightest interest from the village. There is unfortunately no clan spirit in the valley, and the leadership of Mr Law Loi Tak, the Chairman of the Rural Committee, cannot be described as distinguished.

Most of the better paths through the rice field area of the valley have been repaired at the cost of the Buddhists of Ngong Ping and Tei Tong Tsai; and as long as Tung Chung can count on them, and on the wonderful possibility that Mr Brook Bernacchi might utter public criticism of the Government for neglecting the bridge at Tung Chung, the villagers intend to do nothing. The fact is that they are so idle that they will do nothing, even to help themselves. The state of the villages themselves demonstrates this. The resident agricultural officer, when I asked him why certain pigs in the Kadoorie Agricultural Aid Association pens were clearly underfed, replied that the villagers were too poor to feed them properly.

This brings us to another of Tung Chung's problems, which is that wealth is unevenly distributed. In general, it is one of the wealthiest places in the New Territories; but individuals amongst this community are poor and ignorant. There is a daily ferry service to Hong Kong and there is no reason why many cultivators should not have gone over from rice to vegetables years ago. Only since some new settlers from Castle Peak have begun growing vegetables in the valley have the people begun to understand the wisdom of this, and it now seems possible that there may be a slight swing in this direction. The trouble is that the present village tradition is that men go abroad to work, while women look after the fields. Shortage of work in ships is, however, driving some men to work in the fields, lowering as it is to their dignity. The agricultural officer said he had never seen the men doing as much as they are now. If the trade slump continues, Tung Chung as a rural community may revive slightly, but in my view this area should not be considered for any scheme except on a part-Government, part-village basis.

On the beach in front of the Tung Chung group, centrally situated, is a Hau Wong temple of some interest. It seems as if at one time this was a sandbank, with the sea reaching in behind it, possibly as far as the main village. It is now a large, uncultivated sward, used annually for theatre shows. The temple bell and a number of other articles within date from Ch'ien Lung 20th year (1754), when the temple was renovated; and there is a record that this renovation marked the settlement of outstanding disputes between Tung Chung on the one hand, and Sai Chung (Santao) and Keung Shan on the other. It suggests to me that Keung Shan at that time was larger and more important than it is now, and may possibly have stretched over other places, such as the present Luk Wu.

Tung Chung (James Hayes)

The population of the Tung Chung Valley was Cantonese, with very few Hakkas. However, as will be seen, some lineages claimed to have been originally Hakka speakers.[5]

My visits were probably made in late 1957, when poor attendances from most villages were recorded. I recall that early in my tour of duty as District Officer, and very likely about the same time, I was invited to officiate at the opening of a new government primary school inside the Tung Chung Fort. It may be that until this opened its doors, the existing school (perhaps within the same place) was quite inadequate for the number of local children of school age. If so, this would explain the very limited number of children being sent to school. But I remain unsure of the true situation.

San Tao *(San Tau)* (281986) (Austin Coates)

Population 160, including 5 in America, 2 in Borneo, 1 in the United Kingdom, and 1 working as a Hong Kong building contractor, employing 7 other Cantonese. Surnames: Tse, Chan, Chow, Ho Cheng, Wong (water-king 汪). This village is the Sai Chung referred to in the Hau Temple at Tung Chung. The founder family is Tse and the 18th generation is in the first decade, giving a foundation date c. 1550. 170 taochungs, 2 junks, 13 sampans, and one larger transport sampan. Farming, fishing and remittances. Only half the villagers have pigs, and this might be examined by the Agricultural Department. 12 under-employed youths.

The water supply is unsatisfactory, and the Irrigation Engineer has been asked to investigate. It was found that at the time of drought, the new mining company was taking all the water flowing to the village down its main source of supply. This was settled as a matter of urgency by a visit to the mine. This village is poorer than Sha Lo Wan and Tung Chung. It also prefers to be under the wing of Tai O Residents'

Association rather than under Tung Chung Rural Committee. A good deal of young rice has been lost in the drought, and some emergency food supplies may have to be sent later in the year. Hard-working people, very different in character from their indolent neighbours of Tung Chung and Sha Lo Wan.

San Tau (*Gazetteer* p. 77) (James Hayes)

The village comprises San Tau itself, together with the two hamlets of Tin Sham (near the beach) and Seng Chau (near the ancestral hall of the Chau family). Population 160, all Cantonese.

The villagers own 130 taochung of rice fields. All their own rice is exchanged with shops in Tai O, their usual market centre, though they buy a few things from the shops at Ma Wan Chung, in Tung Chung. They sell very few of their own vegetables, and buy only a little.

They have between 40–60 buffaloes; no brown cattle were mentioned. Currently they have five sows, 30 piglets, and 70 porkers.

Very few men from the village are overseas. A few work in Hong Kong, including at least one tailor.

There are 22 children in school, plus another ten of school age not at school because, they said, the existing school was not big enough.

Sheung Ling Pei (299975) (Austin Coates)

Population 151, including 8 abroad. Cantonese. Surnames: Ho, Law, Kam, Chan, Ng (man-five 伍), Kwan, Leen, Tsang, Lee, Cheng, Tai, Lo, Lau. The Ho family, the founders (see above), came from Mui Yuen district. 177 taochungs, owned, 48 rented from other villages, pigs (Kadoorie Agricultural Aid Association pens), vegetables for sale in Hong Kong, one stake-net, one sampan for transport. It seems likely that when this village was founded (or settled by Chinese), it was on or near the coast. The sea is slowly receding in this area, and the rice fields between the village and the sea must have been under water within historical times. The three central villages all derive money from renting fields to other villages.

Sheung Ling Pei (*Gazetteer* p. 76) (James Hayes)

Population 175, of whom 85 are male and 90 are female.[6] The Village representative is Li Sheung. He seems a nice chap but thinks everything is so much trouble.

This is a village of many family names. There are 42 households in all, of which Li and Ho have ten each, Cheng and Tsang have four each, Lo of 3, Lau of one, and others not specified. But all are stated to have lived here from before the war.

Fifteen children of school age are attending school, and another eight do not. [This seems a very low number, and I wonder if they were only speaking of boys?]

17 men are at sea or otherwise overseas, and 10 men are working in Hong Kong or Kowloon.

Currently, they have 70 cows and 70 pigs.

There are 300 taochung of rice fields, and 20 taochung of vegetable land.

They want to have a well, and this has been measured up by staff working for the Kadoorie Agricultural Aid Association, but no news of when the materials may be expected so far. They wish to do the work in the autumn and asked me to enquire. They also want to lay a concrete footpath to Ma Wan Chung (the village where the pier is) but are not sure when they can manage it.

Ha Ling Pei (301976) (Austin Coates)

Population 71, including 1 in New York and 6 working with the Hongkong and Yaumati Ferry Company. Cantonese. Surnames: Hung, Leung. Hung is the older of the two families; they are now in the fourth generation, and came from Ng Wah District.

Ha Ling Pei (*Gazetteer* p. 76) (James Hayes)

Population is 77 in 18 families: again, rather mixed, with 9 families of Hung, two each of Yeung and Chung, a Ho, a Chan, a Kam, a Shiu, and one other. All are said to have been here from before the war, but there are also two newcomer families from post-war. Two of the village families are now living in Hong Kong. The Village Representative is Hung Shue-fuk.

The Hung lineage has no clan record, and does not know when it settled here. However, the first ancestor's grave is at Tai Po (Tung Chung) with an inscription.

One man is in the United States, but there is no one at sea. Two men work for the Star Ferry Company, and three more work in Yaumati, Kowloon.

They reported owning 89 taochung of rice fields, and renting another 26. There is also five taochung of owned vegetable land, with a half taochung rented land.

Education is at a low ebb, with only four children of school age in school, three boys and one girl. If these figures are correct, many more are not being sent.

Lung Tseng Tau (S. H. Peplow)

Lung Tseng Tau. Lung — a dragon, Tseng — a well, Tau — the head. A well near the village where the surrounding hills, have the appearance or shape of a dragon's head. Many hills to the Chinese are shaped like dragons or tigers.

Lung Chung Tao *(Lung Tseng Tau)* (296874) (Austin Coates)

Population 36, including 1 seaman, and 3 working in Hong Kong. Surnames: Chan, Hung. Cantonese. The founder family is Chan. Hung family is in the 18th generation in the first decade. This is a poorer village, renting many of its fields from others; they keep pigs, and one family has a small fishing sampan.

Lung Cheng Tau, Wong Ka Wai (*Gazetteer* p. 76) (James Hayes)

The Village Representative for both these villages is Hung Chiu-cheung.

The combined population is 100 persons in 16 families. In Lung Cheng Tau there are seven families of Chan and two of Hung. Neither can say for how long they have been here. Wong Ka Wai is a single lineage village of Wong of seven families, but they are equally ignorant of length of settlement, saying vaguely that it must be about the same as at Lung Cheng Tau.

One man from Lung Cheng Tau is a seaman, and another from Wong Ka Wai is in the United States.

Lung Cheng Tau has 50 taochung of rice fields and four of vegetable land. Wong Ka Wai has 70 taochung of rice fields and six of vegetable land.

In each village five children of school age are going to school, and two are not. [Again, these figures seem small, and it may be they were only talking about boys.]

Wong Ka Wai (296965) (Austin Coates)

Population 46, including 2 in New York, and 2 working as building contractors in Hong Kong. Cantonese. Surname: Wong. 27 taochungs owned, 41 rented from others. Sweet potatoes.

Lam Tse *(Lam Che)* (288968) (Austin Coates)

Population 14. Hakka. Surname: Lee. The family moved here from Taipo about thirty years ago (1925). They have cultivated marginal land, which they own, and receive remittances from 2 seamen and 1 man working in Hong Kong.

Nim Yuen (*Lim Un*) (288967) (Austin Coates)

Population 14. Cantonese. Surname: Kwan. 4 taochungs; 1 sailing boat. Lam Tse and Nim Yuen are grouped together.

Nim Yuen (with Nam Che) (*Gazetteer* p. 77) (James Hayes)

On the day of our visit, the Village Representative (Kwan Kan-yung) was absent, sick, and had gone to Tai O for an injection.

There are three families of 13 persons and four families of 10 persons respectively living in these hamlets, in Lee and Kwan lineages. The Kwans of Nim Yuen say they have been settled here for 200 years, and the Lee of Nam Che came from Tai Po (mainland New Territories) sixty years ago.

Two men from Nam Che are at sea. No one is working in the urban area.

One child studies, and another of school age is not at school [again, surely only boys are being listed].

I was told in Tung Chung Hang that the only Hakka speakers in the Tung Chung Valley are at Nam Che, "where they have been for thirty years" [the Lees of Nam Che say "sixty"].

Ma Wan Chung (297985) (Austin Coates)

Population 84, including 3 seamen and 2 men working in Hong Kong. Hakka, Surname: Fung. The family came from Wai Chow and evidently arrived shortly before the lease of the New Territories. About 30 of the inhabitants are bright-light fishermen, owning several sampans. They ship their catch direct by ferry for sale in Hong Kong. 31

taochungs. This is the coastal village at which one lands on arrival by ferry. There is also a floating population of about 80.

Ma Wan Chung (*Gazetteer* p. 75) (James Hayes)

I have no entry for this village, which was the landing place for goods as well as being the pier for the Hong Kong–Tai O ferries. It consisted of a few shops and temporary structures, and was also home to some Hoklo boat people's families. There was a concrete bridge linking it with the route to the Fort and nearby villages.

Shan Ha (306975) (Austin Coates)

Population 39, including 1 in London, 1 in Borneo, and 7 working in Hong Kong. Surname: Tang. The male line is originally Hakka, coming from Wai Chow, but due to intermarriage with Tung Chung Cantonese they describe themselves as Cantonese. They evidently settled here around 1855, their first village being the comparatively remote Pa Mei, which the Japanese set fire to in 1943. After this they founded Shan Ha, which is further into the valley and nearer other villages.

Shan Ha (*Gazetteer* p. 76) (James Hayes)

This village is also called Pa Mei San Tuen. This is because Pa Mei was burned by Japanese troops during the second year of the wartime occupation, in retaliation for what they thought was the help given by the villagers to a band of robbers (or guerrilla fighters) whereas, the villagers said, any assistance given had been forced.

There are eleven families of 38 persons here, all of the Tang lineage. They think that their ancestor was one of the Tangs of Kam Tin in the mainland New Territories, but they themselves are Hakka, speaking Cantonese, whereas the Kam Tin Tangs are Cantonese. All rather confusing, and I think this connection to be unlikely [there are, however, Hakka Tangs in the Kam Tin area, living at Wang Toi Shan Village].

The Village Representative's father is in the United States, and two men are working in Hong Kong on what they call water-boats.

They have 90 taochung of rice fields, all in their own ownership, though they also rent some from others. There are 18 brown cattle.

Only three children of school age are attending school.

They say they collect hill tea from tea bushes on Tai Tung Shan Keuk, above the village.

Shek Pik Au (290962) (Austin Coates)

Population 18, a Buddhist settlement on the lowest slope of the steep hills rising to Tei Tong Tsai and Ngong Ping.

Ma Wan (297984) (Austin Coates)

Population 61, including 1 in New York, 1 in Sandakan, and 1 running a grocery store in Western Market, Hong Kong. Cantonese. Surnames: Ho, Fun, Hung. This is the first village inland after crossing the bridge from Ma Wan Chung. A man of Fan of the 7th generation is 47, giving a date of arrival c. 1780. 40 taochungs owned, one stake-net, one transport sampan. Villagers go as far afield as Taipo and Yuen Long to do their twice-yearly rice exchange.

Ma Wan (with Wong Nai Uk) (*Gazetteer* p. 75) (James Hayes)

Population is 79 with 19 families of Fan, two of Hung, and three of Ho. The Village Representative is Fan Fu-ying.

One man is in New York, and eight are working in Hong Kong or Kowloon. No men are at sea.

Only three children of school age attend school. Obviously, a great many are not.

They have 160 taochung of rice fields, of which 108 taochung are stated to be their own, and 34 of vegetable land, all their own. They have 33 brown cattle.

Their potable water comes from the stream, but they would like to have a well in each place, and will indicate where it is best located. They also request materials for an irrigation channel. I shall ask Assistant Inspector of Works Mr Chan to visit and discuss with them.

Wong Nai Uk (300982) (Austin Coates)

Population 14. Hakka, but treated as a subsidiary hamlet of Ma Wan. Surname: Ho. 17 taochungs owned, 17 rented from others. Average rent 12 taos of paddy per taochung.

Shek Lau Po (S. H. Peplow)

Shek Lau Po. Shek Lau — the pomegranate, Lo — a mart. A village where these trees are grown, and the fruit sold.

Shek Lau Po (292969) (Austin Coates)

Population 109, including 14 abroad, several of them in America. This is reputed to be the richest and laziest village in the valley. The Chairman, Law Loi Tak, lives here. Cantonese. Surnames: Law, Tang, Ng. The place was founded by the Law clan, from Fa Yuen District, in 1775. 32 taochungs owned, 103 rented from others, pigs (sold in Tai O), one transport sampan.

Shek Lau Po (*Gazetteer* p. 77) (James Hayes)

There are upper and lower villages. The Village Representative is Law Loi-yau.

Population is 95, with 21 families. There are 17 families of Law, three of Cheung, and one of Tsang. Also, a newcomer family of Ng which came ten years ago.

The Village Representative said that the Laws had been settled here for about 200 years, that is, from around the 1750s, but this was a round figure, and he was not very sure. He also said that the village of Shek Mun Kap (see below) is the oldest settlement in the Tung Chung Valley.

There are seven men in New York, two in Liverpool, one in London, and another four (two Laws and two Cheungs) are working at sea. One man works with the Agricultural Department.

The village families own 57 taochung of rice fields, plus seven taochung of rented fields, and ten taochung of vegetable land. There are 31 brown cattle currently.

There is a two-classroom school in the Tung Chung Fort, taking 45 children in each of its two daily sessions. Six children of school age attend classes, but five do not "because they are not big enough" to walk to school.

Shek Mun Kap (294963) (Austin Coates)

Population 71, including 5 in the United Kingdom, 3 in the United States, and 3 working as assistants in a stationer's in Hong Kong. Cantonese. Surnames: Cheung, Wong, Law, Cheong, Wong (water-king 汪), Tse, Sun. The founder family is Cheung, coming from Fa Yuen District; a man of 47 is of the 4th generation, giving a foundation date c. 1855. 63 taochungs owned. This is near the site of an older village called Fa Yuen, the inhabitants of which have either moved away or died out. Ruins of the former village can be seen.

Shek Mun Kap (*Gazetteer* p. 77) (James Hayes)

Population is 60 persons, in 14 families of many surnames. The Village Representative is Cheung Mun-hei.

There are five families of Law, three of Cheung, two Wongs, and one each of Tse, Wong, and Chong. All have been settled here for a long time, but they do not really know for how long as there are no records available.

There are five men working at sea, and two in the United States. Two men work in a weaving factory in Hong Kong, but none in Kowloon.

They have 60 taochung of rice fields and five of vegetable land. Some families rent from other village families, but no one rents from persons in other local villages.

Six children of school age are attending school, and two do not.

They say they have enough potable water, and have nothing to request by way of Local Public Works.

Mok Ka Tsuen (290966) (Austin Coates)

Population 61, including 3 in the United Kingdom, 3 working in Hong Kong, 1 in Borneo and 1 in Nauru. Cantonese. Surname: Mok. Foundation c. 1855. The clan came from Sha Mei, in Po On District. 26 taochungs owned, pigs, fishing (angling from rocks and sampans).

Mok Ka Tsuen (*Gazetteer* p. 77) (James Hayes)

Population, 144 persons in 18 families, all of the Mok lineage. The Village Representative is Mok Ping.

They say they have been here for almost 200 years.

There is one man in New York, and five at sea. There are another eight men working in Hong Kong or Kowloon, one of them with Taikoo Docks.

They have 140 taochung of their own rice fields, and 30 taochung of vegetable land. They also rent another 11 taochung of fields from the village below them, and some more from Tai O people who own land in this locality.

Eleven children of school age are studying, and four do not. [Again, possibly boys only, as these numbers are low for the reported population.]

They have enough potable water, from a well. They would like cement and iron rods in the autumn, to construct a small footbridge over a stream.

Ngau Au (S. H. Peplow)

Ngau Au. Ngau — cow, Au — a mountain pass. A village near a pass having the shape or appearance of a cow.

Ngau Au (287974) (Austin Coates)

Population 63, including 1 in Borneo. Cantonese. Surnames: Law, Wong, Yeung, Siu, Chow. A village that has grown up since the lease of the New Territories, the inhabitants are all from Tung Kun.

Ngau Au (*Gazetteer* p. 76) (James Hayes)

We sat under a splendid old tree.

Population 60, in 18 families. They comprise eight of Law, four of Yeung, two each of Shiu and Chau, and a Wong family. About the same as pre-war, they said.

They disclaim all knowledge of how long they had been here, and do not seem to have any idea: at least, that is what they say.

There are two parts of the village. A hundred years ago, where the Yeung families live was called Tung Hing, whilst the Laws lived at On Hing. Tung Hing has a surrounding wall, but not On Hing. At present, the village is simply called Ngau Au. There is a Tai Wong shrine.

One man is in New York, one is at sea, and three men work in the urban area.

There are 180 taochung of rice fields and 30 taochung of vegetable land. Last year, they got a Kadoorie orchard, in which they grow pineapples, oranges, peaches, and the like. There are over 30 brown cattle, with five buffaloes, and there are currently two sows.

Seven children of school age go to school, and two do not. [Again, boys only being listed?]

They rely on stream water for their potable supply, but say it gets very dirty. They have already applied to Kadoorie Agricultural Aid Association for materials to build an intake on a stream, and 1,700 feet of piping to bring a clean supply into the village.

Tai Po Tsuen (311990) (Austin Coates)

Population 25, a small coastal village just east of the main Tung Chung valley. Hakka. Surnames: Wong, Tsang, Chung. The village dates from the first decade of the New Territories lease, and was founded by the Wong clan. These people are hard-working and have been most enterprising in developing marginal land. A complete contrast to the people of Tung Chung. Excellent hill tea grows near here.

Tai Po Tsuen (*Gazetteer* p. 76) (James Hayes)

Village Representative is Wong Kam, but he was not at home at the time of the visit.

There are 38 persons here, in four families of Wong, and one each of Tsang and Chung. They are all natives of Ching Yuen County in Kwangtung, and came here over thirty years ago, say around 1910–15. The Wongs came first.

All the men are at home.

They farm 80 taochung of rice fields and five of vegetable land. Most of the fields are rented from Li families in Sheung Ling Pei and Hung families in Ha Ling Pei. They have six brown cattle and four buffaloes.

Six children go to school in Tung Chung, and two others should.

In the dry season, they do not have enough potable water, and have to seek it from higher up the hills behind the village. They think it possible to dig a well here, beside the stream. The Assistant Inspector of Works will visit, and discuss with them.

Tung Chung Hang (*Gazetteer* p. 76) (James Hayes)

Population not stated, but there are 15 families resident. These comprise four or five families of Wong, four of Lai, three of Cheung, and three of Lok. The first three lineages are long settled, here, whereas the Lok families moved here more recently from Nim Yuen on the western side of the Tung Chung valley (see above).

The people here say they are "Hakkas but speaking Cantonese".

Luk Wu (244942) (Austin Coates)

Population 70. A Buddhist settlement about one hour's walk from Tai O, 800 feet above sea level, halfway between Tai O and Ngong Ping. The many foundations here are chiefly nunneries, although some of them have a monk as disciplinary head (who lives free), and most acknowledge as spiritual head the Abbot of one of the principal monasteries, either Po Lin Monastery at Ngong Ping, or Chuk Lam Monastery at Tsun Wan. The oldest foundation is Tai Luk Wu, former Taoist hermitage which was deserted by its founders and reconstituted by a single remaining disciple on Buddhist lines in Kuang Hsu 10th year (1885). The present leading personality here is the Abbess of Chuk Yuen Nunnery, a hospitable and kindly person. It seems to me that, just as the nunneries of Luk Wu have succeeded in buying up fields belonging to the Wans of Keung Shan, and may one day absorb that village, so they may have earlier absorbed other agricultural land of which there is today no record. Most of the Buddhist development at Luk Wu dates from 1912 onwards (the foundation of the Republic and the prospect that Buddhist institutions in China would lose their last vestiges of patronage).

Pa Mei (S. H. Peplow)

Pa Mei. Pa — an embankment, Mei — the end. A village at the end of an embankment.

PAK MONG GROUP OF VILLAGES

Pak Mong (336997) (Austin Coates)

Population 67. Hakka. Surnames: Kwok, Cheung. 1 man working in the United States, one in North Borneo, and 3 in restaurants in Hong Kong and Kowloon. The village was founded by the Kwok clan, who came from Wai Chow to Castle Peak, and crossed from there to Pak Mong by bamboo raft. A man of 36 is of the 17th generation, giving a foundation date c. 1545, or, at the 27-year rate, 1511.

About 70 taochungs of rice fields and some vegetable fields. Vegetables are carried over the hills to Silvermine Bay, by ferry to Cheung Chau and sold there. Sweet potatoes are grown, and there was a heavy loss this year due to frost. Stake-net fishing. One sampan. The village has not enough pigs; this is being brought to the attention of the Agricultural Department, with a view to starting loans. Loans are also required for bringing more land under cultivation, for onions.

A new school building was completed this year for the three main villages in the group.

Water supplies and irrigation are bad, and the Irrigation Engineer is being asked to advise. The size of the improvements needed indicates a Colonial Development and Welfare application. Repairs are needed to two small bridges near Ngau Ku Long, and to the main track to Silvermine Bay.

Pak Mong (*Gazetteer* p. 85) (James Hayes)

The Village Representative is Kwok Hing-chuen. There are 81 persons in the village, all named Kwok, save for some families of Cheung from Tai Ho.

The Kwoks came from Wai Chow, and claim 16 generations of residence here.

The village has 80 taochung of paddy fields, and 6 of vegetable land. There are 20 brown cattle and 25 pigs, but currently no sows.

The village carries on a little fishing, selling part of the catch. So do the Ngau Kwu Long people, but not those at Tai Ho. Fishing is mainly by stake-net, and not from boats.

Five men are working in Hong Kong or Kowloon, 3 are in the United States, and 3–4 are seamen.

There are 14 children of school age, presumably all attending the Pak Mong village school.

This village has a rubble wall and a stone watch-tower. This is another of the watch-towers built in these villages during the Japanese occupation, and there is a third at Tai Ho. They were erected by the villagers to protect themselves and their livestock in troubled times, when robbers and bandits were also a threat.

The village wall, like many of the houses, is substantial, and clearly old, but no one can say when it was built. The houses are built to various designs. Some have stone railings on top, and others are decorated in traditional style. They include the Kwok's ancestral hall.

They showed me a letter from the Registrar of Births and Deaths, dated in March 1956. I did not take the details, but in hindsight it would be interesting to know how much attention the Registrar was paying to village births at this time, and whether the letter related to a new birth or a post-registration.

Ngau Ku Long (*Ngau Ka*) (343992) (Austin Coates)

Population 62, including 3 in the United States, and 2 seamen. Cantonese, but intermarried with Hakka. Surname: Lam. The clan came from Tung Kun District and settled here a few years after the foundation of Pak Mong. About 100 taochungs. Vegetables, and a good herd of cattle. 2 stake-nets and 3 sampans. 8 under-employed youths.

Ngau Ku Long (*Gazetteer* p. 85) (James Hayes)

Village Representative: Lam Kau, aged 74, "a nice old boy". He has a very good new house.

There are 78 people, all named Lam. They are Hakkas and have been here for over ten generations. However, they are not very sure for just how long, because the clan record was lost when 24 houses were burned by Japanese troops during the wartime occupation. They said they came from near Shek Lung in Tung Kwun, but Chai Lun in Wai Chow was also mentioned.

There are 7 men working in Hong Kong and Kowloon, and another 5–6 are abroad. As at Tai Ho, they want employment for their young men, meaning work outside the village.

They would like Local Public Works materials for a bridge and well, now if possible.

The village has 18 taochung of paddy fields, and over 10 of vegetable land, all their own. There are 27 brown cattle, and 24 pigs.

We saw a single telegraph pole, a relic of the pre-war system which (they said) linked the police stations at Cheung Chau, Tai O, and Kap Shui Mun (Ma Wan). During the occupation the Japanese had got the villagers to take down the poles in the valley and carry them to the beach, to be taken away.

There is a duck farm at the bottom of the valley, operated by a Ho from Tung Kwun, who came there with his wife just after the war. They now have a son.

Tai Ho (345985) (Austin Coates)

Population 57, including 1 in Sydney, 2 in North Borneo and 1 in Nauru. Hakka. Surnames: Kwok, Cheung, Ho, Chow. This is an offshoot of Pak Mong, the families being the same. 57 taochungs; a good herd of cattle. No fishing.

Tai Ho (*Gazetteer* p. 85) (James Hayes)

Village Representative: Cheung Au. There are 69 people in 14 families: 5 of Kwok, 4 of Chau, 3 of Ho, and two of Cheung.

The settlement is now a straggling village along the footpath, but about 60 years ago it was located higher up. At that time, I was told, there were only ten families left of about 60 or 70 at an earlier time, with few births, and so they abandoned the place and moved to their present location.

All families listed had previously occupied the old settlement. After the move, they lived at first in huts, and then erected their present houses.

The village is a pretty miserable place, which is hardly surprising. It has a watch-tower built during the Japanese occupation, with funds provided by Yuen Wah Chiu, the present Chairman of the Mui Wo Rural Committee.

No men are abroad. The rest are at home, apart from several working in Kowloon. Twelve young men have no employment.

There are 12 children of school age, all of whom go to Pak Mong School.

The villagers have 70 taochung of paddy fields and over 10 of vegetable land. They have 15 brown cattle, no sows, but 20 pigs.

There is enough water to drink from nearby streams, but they would prefer to have a well or a water tank, adding that, since they would have to dig 15 feet or so before they can obtain water, they would prefer the tank, with water piped from a clean supply.

Tin Liu (346992) (Austin Coates)

Population 15. One family of Yeung (7 persons) moved here in 1951 and owns 2 stake-nets. They formerly lived in Ngau Ku Long. Late in 1954 a group of 8 refugees, who had formerly been working in Yuen Long, moved here; they are at present rearing ducks, which are thriving, and have applied for some of the disused land, formerly cultivated, in the area.

Tin Liu (*Gazetteer* p. 85) (James Hayes)

This is a hamlet located below Tai Ho, occupied by several families of Kwok. It is considered part of Tai Ho.

LANTAO SOUTH COAST: SHEK PIK TO PUI O

Shui Hau (274906) (Austin Coates)

Population 126, including 2 in the United States, 1 seaman, and 2 in the Urban Services Department, working in cemeteries. Surnames: Chi, Fung, Chan. The first settlers are said to be the Chi family, who moved over from Shek Pik; This clan describes itself at Shek Pik as Cantonese, but at Shui Hau as Hakka. A man of 53 is of the 13th generation, giving a foundation date c. 1625. The Fungs are Cantonese, a branch of the same clan as at Fan Pui and Shek Pik; they evidently moved from one or the other some time after the Chi. The clan family is Cantonese. The villagers claim they have no clan records, but the Village Representative of Fan Pui says this is nonsense, and that the Village Representative of Shui Hau, a Fung, has a record. Further enquiry is necessary.

A new school building was completed in 1953.

200 taochungs, 3 large sampans, one of which is motorised and provides a ferry service between Shek Pik, Shui Hau, Tong Fuk and Cheung Chau. The Police recently arrested the ferry sampan for carrying a full cargo and 57 passengers. Pigs, grass and wood are all sold in Cheung Chau.

The bay in front of the village is very shallow, and the only satisfactory way of improving communications is by the construction of a road from Silvermine Bay, which is already under consideration.

Shui Hau (*Gazetteer* p. 78) (James Hayes)

Population 142, all Cantonese, belonging to the Fung, Chan and Chi lineages.[7] The first two say they have clan records. The Fungs and Chis are of the same lineages as those in Shek Pik. Marriages with women from Shek Pik, Tong Fuk and Pui O are the most common, and Shui Hau girls mostly marry out to these places.

There are about 300 taochung of fields (wet and dry cultivation), but 100 of them are out of use: but I did not record the reason. They grow vegetables also, but only for their own use. They have 70–80 brown cattle, and around 150 pigs at present, 20 of them being sows. Porkers are selling on Cheung Chau for $160–170 per picul.

They take 90 per cent of their own rice to the Cheung Chau shops to exchange for poorer quality imported rice, described to me as "broken rice". Their rice sells for around $40 per picul, and they get roughly twice the amount in return.

They are taken on a kind of local ferry service run by fishermen's motorized junks. This leaves Shek Pik around 7.30 a.m., they spoke of "a race from Shek Pik", indicating there was more than one boat involved, and of the boats "being overloaded with people and goods". The return journeys are made in the afternoon.

Ten men are away from the village. Five are working on ships as stokers and fitters. Two are working in Pokfulam, and their families are with them. This suggests long-term employment, perhaps with the Dairy Farm, but I did not ask. Three more are working in casual employment in Hong Kong. A further five are working on Cheung Chau, but these are not counted as being "away from the village", as they are still based at home, coming back every two months or so to visit their families. Currently, 15 men are working on the extension to the South Lantao Road (five of them had come to see me the day before, but I had not recorded what this was about).

Besides farming, there is coastal fishing between the third to the sixth lunar months, for shrimps and pomfret. Eight sampans are used in this work.

The village school operates primary classes 1–4. The times are from 9–12 and 1.30 to 3.50 p.m. The school began in 1953, prior to which children went first to the Shek Pik school, and then to the Tong Fuk school (which was nearer) when a school was built there. I think this refers to government subsidized one- or two-classroom buildings.

There is only one teacher, a Mr Wong, an "outsider" who lives in the village. He had come here first for a month last year, to relieve the then schoolmaster. He is unmarried and goes back to his home in Sham Shui Po, Kowloon, once a fortnight. He is a graduate of a middle school in Canton, and after coming to Hong Kong studied for two more years at Chu Hoi College in Mong Kok. He began teaching in 1951.

During the visit, we saw three men at work on repairs to a house. They belonged to a building contractor's firm from Cheung Chau.

Despite the employment position, there are 4–5 young men aged around 19 who are looking for other work. I said I would write a letter for them to the Labour Department's employment service.

On the face of things, it would seem that the women and girls do most of the farm work.

Local Public Works: they want cement and other materials for an all-weather footpath to the line of the South Lantao Road. Repairs to an embankment were also mentioned. Our Assistant Inspector of Works, Mr Chan, will visit.

Tong Fuk (Paul Tsui)

Tong Fuk has a total population of approximately 120, made up of 38 families, with the Tangs (鄧) and the Chans (陳) claiming to have an equal strength in manpower (about 45 each). The Mans (文) have 12, the Lais (賴) have 6. The Yuens have 5 and the Tsangs have only 2. They all live intermixed in two groups of houses built close to each other.

Land. The total acreage of cultivable land is said to be over 200 mu, all producing two crops of paddy. Details of present holdings are as follows:

Tang Kwai Hung	10 mu	
Tang Man	3	
Tang Kwai Po	2	
Tang Wai Kin	2	
Tang Wai Laan	2	
Tang Shu Yung	2	
Tang Fuk Hing	8	
Tang Loi Po	3	
Tang Loi Shing	2	
Tang Yau Wan	3	
Tang Fung Shi	3	
Tabg Lin Yau	7	
Tang Kiu Tsai	8	
Tang Yung Kan	7½	
Tang Tso	15	
		77½ mu
Chan Wai Fat	11 mu	
Chan Chung Low	10	
Chan Muk Tsai	4	
Chan Fuk	10	
Chan Koon Shing	7	
Chan Fuk Kwai	5½	
Chan Shu Fat	3	
Chan Man Yau	3½	
Chan Tan Yau	3½	
Chan Yiu Kan	5	
Chan Shek Fat	4	
Chan Tso	nil	
		66½ mu
Man Wan	11 mu	
Man Chu	12	
Man Yung Kan	4	
Man Tso	nil	
		27 mu
Lai King Cheung	5 mu	
Lai Tai Cheung	3	
Lai Tso	4	
		12 mu
Tsang Hei Yung	6 mu	
Yuen Koon	6	
		12 mu
Grand Total		195 mu

The average yield per mu is said to be 1.5 piculs of kuk *(unhusked rice)* per crop, and all land is said to be producing two crops of paddy plus a winter crop of onion or sweet potatoes in some cases where spare labour could be made available. Those who own less than 5 mu are part-tenants of their own cousins who own more than they require, at an annual rental on a 50:50 basis of the paddy produce. Any winter crop made available goes entirely to the tenants. Not a single piece of land is said to have yet been alienated to outsiders, nor has anyone been known who has mortgaged his land to any outsider. The smallness of their present holdings in most cases is due to divisions amongst members of the same family when the older generation die.

Like Pui O and Ham Tin, all men are trained to handle boats and fishing gear. They do not own any square trap nets (though someone tried not very successfully pre-war) but they go for shrimp, pomfret and cuttlefish with intensity. The village community has a total of three shrimp boats and 14 small sampans. The sampans are suitable for pomfret and cuttlefish catching. The cuttlefish season occupies the 2nd and 3rd moon (March/April); the pomfret season occupies the 4th and 5th moon, and shrimps the 5th and 6th moon.

A good cuttlefish season may bring a maximum of 180 catties a day, varying to a minimum of 20 catties a day. An average of 8 to 12 piculs of live cuttlefish per boat per season is usual. Pomfret catching is unreliable. They have had bad years for three successive years since 1947; none of the villagers claimed to have sold more than $50 worth during the last three years, but they had very good years of pomfret catching throughout the Japanese occupation. Shrimp catching has always been steady and reliable and an average of 50 to 70 piculs per boat per season is usual.

Estimate for the cost of equipping a shrimp catching team is as follows:

1 shrimp boat	$480
2 nets (shrimp thick type)	$70 ($35 each)
1 puller	$45
1 shaft	$25
1 篙尾 (long pole?)	$35
1 set of oars	$50
20 catties of rope	$40
25 catties of yam fibre	$40
1 anchor	$27
Miscellaneous items	$10
Approximate total	$822

A shrimp catching team consists of 3 persons, who require wages as well as food and clothing. A Cheung Chau shrimp dealer has established a shrimp collection depot right at Tong Fuk. This collection depot is made up of a small hut housing 4 fokis, one being an expert treater. The main equipment of this collection depot consists of three large-size wooden buckets, each with a capacity of approximately 50 piculs. Shrimps when caught are brought to and sold at this collection depot. The depot fokis pour them into these wooden buckets, and have them smashed into paste (by having men standing and stamping on the shrimps, smashing them with their feet). Salt is then added as a preserving and curative agent. The shrimps are then despatched to Cheung Chau for fermentation and maturation.

The annual household budget for a family of 5 was given as follows:

Rice	1,300 catties
Oil	3 catties
Salt	300 catties
Pork	36 catties
Sugar	150 catties
Fish (fresh)	300 catties
Fish (preserved)	180 catties
Vegetable (green)	300 catties
Vegetable (preserved)	30 catties
Sweet potatoes	120 catties
Taro	30 catties
Fruit	30 catties
Eggs	70 eggs
Chickens	6 birds
Rent	nil

9 young men in the village had succeeded in finding employment in town. 5 of them work as coolies in the Health Department and 4 work as forestry guards in the Forestry Department. None is said to have gone abroad; nor anyone known to have been away as a seaman. There is one small shop in the village selling candies and other odd items. The shopkeeper claims to have done only $7 of business each day during the last 3 days, but admits that his average business is approximately $100 per month.

Only the men go out fishing. The women confine themselves to work in the fields as well as in the homes. When they are free from farm work, they will also go up the hill to cut grass or fire wood. The men also share this work in the field when required.

Every household rears 6 or 8 pigs per year. No one complains of not having any cattle; in fact most of them have more than one cow, which is what they need as a draft animal. Spare cattle serve as additional sources of income for the family. Every housewife keeps a few dozen chickens.

Tong Fuk (294914) (Austin Coates)

Population 130. Cantonese. Surnames: Tong, Chan, Man, Yuen, Lai, Chiu, Wong Tsang. A large number of under-employed. Several formerly had coolie jobs in Hong Kong, and some still have, but most of them have returned to the village. They are rather a difficult type to find work for, too intelligent for the lowest jobs, and not quite intelligent enough for clerical jobs. A village of fine physique; the men are unusually tall for Cantonese. All clan records were taken away by the Japanese, and the age of the village is not known. The founder family is Tang, and they are a branch of the Tangs of Kam Tin, Yuen Long District. It should therefore be possible to find out from Kam Tin when this branch split off.

180 taochungs. 3 transport sampans, 18 small sampans used for bright-light fishing. The principal occupations are farming, fishing, grass-cutting (for sale in Cheung Chau), and some pigs. Transport difficulties here are due to the frequent danger of landing from the sea. This village would benefit greatly from the proposed Silvermine Bay Road. There is a school housed in a post-war building. A small temple near the landing cove has a bell in it dated Chia Ching 7th year (1799).

Tong Fuk (*Gazetteer* p. 78) (James Hayes)

Population 134, all Cantonese. Village Representative is Chan Man-yau. The families belong to the Tang, Chan, Yuen, Man, Lai, Chiu, Tsang, Liu and Leung surnames. The Tangs belong to the main Tang clan in the mainland New Territories, and have a genealogy. The Chans think they may have been here for over ten generations.

Around 20 children of school age, of whom 15 attend school. The rest are said to be "too poor".

Several men are working in the United Kingdom, but there are none at sea. A few men are said to be working in the Government's Sanitary Department.

The village has about 180 taochung of rice fields, with some vegetable land behind the village. They have around 70 brown cattle.

Their forbears used to grow sugar cane and make brown sugar, but this stopped a long time ago.

They exchange their own rice for imported rice of lesser quality at the Cheung Chau shops. On average, they have enough rice for a 6-month supply, sometimes as much as 8 months, and sometimes less.

They have no forestry lot under licence from the District Office, but all this means is that they use the hillsides anyway, as they have always done, but without having to pay for the licence.

We were given some hill tea to drink. They say they use it a lot. It is grown on bushes at Fung Wong Shan, the high peak behind the village. The leaves are plucked in the 2nd and 3rd lunar months.

There are two temples in the village. The Kwan Tai temple is in the village, and the Hung Shing temple is near the landing place down by the shore. There are also two earth shrines, styled Baak Kung, one at each end of the village.

In answer to a question about widows in the village, some were reported, but no assistance was requested.

In regard to Local Public Works, they reported that the embankment question is the big one, owing to earth and stones washed down by the recent floods. My note is too brief, and I am not quite sure what was meant. But I had said that Assistant Inspector of Works Mr Chan would visit them to inspect and discuss.

During this visit, we saw that a blacksmith was at work beside the large shrub-covered boulder in the space between the rubble wall in front of the village and the first row of houses. I think we also saw him there the following winter. This was an outsider, an itinerant workman from the Mainland, who had been coming to Tong Fuk and other Lantao villages in this season for decades, and so the elders told me, his father before him. It was the end of a long connection, from a time when itinerant artisans like blacksmiths and weavers came regularly in the winter season, and been a familiar and accustomed part of rural life.[8]

Cheung Sha (S. H. Peplow)

Cheung Sha Wan. There are several villages bearing this name. One or two on Lantao Island and one near Sham Shui Po. Cheung — long, Sha — sand, Wan — bay. A village near a long sandy bay.

Cheung Sha (325924) (Austin Coates)

Population 49. Surnames: Lau, Tsang, Cheung, Fung, Chung, Chow. Hakka. All clan records, if there ever were any, have been lost, and the age of the village is not known. All that is remembered is that Lau is the founder family. A comparatively poor village, with no remittances. One boy working as a baker's delivery man in Hong Kong. Occupations: rice growing, fishing (7 fishing boats, bright-light method used), grass-cutting (for sale in Cheung Chau), pigs. 2 transport sampans, and the usual trouble with the Police for overloading. Considerable improvements to irrigation could be made here, but it will have to be a contractor's job; it is too big and expensive for the village. Mr Kadoorie is being approached about this. Other smaller jobs on wells, etc., can be done by the villagers.

Cheung Sha Lower Village (*Cheung Sha Ha Tsuen*) (*Gazetteer* p. 80) (James Hayes)

A post-war village, settled by people coming here from Po On and from Outer Ling Ting and other islands lying south of Lantao. Population now about 50 persons.

There are 14 children of school age. Of the 22 children who should be at school from the two villages, only 12 travel to the school at Pui O. Among the rest are some who are described as being "too small". For others, the bus fare to Pui O (at 20 cents return) is said to be "too expensive".

They own only 7–8 taochung of paddy fields, and rent another 20 taochung from people at Pui O who own land here.

They have 2 sows at present, with 20–30 porkers.

A seawall here is reported to have been damaged. Assistant Inspector of Works Mr Chan will visit and report back.

Like persons from the Upper Village, it is likely that some of the villagers do some fishing in season.

Having seen cultivation on Crown land without a permit, I jotted down a reminder to re-issue our public notices stating that no compensation will be paid for any crops affected by the ongoing road works unless a permit is held by the cultivators.

This brought on a discussion about fertiliser (night soil) being in short supply. It is purchased from Cheung Chau, and the shortage has led to difficulties in growing vegetables.

Cheung Sha Ha Tsuen (323921) (Austin Coates)

Population 58, including 2 working in Hong Kong as earth coolies. Hakka. Surnames: Chan, Ng, Ho, Lai, Wong, Yeung, Lee. The village was founded in 1905 by the father of the present Village Representative, Chan Tsing. The Chan came from Po Kat. They have had a difficult time getting on with the villagers of Cheung Sha, who still resent their intrusion and describe them as if they had arrived last year. They own 5 taochungs of rice land, and rent 6 more from the Pui O people who own land here (the Cheung and Law families). They have 3 transport sampans and 4 fishing sampans. Fishing and grass-cutting are the most important means of livelihood, the place of sale being Cheung Chau. The Silvermine Bay Road may be expected to benefit both these villages.

The long beaches of Cheung Sha are the largest in the colony. Turtles come here to lay their eggs at the end of the 5th and beginning of the 6th moon.

There have been serious disputes about grass-cutting rights between the two villages, with acts of violence in 1953 and 1954. After the second I sent a very stern letter to both villages and caused a full Police enquiry to be held. Recent months have been quieter, but there is still the possibility of recurrence.

Requests for a school need not be listened to; the future road will bring this place into closer touch with Pui O, where some of the children already go as it is.

Cheung Sha Upper Village (*Gazetteer* p. 80) (James Hayes)

An old village, with 50 inhabitants, all of the Lau lineage.

There are 8 children of school age: see under Cheung Sha Lower Village, following.

They have 70 taochung of paddy fields. These are their own, and there is very little rented or on permit.

They are currently rearing about 20 pigs, but have no sows.

They grow some vegetables for their own use, but also sell sweet potatoes and Chinese onions.

They have some fishing sampans, and catch shrimps and small fish in season (March to August). There is not much fish available at other times.

Water in an irrigation channel here is reported to be not flowing properly. I said I would ask our Assistant Inspector of Works Mr Chan to visit and discuss with them.

Shan Sek Wan (*San Shek Wan*) (Paul Tsui)

To get to San Shek Wan takes about half an hour's walk west along the sea coast from Pui O. It is a collection of huts and hamlets. The Village Representative is known by the name of Mo Kau（毛九）, who owns a mud house which is kept very clean. He is related to the Cheung's of Shap Long. He has a total acreage of approximately 16 mu. He also possesses a full set of equipment for catching pomfret as well as cuttlefish. He has a large family of 8 including 2 married sons and 2 younger sons. His younger brother, who lives separately, also works on 6 mu of land in addition to fishing. Mo owns 3 small boats, but no square trap net.

When I commented on his cleanliness, he remarked that unlike the people in Ham Tin he keeps all his pigs indoors. The floor in his house is properly surfaced with cement, and he keeps his odds and ends much tidier than the villagers in Ham Tin or Pui O. He also told me that no outsiders own any land in his village.

There is one fairly decent house being left almost deserted, and I was told that the owner went abroad some years ago, and has not been heard of for some time. It is quite possible that he may have died.

Shan Shek Wan (336928) (Austin Coates)

Population 34. Hakka. Surnames: Mo, Tsang. A small village on the coast west of Pui O. On the track to Cheung Sha. A man of 55 is of the 5th generation, giving a date of settlement c. 1825. The Mo family came direct from Caichow. 60 taochungs, with bad water supply. 5 small fishing sampans and one larger transport sampan. Some pigs.

Shan Shek Wan (*Gazetteer* p. 80) (James Hayes)

There 30 persons in the village, belonging to the Mo lineage, eight families in all. They have been here for 6–7 generations, 200 years they said.

Five children of school age go to the Pui O school. A few others do not, either girls, or small boys who have to look after the cows.

Several men are working in Hong Kong. The Village Representative's sons are apprentices, one of them in a printer's shop. There are others too, including one in a weaving factory in Hung Hom, Kowloon.

They have 40 to 50 taochung of rice fields, but very little vegetable ground, only growing enough for their own use. They have a new drainage channel project in mind, but have no manpower to do it at present.

They have 12 brown cattle, one sow, and 10 porkers.

They say they eat most of their rice themselves, taking the rest to the Cheung Chau shops for exchange for a larger quantity of broken imported rice.

They have a few small sampans for fishing.

PUI O GROUP OF VILLAGES[9] (*Gazetteer* p. 80) (James Hayes)

There are some small Cantonese lineages here, but the great majority of villagers are Hakkas.[10]

The school for the area is located at Pui O, and I met the local leaders there.

There are currently 70 children in the school, in classes 1–5. Because it is a wholly rural district, connected (apart from the new South Lantao Road from Mui Wo, running across the Pui O valley and along the coast) only by village footpaths and country tracks, many of the younger children of school age are kept at home and do not yet attend school.

There used to be a primary 6 class, but there is now no room for it. Parents who want their children to progress further must send them to Cheung Chau or Hong Kong.

There is a request for the ten villages within the sub-district to run a lorry for their own purposes, moving people and their goods and produce, for a start. They need this to make better use of the new South Lantao Road, completed in 1956.

The sides of the main stream at Pui O need strengthening, they said. Assistant Inspector of Works Chan will visit, discuss with them, and report.

Pui O (Paul Tsui)

Pui O consists of two separate villages, namely Lo Wai (old village) and Sun Tsuen (new village). The former has a population of 140 making up a total of 32 families; the latter has 86 and 24 families. Out of the total 226 individuals, 16 exceed the age of 60 while 28 had not yet attained the age of 15. Not all the families here are of one surname, but the most successful ones have been the Hos. The Chungs here are not related to their neighbours in Ham Tin. There are also Chans. The Laws occupy a separate hamlet of 6 houses situated at about 100 yards towards the west of Sun Tsuen.

Pui O has a larger acreage of land than Ham Tin, but their land is less fertile than that of the latter. The produce of the salt fields of Ham Tin is superior both in quality and in quantity than that obtainable from all land possessed by the Pui O people. Out of the total of 56 families in Pui O, 4 families possess no land, while 2 own more than 10 mu each (one owns 12 mu and one 20 mu). Half own less than 5 mu; the remainder owns an average of 6 to 7 mu each. With the exception of 2 families who have sold their land to Cheung Chau people due to their failure in business, none of the remaining families have yet alienated their land to outsiders.

Like the inhabitants of Ham Tin, villagers in Pui O live partly on cultivating the land and partly on catching fish at sea. The Pui O community owns one big boat (30 picul size) and 10 small sampans. The sampans are used mainly for catching pomfret and cuttlefish, but they also serve for other odd jobs. All men are trained to handle boats, nets as well as fishing lines, but women restrict themselves to agricultural work in the fields as well as firewood collection on the hills. Rearing of pigs and other domestic work is exclusively the duty of the women. Though the villagers used to operate a number of square trap nets in pre-war days, none of them has any these days.

5 villagers have now gone abroad. 1 operates a shop in Cheung Chau but is now at a point near bankruptcy. One was fairly successful in business in Siam, but no news of his latest achievements has yet been heard. It is doubtful whether he is faring well at all lately, for no-one has heard of his sending any money home to his folks.

Pui O (Austin Coates)

The most important matter affecting Pui O is the maintenance of the sea-walls on either side of the large inlet which comes up into the valley, with extensive rice land on both sides. For several years the District Office has provided cement for piecemeal repairs, and the village has spent a good deal of money on the labour involved. The other day I visited the place at the highest tide, and was shocked to see that on a calm day the wall is only three inches above water level. If a typhoon occurred at a high tide, which providentially has not happened for several years, there would be a very serious inundation involving ¾ of all the fields cultivated. There seems to be no alternative but to put this forward urgently as a Colonial Development and Welfare scheme. Combined with raising the heights of the walls might go the construction of a small path on one of them, giving dry connection in all weather between Ham Tin and the Pui O school.

Pui O Lo Wai (346936) (Austin Coates)

Population 134. Surnames: Cheung, Wan (both Hakka), Law (Cantonese), 6 men in Singapore (4 Laws and 2 Wans), and 1 Cheung is a seaman at present in the United Kingdom. The Cheung family is the oldest in this part of Lantao Island. A man of 51 is of the 15th generation, giving a foundation date c. 1580, or by the 27-year period, 1550. They came from Tamshui, and the story is that the founder came first alone, saw the land was good, but was afraid of bandits. He therefore returned to his village, collected some of his clansmen, and a group of them returned to Pui O, where in the course of time they fathered a number of villages. The family tradition is that when they first arrived the area was uninhabited. Lo Wai people make a certain amount of their income by renting out fields to later arrivals; they own more land than any of the others and are a diminishing village. The Law clan speak Hakka and have evidently intermarried, though they still call themselves Cantonese. Nearly every family has a fishing sampan, the bright-light method being used, rice, fish, pigs. Their nearest convenient market is Cheung Chau. The village owns one transport sampan.

Pui O Lo Wai (*Gazetteer* p. 80) (James Hayes)

Population 130. There are 24 families of Cheung, five of Wan, and four of Law.[11] This is the oldest village in the group of villages at Pui O.

The Cheung lineage is the major one in the sub-district. They say they have been settled here for 14 generations.

There is one old man overseas, but no young ones. Two men work in shops or restaurants in Hong Kong, and six men are currently employed in quarries.

Ham Tin (350931) (Austin Coates)

Population 93, including 1 in North Borneo, and 1 working with the Kowloon Motor Bus Company. Hakka. Surname: Cheung. A branch of the founder family of Lo Wai. Ham Tin seems to have been founded shortly after Lo Wai, as suggested above. 47 taochungs, a low amount per mouth, even by New Territories standards. Some of the fields are rented from Lo Wai. Bright-light fishing is most important to this village; there are 8 fishing sampans, and 2 large transport sampans. Fish and pigs are sold to Cheung Chau.

Ham Tin (*Gazetteer* p. 80) (James Hayes)

Population 110. There are over 20 families, all Cheungs, save for one family of Chan. The Chan family is the remnant of a lineage settled here for centuries.

Pui O Sun Tsuen (343936) (Austin Coates)

Population 85, including 1 in the United States and 1 in North Borneo. Hakka. Surname: Ho. A man of 70 is of the 7th generation, giving a date of settlement c. 1760. Farming and fishing. 3 sampans and 1 larger transport sampan. Some pigs.

Pui O Sun Tsuen (*Gazetteer* p. 80) (James Hayes)

Population 110. There are 23 families, all of the Ho lineage.

Several men work overseas. Another five are working in shops and businesses in Hong Kong.

Lo Uk (*Law Uk*) (344934) (Austin Coates)

Population 28. Cantonese. Surnames: Law, Fung, Tse. The village seems to have been founded c. 1865. At first the Law family lived entirely in their own village, but gradually some of them began to move down into Lo Wai. This is a poor village. It owns no fields and no sampans, and has no people working outside.

Lo Uk (*Gazetteer* p. 80) (James Hayes)

Population 40. There are five families of Law, and one each of Tse and Fung.

Three men are working in Hong Kong, leaving only one able-bodied man in the village. Two older men have been in Singapore for over twenty years.

SHAP LONG GROUP OF VILLAGES

Shap Long (353906) (Paul Tsui)

Shap Long (十塱) is situated at a bay adjacent to the south of Silvermine Bay on Lantao Island. It is approximately 12 miles (on a direct line) from Tsim Sha Tsui in Kowloon. The bay is facing almost due east, but is screened at a distance (approximately 2½ miles) by the island of Ni Koo Chau *(Hei Ling Chau)*. It is just about 4 miles (by sea route) to Cheung Chau. A short walk across a low ridge (250 ft) (approximately 1 mile) due west will lead you to Ham Tin village at Pui O bay. The northern arm of the bay is a mass of hills separating this valley from Mui Wo (Silvermine Bay) on the other side of the slope. The southern arm consists of another mass of hills preventing you getting a direct view of Cheung Chau and the wide seas.

Shap Long is a small valley with approximately 200 mu of cultivated paddy land. The old village of Shap Long was built at the end of the plain where the slope starts greatly upwards towards west; but this old village has been abandoned, leaving only one house now still inhabited this day. The village was noted for its fung shui in the old days. It derives its name from the resemblance of its hills to the shape of ten wolves（十狼）, hence the name, Shap Long. But some prefer to say the hills look like twin white tigers. But the wolf or the tiger has been driven away by smoke when someone put up a lime kiln at the tip of Ni Koo Chau, which is directly opposite to the village. Ever since then the young men in the village began to die at unusual speed, with the dying rate highest during the Japanese occupation (probably due to malnutrition). The surviving villagers have since moved to a new village site near the northern end of the beach, but they have never yet regained their original prosperity. A stream runs down the valley flowing to the sea at the southern end of the bay. This village is occupied by a separate community who choose to name their group of houses Chung Hau（涌口）(mouth of a stream). The occupants of Shap Long village area, all Cheungs（張）who are related to the Cheungs at Ham Tin of Pui O, but the inhabitants at Chung Hau are Chans（陳）. They were contemporary settlers.

The total population of Shap Long valley (including inhabitants of Wang Tong) consists of 23 families of 89 mouths at the time when the Japanese surrendered. Since then, one died of beri-beri, but 10 more babies (6 males and 4 females) were born in these families.

8 families (all Cheungs) now occupy the new village site of Shap Long. The elder is known by the name of Cheung Hoi Shing（張開盛）, but his cousin Cheung Koon Shing（張觀盛）owns the largest holding of land (22 mu). The other Cheung families are Cheung Ma Yau, Cheung Si Kong, Cheung Chan Hei, Cheung Fook and Cheung Yau Sang. Cheung Mun, however, still lives at the old abandoned village site.

People living in Shap Long do their marketing in Cheung Chau. They all patronise the well-established firm of Lau Shing Cheong at Cheung Chau. Any letters to be sent to them may be addressed c/o the Lau Shing Cheong, Cheung Chau.

One Cheung family, now headed by Cheung Shu Kan, who used to own most land in the valley, has practically sold all the land his father used to possess. His total holding used to be 15 mu, but at least 6 out of the 15 were sold to Lau Shing Cheong during the Japanese occupation; the rest were sold to other fellow-villagers before the war. His present holding consists only of 0.2 mu of dry fields, and a small hut up the hill slope. His neighbour remarked to me (confidentially) that Cheung Shu Kan has spent all his wealth through his own follies. He was the headman of the village in pre-war days, but he used to indulge himself in gambling and prostitutes which got him in debt. He has now reduced himself to a mere tenant farmer. Lau Shing Cheong has been kind enough to let him go back to work on the land he formerly owned for which he pays 4 piculs *(400 catties, 60.5 kg)* of kuk *(unhusked rice)* for a nominal area of 6 mu which measures in fact only 3½ mu. He sold the land for $400 in 1942. He has a younger brother assisting him in his work.

Only two families own less than 5 mu of land, but 3 own more than 10 mu. Cheung Kam Shing 22 mu, Cheung Hoi Shing 20 mu, Cheung Mun 12 mu. Each of the rest has an average of 6 to 7 mu. With the exception of the 6 mu, which were sold to Lau Shing Cheong during the Japanese occupation, no outsider owns any land in Shap Long. There is a total of approximately 5 mu of unirrigated dry fields.

Chan Kam Hei has a family of 8 persons (self, wife, elder son, daughter-in-law, 2 younger sons and 2 grandchildren). His father used to own over 20 mu of land, but the title of this land consists mainly of Chinese-Customary Mortgage ownership. When the Japanese came, all former owners redeemed their land, leaving him owning only 3 mu on this day.

The land in Shap Long is said to be not very fertile, and fertiliser cannot easily be made available. The farmers have to go to Cheung Chau to buy their fertiliser, and the chief fertiliser they buy has been human urine collected in kongs. Such urine has to be carried back in narrow neck jars (changs) of the wine-dealers type. They have to pay 50 cents per every 2 jars and have them transported back from Cheung Chau by sampan. (This practice is common amongst most of the villages in southern Lantao.) They envy Silvermine Bay for having the Agricultural Department arranging the delivery of night soil.

The original settlers came here as coastal fishermen. In the good old days they had a number of big trap nets (those big square trap nets with the fishermen sitting on a rock watching the fish coming into their trap nets). Their secondary form of fishing used to be catching pomfret and cuttlefish; shrimps have always been their third choice.

The way to catch pomfret is to have a number of dummy wooden fish tied up in strings attached to a small boat so that the live fish follow these dummies and run into a spot where nets have been laid round forming a trap enclosing the whole regiment of pomfret fish. The way to catch cuttlefish is to have a pair of live cuttlefish (1 male and 1 female) securely attached to strings led by the fisherman onshore. The female will attract the males, and male attracts the females. As they come near to the rocky beach, a small net attached to a long pole will be used just to lift up the sexually hungry cuttlefish. For shrimp catching, big and thickly woven nets are used by pairs of longish sampans. When the pomfret season comes, villagers of Shap Long go across the valley and try to catch them in Pui O. Shrimps used to fetch $3,000 per boat in 1945, but only $600 last year.

Grass-cutting forms the main source of subsidiary income for the inhabitants of Shap Long. They complained that a day's efforts would not even fetch them $2 a day. It is true that shipyard people resell their grass to the junk people at 25¢ a catty, but these middlemen pay only 6¢ to 10¢ to the village grass-cutters. Besides, when grass is being sold by the villagers to the shipyard middlemen, a picul would weigh 120 catties, but when the middlemen resell the grass to the junk people their picul would weigh only 75 catties. The middlemen in Cheung Chau use two different weighing scales in their business.

A shipyard middleman in Cheung Chau (KSC) charges $40 for the use of his slipway for decoralling. In addition, he would only admit a junk on his slipway on condition that the junk would buy from him not less than 2 piculs of grass for the purpose. Should the junk fail to use up the whole of the two piculs of grass, he has to pay for the 2 piculs. Should the junk, however, use more than 2 piculs of grass, additional charges will be made. Apart from grass, the junk is also expected to buy from the slipway a certain quantity of Tung Oil and other odd items. Every time a junk is sent up to a slipway it would cost the owner not less $100, and each junk would normally do this twice a month unless that particular month happens to be a bad one.

Every family claims to rear at least one pig. Some, of course, rear more than one. Poultry is kept by every family, usually 15 to 30-odd chickens of varying size and age. None of them complain of lack of cattle as draft animals. Although not every family can afford to own an ox or a cow, nevertheless there are more than enough to go round to meet the requirement of everybody in the village.

Shap Long (365929) (Austin Coates)

Population 117. This village was originally situated further in from the sea than at present, on the south side of the valley. It was founded by a branch of the Cheung family of Pui O, who are Hakka. A Cheung of 63 is of the 13th generation, giving a date of foundation c. 1615. Around 1935 they moved to Sun Tsuen (362933), a new village on the north side of the village. About 1875 a Cantonese family of Chan, from Wai Chow, settled in the present southern section of the village, where the school is. They also founded the subsidiary village of Wan Tong Tsai (see below). Since the Second World War a third hamlet has been founded at Ngau Ku Wan (371941), most of them Hoklo living in huts on the beach. Their names are Ip, Chow, Chan and Lam, and their number is 14. They came from Hoifong. In 1952 they were joined by 3 families of Lam, who were moved from Hayling Island when it was made over to the Mission to Lepers. The Lams have built proper village houses.

Shap Long has a few men working in Borneo, and 1 working in Cheung Chau for a building contractor. Vegetable production is the chief source of livelihood, with grass-cutting and pig-breeding. Produce is sold to Cheung Chau. There are 5 sampans at Shap Long and 3 at Ngau Ku Wan. A large settlement for the disabled and destitute, run by the Social Welfare Office and originally intended to house former soldiers, etc., from Rennie's Mill (at a time when there appeared to be no likelihood of Taiwan taking any of them), is being established not far from Shap Long, and in the hills beyond this camp is the new reservoir, now under construction, from which Cheung Chau's water supply is to come.

A hard-working group of villages, in which the later arrivals have apparently got the better of the much older Hakka inhabitants. The present Village Representative, a Chan, demonstrates an unusual and creditable interest in all the inhabitants, including the Hoklo refugees.

Shap Long (*Gazetteer* p. 81) (James Hayes)

There are two settlements here: the Old Village, at the head of the valley, and the New Village, down by the shore and on the way to HM Prison at Chi Ma Wan.

Population 117 persons, in 30 families, mostly of the Cheung lineage and related to those at Pui O. The Village Representative, Mr Chan ("a nice chap"), who lives in the New Village, is a building contractor.

A few men work in Hong Kong, including the Village Representative's son.

24 children are in the local school, situated in the New Village. All come from the locality, including 5 from nearby Cheung Sha Lan. The school building is old, and they have requested glass windows for it.

They have 140 taochung of paddy fields, but little vegetable land. They eat their own vegetables, and take some to Cheung Chau to sell. They exchange their rice in the Cheung Chau shops, value for value, i.e., getting back about double the quantity they take in.

There are 60 brown cattle, but few pigs currently.

There are a few boats in the village, and there is some local fishing. The people use these to go to Cheung Chau, taking their livestock and produce, and to do their marketing.

Some Tanka boat people use the small bay, coming from outside, but the villagers say they have little contact with them.

Cheung Sha Lan is a post-war settlement of vegetable cultivators, mainly of the Fu lineage, who have links with Cheung Chau and its Vegetable Growers Association.

Ngau Ku Wan (S. H. Peplow)

Ngau Ku Wan. Ngau Ku — a bull, Wan — a bay. A breeding place for these beasts.

Chung Hau *(Shap Long Chung Hau)* (Paul Tsui)

Chung Hau is occupied by 5 families. Chan Kam Hei acts as headman of the village. He has Jim Loy Fat (a carpenter-cum-mason) living to his right and Chan Shek Kin to his left. Chan Tim (also a mason) and Tsang Pak Tai live in small houses at the back of his house. A small school attended by less than 20 children was built under a big tree at Chung Hau.

The Chung Hau inhabitants live mainly from cutting grass for the boat people who come along twice monthly, on 1st and 15th day of each moon, to have their boats decoralled. The stream at Chung Hau, with a sandbank beach serving as a dyke, provides enough space to shelter about 100 small boats during typhoons. 2 families here also own a shrimp boat each, with which they go out to catch shrimps at certain times of the year. The Shap Long villagers, however, own no boats of their own these days. They live mainly on land. The Cheungs have settled in this part of Lantao for over 10 generations. They don't seem to have a clear idea as to which part of China their grandfathers might have come from. They believe that they must have come from somewhere in eastern Kwangtung. They are Hakkas, but their Hakka has been mixed up with strong and notable Punti accents. The Chans also were of Hakka origin, though they speak Punti fluently. Jim Loy Fat, the carpenter, however, told me that he came to live here only since his father's day. He is now 62 years of age, and his family came from Chik Ngai Kon of Shatin on the mainland in the New Territories.

Wong Tong Tsai *(Wang Tong)* (367937) (Austin Coates)

A subsidiary hamlet of Shap Long, the population included in that of the parent village. Surname: Chan. Cantonese.

CHI MA WAN PENINSULA

Mong Tung Wan (352912) (Austin Coates)

A small hamlet seldom visited. Independent of its neighbours, Tai Long on the southeast and Pui O to the north, it has little contact with the District Office. There are white clay workings in the hills above it which have been worked spasmodically for several years; at present idle.

Mong Tung Wan (*Gazetteer* p. 80) (James Hayes)

Population 48, in ten families all named Fan, and Cantonese. They came here from Nam Tau in San On County, and the Village Representative is of the 4th generation. All speak Punti, but with a very odd, presumably Nam Tau, accent.

There are 15 children of school age, thirteen of whom attend the small school in this village. Another goes to the Pui O school, and stays with relatives there.

They have ten taochung of rice fields locally, and another four or five taochung at Pui O. Also, some vegetable land.

There are currently 20 brown cattle, seven sows and 30 pigs.

All the men work at home, on the farm and in coastal fishing, about 12 of them.

There are several widows, but no assistance was requested for them.

Local Public Works: they have in mind concreting a footpath down to the beach, and a small footbridge.

They are interested in having a Kadoorie orchard, and I shall ask the Director of Agriculture, Fisheries and Forestry to look into this with them.

Tai Long (369907) (Austin Coates)

Population 80. Surnames: Cheung, Ho (both Hakka), Chan (Cantonese). 1 man in the United States, 2 working in factories in Cheung Chau (soya and leather). 130 taochungs, 10 fishing sampans and 3 larger ones used for transport. Grass-cutting for sale in Cheung Chau. Pigs. The village is almost entirely illiterate. The District Commissioner visited it early in 1955, and as a result a school is to be started this year. The village is an offshoot of the Cheung villages of Pui O. Irrigation needs improving, but the Engineer advises that it will need a contractor to do the work, which is beyond the competence or funds of the village. I am approaching Mr Kadoorie about this.

Tai Long (*Gazetteer* p. 81) (James Hayes)

There are 90 persons here, in 19 families. These include 7 of Cheung, 7 of Chan, and 6 of Ho, all branched off from the lineages at Pui O.

There are 14 pupils in the small village school, all from here. 3–4 men are working on Cheung Chau, and two are at sea.

They have over 100 taochung of paddy fields and 16 of vegetable land. There is not enough water for the fields, and Assistant Inspector of Works Mr Chan will visit in two weeks to discuss the problem with them.

They engage in coastal fishing in the 1st, 6th and 8th lunar months.

They have 6 sows at present. The sale price of pork is currently $140 a picul.

Lung Mei (S. H. Peplow)

Lung Mei. Lung — a dragon, Mei — tail or end. A village near hills which have the appearance of a dragon's tail.

Lung Mei (371913) (Austin Coates)

Population 12. Surname: Cheung. A subsidiary hamlet of Tai Long, situated in the hills just above it. It has a small but well-watered area of rice land, which could be slightly extended. 2 men working in Cheung Chau (leather factory), and others living there doing odd jobs.

Lung Mei (*Gazetteer* p. 81) (James Hayes)

I have no separate note. Included with Tai Long.

MUI WO (SILVERMINE BAY) GROUP OF VILLAGES

Mui Wo (Eric Hamilton)

I remember the trip to Silvermine Bay when I took out the Governor and Colonial Secretary. They came because some bloody fool of an Unofficial Member (English) had expressed the desire that "the smooth green slopes of Lantao" should be covered by a myriad growth of cabbages. He had seen them from about 5 miles away when going round to Deep Water Bay in his launch. I don't know what the slopes are like now, but then they were brushwood and ravines. Wherever there was a bit of flat ground, usually at the outfall of a mountain rivulet, you could find a patch or two of paddy.

I once persuaded the Governor (Stubbs) and the Colonial Secretary (Claude Severn) to come along with me plus his aide-de-camp. There was some question of taking building sand in too large quantities from the Silvermine Bay beach. There is a barrier of high sand dunes at the back with only a small percentage of salt in the sand. I was afraid they would dig this away and let the sea into the swampy low land behind near Mui Wo. We had a great time, an excellent bathe and I remember the ADC and I took $5 off Stubbs and Claude at Auction Bridge. With this precedent I was thereafter on a good wicket.

Mui Wo (S. H. Peplow)

Mui Wo. Mui — a plum, Wo — a lonely peaceful dwelling. Thus we get a lonely peaceful village where plum trees abound. The place is well named. Many people will remember it by its beautiful long sandy bay, a favourite place for swimming picnics.

Mui Wo (*Silvermine Bay*) (Paul Tsui)

The most important and influential individual living in Mui Wo today is Yuen Wah Chiu (袁華照). He is better known by his nickname "Yuen Hai Kao", though people who know him usually call him (九爺) "Kao Yeh", just to show him due respect. He has lived in Mui Wo for nearly 20 years. His original native village was in Tseng Shing District (East of Canton and north of Tung Koon).

He was a smuggler by profession. He is well known and well respected by all those who earn their living in underground or semi-underground way in the whole of the Pearl River Delta. He is famous for being a crack shot with a revolver. The legendary story told about his skill with a revolver has been that he could hit a wire within the distance of a hundred feet without ever missing. He has also earned himself a reputation as a heavy opium smoker. But the real goodness in him, for which he commands respect, has been his unrivalled generosity in never refusing any help to those who are in distress, and that he always keeps his promise once he has undertaken to do a job for his clients.

Mui Wo first attracted his interest when he was actively engaged in his smuggling business. He used to use Cheung Chau as one of his advanced field bases for his operations (in these days he still holds considerable interest in various investments on the island of Cheung Chau). He first set foot on Silvermine Bay in the course of his many operations in connection with smuggling. It was through such contacts that he first developed a liking of the place. He has since built himself a house, fenced it with brick walls all round and has since bought himself a fair amount of agricultural land (including hillsides and paddy). At this moment he has an orchard consisting of approximately 5 acres employing about 20 farmhands. His orchard mostly grows laichees and guavas, with odd bits of other varieties. His son is doing most of the donkey work for him, while he himself considers he is living a life of retirement. Unless you know him well enough, you would easily overlook his presence, as he normally makes his appearance in the simple habits of a country villager.

He has a house in town (in Yaumatei). He told me that at the time when he came to settle in Mui Wo, there was only one miserable-looking hut at the beach where now stands the prosperous village of Chung Hau (涌口). That hut belonged to Tsang Sam Lee (曾三利) whose occupation was to clean coral from the boat bottoms on the beach. If he doesn't own half of the built-up area of Chung Hau village today, he has at least backed it with his finance and influence so much so that the present so-called village representative of Mui Wo is one of his fokis.

It was he who got all the inhabitants on the eastern coast of Lantao Island organised during the Japanese occupation. It was he who first thought of building all the lookouts in defence against bandits now dotted all over the inhabited districts of eastern Lantao (from Pui O to Mui Wo to Pak Mong). I suspect that the Japanese massacre of villagers in August 1945 was indirectly caused by their being too well organised in this respect.

Next high in the rank of importance among the squires in Mui Wo is Yeung Chim Tat, the Secretary of Hong Kong Yaumati Ferry Co., the managing director of United Deliveries Ltd. and the managing director of Hong Kong Shipyard Ltd. For practical purposes, Yeung's influence, as an individual, is no less (and maybe more) than that of Yuen Wah Chui. In contrast to Yuen, Yeung belongs to the other school. Yeung is a successful business man in the open, and very well educated in the Western way.

Yeung's father was a native of Pun Yu (Canton) but he emigrated to Australia in his young days. Yeung Chiu Dart and his brothers (5 of them in all surviving) were born in Australia. When Yeung first came back to Hong Kong he worked in a local solicitor's firm in a humble capacity. He soon gave up that job, and joined Hong Kong Shipyard. On one occasion he was going to get the sack by his boss, but being acquainted with the law, he fought his case and won it. He soon found himself made the managing secretary. This shipyard is under the same management as the Hongkong and Yaumati Ferry Co Ltd. When the ferry company got a monopoly he was made secretary. He has since been the de facto manager of the ferry company, and it was his brain work which led the ferry company to its present prosperity. He has done a marvellous job rehabilitating the ferry company since the war, but he never signs his name on any official correspondence of the company; rather he always makes Lau Tak Po, the manager of the company, sign official letters. He was the founder of United Deliveries Ltd., which started with only 2 lorries about 15 years ago, and the firm is now entirely in his hands, with the assistance of his capable brothers and cousins. His partner (a Hakka man) is gradually being pushed to the back stage behind the scenes. Yeung is considered by the business public to be a clever, capable but rather mean type of a businessman.

Yeung's interest in Mui Wo consists of a country bungalow with an attached farm. In addition, he has also bought up a fair acreage of paddy, and is now expanding on the hillside by terracing a fair amount of "unwanted" land. To please himself and his neighbour in Mui Wo he has found a very convenient but good excuse to divert the Cheung Chau ferry to call at Silvermine Bay twice daily, both ways. He also offered an "adequate" number of employments to a number of his neighbours in his establishments. People in Cheung Chau criticise Yeung for his special treatment in favour of Mui Wo at their expense. Any and every time you mention the name of Yeung Chiu Dart to the residents of Cheung Chau you will notice the feeling of resentment immediately shown in their gestures.

Without reference to the individual goodness or badness of either of them, the active interest that both Yau and Yeung have shown towards the development of Mui Wo has done a great deal for the prosperity of Mui Wo today.

Mui Wo Group of Villages (Austin Coates)

Pak Ngan Heung (also called Old Village), Luk Tei Tong, Tai Tei Tong and Chung Hau (Creek Mouth) are the principal villages of which the Mui Wo group is composed. No one in the valley possesses clan records, and there is little reliable material available therefore concerning the background of the area. The first village to be settled was Pak Ngan Heung, and the earliest families here were Tsui and So. When they came, most of the present cultivated area of the valley was under the sea, Pak Ngan Heung was situated on the coast, and the little hill in the middle of the valley was an island. The first two families apparently have no knowledge of their origin, but the third family to arrive, name To, understand that their ancestors were soldiers who, on being demobilised, were allowed to settle wherever they could find untilled land and live there tax free. The lanterns which their family displays at appropriate festivals have characters written in black on one side and in red on the other, which, I am told, suggests definitely that this tradition of having been soldiers is correct. The present Village Representative of Pak Ngan Heung is a To, aged 71, of the 14th generation, which puts their date of arrival c. 1585, which accords with the fact that they believe their ancestors to have been soldiers under the Ming. The family has an ancestral hall, but it is not in the centre of the village, which bears out their statement that the village was already in existence when they came. The sea in front of the village began to recede, and the next village to be founded, by a family of Kam, was Luk Tei Tong, which was at that time on the seashore. The process continued, until by the time

Tai Tei Tong was founded (by a Wong, of whom the 6th generation is 33) around 1820 the central hill was no longer an island, and access to the sea for the new village was along a creek which is now no longer navigable.

Chung Hau, which is today the main village and market for the valley (and the largest market in eastern Lantao), was founded at the beginning of this century by Yuen Wah Chiu, once a famous Pearl River bandit, and now an old man, Chairman of the Mui Wo Rural Committee. Chung Hau still has a precarious channel to the sea, and the recent dredging of the creek mouth by the Sand Monopoly has probably improved it. The sea is, however, still receding.

The preoccupation with the problem of access to the sea suggests that most of the villages were founded, as Chung Hau was, by bandits in the process of domestication; and I think we may say with reasonable certainty that ever since the Ming dynasty, Silvermine Bay has been a quiet place of retirement for rogues of one kind and another, possibly not always quiet either.

The area derives its English name from a silver mine which was operated here in the last decade before the lease of the New Territories by a group of Hong Kong Portuguese and Chinese businessmen. The mine still exists (359971) together with the last battered remains of a pier which the group also built on the north side of the bay. Not much silver was found, and the project failed.

In the 1930s a number of Hong Kong Chinese started building European houses in the valley, as weekend homes, and during the Japanese occupation several of these families took refuge here. Since the war this development has continued, and in addition Silvermine Bay has become one of the most popular places in the Colony for swimming, picnics and other holiday excursions, particularly with the Chinese community in Hong Kong and Kowloon. Noone did more to encourage this than the late Young Tsun Dart, a New Zealand-born Chinese who was for many years Secretary of the Hongkong and Yaumati Ferry Company. Under his auspices, and with his tactful handling of the older village residents, Silvermine Bay has been improved out of all recognition in the last 20 years. Young Tsun Dart was an exceptionally far-seeing and public-spirited person who, while intent on promoting his Company's interests, genuinely wanted to see more of Hong Kong's urban population having the opportunity to get out of town on a Sunday and go swimming or hiking in such places as Lantao. Seaside House, the first modern restaurant, with a residential hotel above, to be built on Lantao Island, was started by the Ferry

Company at his instigation; and there are now two similar places, one a boarding house, the other a restaurant with large changing rooms for swimmers. He was again the moving force behind the small Health Centre, which was presented to the Government by the people of Silvermine Bay in 1951, built under Government supervision, and now staffed and maintained by the Medical Department. Young Tsun Dart died in 1954, very shortly after his two sons, the elder of whom was killed in an aircrash near Rome, the younger dying in hospital in New Zealand. The ashes of the father and his two sons are interred on the top of the central hill, just above his house.

Silvermine Bay is the only area in the Southern District, apart from the Tsun Wan region, to be seriously affected by the post-war influx of refugees. Most of them were cultivators, and during 1951 and 1952 they were coming in fairly steadily from New Kowloon. All cultivable land being privately owned, the refugees became tenant cultivators, and with their migratory habits and quick wits there was a good deal of rent owing, which brought the troublesome Planters Association on the scene. By 1953 the Silvermine Bay people had taken stock of the situation, and the Rural Committee asked for approval to hold a census, which was completed by the end of 1953 and served to scare a number of troublemakers out of the valley. Yuen Wah Chiu, the aged Chairman, has here been dealing with a problem which he understands from top to bottom, and by outwardly tranquil methods of intimidation he and the Rural Committee have brought the refugee situation under control. Although there are still a fairly large number in the area, the troublesome ones have either gone or remained quiet. I also found that some of them were being encouraged (whether intentionally or not I do not know) by the Chinese Agricultural Officer in charge of the Government Agricultural Station. At my request, this officer was transferred.

Silvermine Bay is a smuggling anchorage, like Cheung Chau and Ping Chau, and it is almost certain that there is a well-organised drug traffic through it, although the people operating it are clever and no important seizures have been made. My private information on this, from a source within the valley, is precise as to its existence.

In 1951 the Government built a ferry pier at Silvermine Bay and in 1953 a small Police post was opened, in a private rented house, coming under Cheung Chau Police Station. A Government nurse/midwife resides at the Health Centre, and doctors from St John Hospital, Cheung Chau, visit twice a week. For emergency cases police launches are often used to transport the sick to Cheung Chau. A Government

Agricultural Station gives advice on cultivation, supplies seeds and chemical fertilisers and looks after animal health. The people are at present raising subscriptions to build a first-rate school; the condition of the present school is deplorable, and the teachers live in conditions no better than squatters.

Although hit by the full impact of modern urban Hong Kong, Silvermine Bay has reacted far less socially than Cheung Chau has under similar circumstances. The older generation, opium, gambling and smuggling, predominate and the younger generation looks as though it may be following in its father's footsteps. For example, few local men swim as a recreation, and sports are confined to the games of small children. Education is largely responsible. There has been trouble of one kind and another with the school; no teacher who in the slightest way defies the established ideas of the ruling clique, under Yuen Wah Chiu, can expect anything but trouble. The Catholic Church applied in 1954 for land to build a large school and I think this might render a great service to the community. The application has, however, been completely held up because of planning arrangements. The Public Works Department has prepared a plan of future development at Silvermine Bay which is unpractical and although practical amendments have been suggested by the District Office, the Planning Officer will not accept them. All decent housing development in the central (Chung Hau) area has thus been suspended for three years. Another approach to the Catholic Church will shortly be made, and if they are still interested in obtaining land, I propose that the land should be granted in disregard of the plan.

There is a vegetable cooperative for the whole valley, and extra supplies of night soil are obtained by junk from the Marketing Department's depot near Tsun Wan. At the end of the bay is a boys' camp run by the Social Welfare Office and donated to the Government by the Rotary Club of Hong Kong. The aim of the camp is to train and discipline boys picked up off the streets of Hong Kong and Kowloon and who are potential delinquents, by giving them a complete change of environment and some idea of community living and of service to society as a community. It is rather starry-eyed, but probably does a certain amount of good. The children thoroughly enjoy themselves.

Seaside House and various other houses have private generators, but there is no public supply of electric light. There is a radio telephone at Seaside House, connecting with Hong Kong, but it is kept up at a severe loss; the police post has wireless connection with Cheung Chau. The place is not yet large enough for the installation of main water supplies, but under Colonial Development and Welfare definite irrigation improvements have been made, distributing the existing sources of water, which are fairly good, over a wider area than formerly.

Ferry communications are good. There are 4 services daily between Hong Kong and Silvermine Bay, via Ping Chau. The journey takes 1½ hours and the first-class fare is 80 cents. There are also 4 daily services to Cheung Chau, the journey taking 30 minutes. At weekends sometimes several extra ferries are laid on to bring over holiday-makers, often over 3,000 on a Sunday, and most of these go direct to Hong Kong on the return trip, without calling at Ping Chau. During the summer school holidays there are special daily ferries for schoolchildren, about 1,200 per day.

The inlet which is all that now remains of Mui Wo's once well-protected harbour is a good typhoon shelter, used by quite a large number of boats. The Rural Committee charges a small toll from all people crossing the bridge over the creek mouth, the planks of the bridge having to be withdrawn every time a large boat passes in or out.

Silvermine Bay (*Mui Wo Group of Villages*) (James Hayes)

The local population is part-Cantonese, part-Hakka, but now made more motley with the post-war arrival of many immigrant vegetable farmers and livestock breeders.

First visit to the Mui Wo Rural Committee: Miscellaneous notes include the information that the vice-chairman, Yip Wai (like the chairman, not a native of Mui Wo), had been a member of the District Watch Force (Secretariat for Chinese Affairs) pre-war, but during the Japanese occupation had invested funds in cinemas in Queen's Road Central, and was now the joint owner of the Central Cinema.

There was information about the school extension, now in hand. When, as is the practice, the District Office passed the tenders to the Public Works Department for checking, it was found that the specifications had not matched the plans. The architect was asked to make the necessary adjustments, and fresh tenders were called. The government subsidy is on a $ for $ basis, with the local community matching the government dollar. The cost is now expected to be higher, but the successful tenderer has said he doesn't mind building the work for the original sum.

The Rural Committee confirmed that it does not have any regular revenue save for the tolls levied on the wooden swing bridge (centre portion up and down to allow boats to pass in and out of the creek). Noted that "there is, apparently, a confidential file on the revenues of rural committees in the Office".

Local Public Works materials for various works have been requested from the District Office and the Kadoorie Agricultural Aid Association, again through the agricultural station. Kadoorie Agricultural Aid Association assistance is said to include some cash, but the District Officer Local Public Works vote does not.

The projects they have in mind include strengthening the embankments at the main stream, which overflowed in last year's rainstorms, and a wide "public road", 10 feet wide, to run from the agricultural station, with narrower ones branching off to the several villages of the area. Both seem quite major jobs.

My staff say the "public road" idea has never been broached before, and wonder how it might fit in with any development planning schemes for the Mui Wo area. I seem to have queried whether the local people were willing to do the work, or whether this was being suggested as something for the government to provide, but this was left up in the air. Indeed the whole thing needs a proper discussion, and on the ground.

Visits to the individual villages of the Mui Wo valley were made on a subsequent occasion, as described below.

Additional information about Mui Wo in 1957: *Gazetteer* p. 82, gives a total population of 1,480 for Mui Wo in 1960, but does not provide a breakdown by villages.

Besides the native inhabitants living in the villages, and the resident immigrant cultivators and livestock breeders scattered in huts and temporary structures near their fields, there were also boat people, not so much as given a mention by the rural committee or the village heads.

There were a good many of them, from memory, perhaps up to several hundred, living in huts and boats drawn up on the foreshore in the creek at Chung Hau. There were working boats, too, as some of them still fished the surrounding waters, as well as a few visiting craft.

The wooden bridge at the entrance to the creek was a toll bridge. I do not know what was paid by boat people to bring their craft in and out; it must have been something, as labour was needed to watch out for, raise and lower its centre section, but there was a notice in Chinese and English for foot-travellers, indicating that, for them, the charge was (in English), "Five cents cross, ten cents double-cross"!

Pak Ngan Heung (S. H. Peplow)

Pak Ngan Heung. Pak Ngan — white metal or silver, Heung — a village. The village is so named from the fact that silver was once found there. The old mine and crushing ground are still to be seen. The mine is not worked now, the quantity of silver not being sufficient to pay expenses. Anyone wishing to explore the place should take care to go provided with torches, the Chinese idea of a mine being totally different from that of foreigners. Tunnels run in any and all directions, and last but not least, the place is full of bats. If these are disturbed they are apt to leave the explorer covered with a mass of unwelcome visitors.

Pak Ngan Heung (358968) (Austin Coates)

Cantonese. Surnames: Tsui, So, To, Wong (king 王), Chow. Incorrectly marked on maps as Mui Wo.

Pak Ngan Heung (*Gazetteer* p. 82) (James Hayes)

150 persons. The Village Representative is To Wo. There are 9 families of To, 7 of Wong, 5 of Chau, 2 of Tang, 2 of Fan, and one each of Ho, Kong, and Poon. Largely Cantonese. The So lineage was the oldest in this village, but it died out recently. In the Fan family, only a woman in her 30s is left.

The Tos are already in their 17th generation here, from Nam Tau in Po On County, previously known as San On County. The So and Faan ancestors had come before them, say 3–4 generations before.

They have 200 taochung of paddy fields, but only 10 taochung of vegetable land. Most of the land is their own, but some rent from Cheung Chau people who (surprisingly enough) did not buy it during the Japanese occupation but had done so before the British took over the New Territories in 1898–99.

There are over 40 children of school age, of whom 20 attend classes.

No water buffaloes here, but 65 brown cattle. Only 4–5 sows currently.

The village has a Kadoorie orchard. It was explained that they had held a meeting to discuss this, and all were agreed. They cleared a small area behind the village (emphasizing "but only so much"), and opened the orchard last year.

Asked about the government's agricultural post, they said the service was OK. However, there had been a vet there, but not now, and they asked for a replacement to be sent out. They had applied to the station for money and cement from the Kadoorie Agricultural Aid Association to repair some of their houses.

They have a grievance over the government's sale of part of their former forestry lot to the Yeung family (Hongkong and Yaumati Ferry Company) who had built a country house here pre-war. The sale of the forestry land to the Yeung descendants was quite recent. They want another area of Crown land from the District Office to replace it, and asked me to write to the Agricultural and Forestry Department for trees to plant out, if and when they get it. They mentioned 43 acres, which seems a very large area if correctly stated.

Like the other villages, they want to make a wide concrete footpath to Chung Hau, and say they will carry out the work in the autumn.

Unemployment. They say there are 10 young men without jobs in this village. A side note states there are another 10 each in Luk Tei Tong and Tai Tei Tong, but very few in Chung Hau.

There is an old Kwan Tai Temple in this village, but it is in need of renovation. There is no commemorative tablet or dated bell inside the building, to help indicate its age.

Luk Tei Tong (S. H. Peplow)

Luk Tei Tong. Luk — a deer, Tei — ground, Tong — a pool. A village near a feeding ground and pool for deer. Deer are plentiful all over Lantao.

Luk Tei Tong (360957) (Austin Coates)

Cantonese. Surnames: Kam, Wan, Lan.

Luk Tei Tong (*Gazetteer* p. 82) (James Hayes)

About 200 inhabitants. The Village Representative is Tsang Lin. There are 8 families of Tsang, 5 of Kan, 3 of Wan, 3 of Lam, 2 of Wong, and 1 of Kam. The Lams and Kams are the longest settled here, for over 10 generations, they say. The Lams came from Fukien and the Kams from Wai Chau. The Tsangs count 8 generations, from Ying Wah.

This is apparently a mixed Hakka-Punti village. The Tsangs told me, "We were originally Hakkas, but now we are Punti." Most families in the village are Hakkas.

There are 40 children [of school age?] of whom 18 attend school in Mui Wo.

There are 160 taochung of paddy fields, and about 40 of vegetable land. The latter are cultivated all year round. The village has a forestry lot of over 20 acres.

There are 15 water buffaloes and 30 brown cattle, with 20 sows at present.

Three men are working in Cheung Chau, and some are working as seamen. There were enquiries about entry to the Hong Kong Sea School for young men (I said I would make enquiries and let them know). Some men also want introductions for labouring jobs on the South Lantao Road, now being extended beyond Cheung Sha.

They have a list of Local Public Works projects they would like to carry out. First, there is a connecting footpath of 6-foot width. Then, there is a dam to provide more water for irrigation, and they also want to strengthen the nearby stream embankments on each side. They say all this work can be done in the autumn, and I promised them the necessary materials for that time.

Tai Tei Tong (360963) (Austin Coates)

Cantonese. Surnames: Wong, Lam. 500 persons. The Village Representative is Wong Chau-fuk.

Mixed Hakka-Punti. The families include Wong (3), Lam (4), Lo (1), Laam (1), Tsang (3), Yau (2), Ch'ak, and are mostly Punti. The Yaus say they are Hakka. The Laams say they are of the 10th or more generation. All are reported to have come from Fukien, but some didn't really know.

The population figure is too high for this number of native families and many of the residents of the area are likely to be recent immigrants.

There are 150 taochung of rice fields, and 20 taochung of vegetable land. Only a small number of native villagers rent Crown land on permit. They say this is also the case at Luk Tei Tong and Chung Hau, with most native families owning their own land.

No water buffaloes, but 30 brown cattle, with currently 11 sows.

Over 10 men work outside the village, and one man is at sea.

There are over 50 children of school age, of whom 30 attend school.

Local Public Works projects include the "road" connection to Chung Hau, also suggested by Yip Wai, vice-chairman of the Mui Wo Rural Committee, during my meeting with them.

Chung Hau (S. H. Peplow)

Chung Hau. Chung — stream, Hau — mouth. A village near the mouth of a stream or river.

Chung Hau (366964) (Austin Coates)

Multi-clan.

Chung Hau (*Gazetteer* p. 82) (James Hayes)

This is the market village nearest the pier and the creek. The chairman, Yuen Wah-chiu, has a farm here, built in the 1930s, after he came here from the East River district of Kwangtung, purchased much land, and planted a large laichee orchard. It has very solid stone buildings and a watch-tower, and looks older than it is.

There are 45 buffaloes here (20 of them belonging to Mr Yuen) and 40 brown cattle. Currently there are between 200–400 pigs, and sometimes many more.

There is no pig cooperative society at present, though a vegetable cooperative is already in operation. Its membership includes both native and immigrant cultivators. The general feeling is that it is working well, and forming one for the pig-breeders as well has been suggested.

In regard to education, there is a government subsidised school which provides for instruction in primary classes 1–6. It operates in two sessions, and there are over 90 children in each session. They come from all around. In addition, 8–10 children from Mui Wo travel daily by ferry to schools in Cheung Chau and another 15 attend higher classes in Hong Kong. It also appears that a Roman Catholic priest is running some classes here.

Notwithstanding, around 150 children are reported to have no schooling at present, though the majority of their parents can afford to send them. Presumably, this unsatisfactory situation will be resolved when the extension is built.

Wong Chuk Long (364984) (Austin Coates)

(The editor has not been able to locate this village on the map.)

Population 20. Surname: Chan. This is the largest of the hill settlements situated on the high ridge that separates the Silvermine Bay valley from the Pak Mong group of villages. It was founded slightly over 20 years ago. The senior member of the family works in Hong Kong. A remote and rather more squalid place than Hung Shui. About 20 taochungs of rice land.

Wang Tong (*Gazetteer* p. 82 has "Mang Tong") (James Hayes)

Not recorded. A small hamlet behind the north end of the main beach.

Mang Tong *(Wang Tong)* (S. H. Peplow)

Mang Tong. Mang — blind, Tong — a pool. A village near a hidden lake. The place is well named, the stretch of water is surrounded by thick bushes, and one almost walks into it before discovering the place.

LANTAO SOUTH COAST: EAST OF MUI WO

Mang Kok Tsui (*Man Kok Tsui*) (395965) (Austin Coates)

Population 32. Surnames: Cheung, Wong, Lee, Kwok, Yung. A refugee settlement of vegetable growers, tenants of the Yuens of Chung Hau. The first of the present cultivators arrived 3 years ago. Cantonese; one family of Hoklo. 3 sampans, 2 stake-nets. They sell their vegetables through Ping Chau for onward transport to Hong Kong.

Man Kok Tsui (*Gazetteer* p. 82) (James Hayes)

This is an outlying location, on the left extremity of the Bay as you look towards Hong Kong. There is a post-war settlement here, of over 30 people, of mixed origin, with households of Wong, Li, Ko, Cheung, Leung, Lam, Chang, and a Hoklo one of Ng, a fisherman. Mostly Hakkas from Wai Chau.

Includes five children below 6 years of age. The two of school age do not attend school.

They catch fish and sell vegetables through Peng Chau. The cultivators all belong to the vegetable guild there. They rent land from Mr Yuen of Mui Wo (the Rural Committee chairman) and have applied for permits for their Crown land fields.

Kau Sat Wan (Dog Flea Bay) (393970) (Austin Coates)

Population 14. Surnames: Lee, Wong. A group of refugees from Chungshan District. They first settled at Ping Chau, then moved here in 1953. Their land is rented from Kwong Fuk Cheung of Ping Chau. Vegetable cultivators, with one sampan, selling their produce at Ping Chau.

Trappist Haven (390980)

Hung Shui (381985) (Austin Coates)

(The editor has not been able to locate this village on the map.)

Population 8. Surname: Lee. A hill village above and not far from the Trappist Monastery. Some well watered rice fields, sufficient to feed the two families who live here. The present inhabitants arrived in 1945, a few weeks after the end of the war. One child goes to school in Silvermine Bay, about a 45-minute walk. Since the monastery brought large cows to Lantao and began making cheese, the Lee families make some extra money selling hay for cow fodder. Strong, sensible people, with plenty of self-respect.

This village was probably on the site of what is now the Discovery Bay Golf Course.

Hang Tse (377997) (Austin Coates)

(The editor has not been able to locate this village on the map.)

Population approximately 10, settled since the war. There is a blind young man here who needs Social Welfare Office attention. The village is situated in the midst of the Lantao Development Co's forestry lot, but its grass-cutting area is specially excluded from the Company's zone.

This village was probably on the site of what is now Discovery Bay.

Cheung Sha Lan (391990) (Austin Coates)

Population 15. Surnames: Fung, Lau. An old village whose inhabitants, by the end of the war, had either died out or moved away. The present occupants arrived in 1946 and have since lived by growing vegetables on a small strip of shore land, containing a good deal of sand. They sell their produce at Ping Chau. The people are of mixed Cantonese and Hakka descent known as Pun Kong Cham (half-filled pitcher). Although older than Nim Shu Wan by one year, Cheung Sha Lan is tending to become a subsidiary of the younger place, and the two can be dealt with administratively together.

This village was probably on the site of what is now Discovery Bay.

Nim Shu Wan (S. H. Peplow)

Nim Shu Wan. Nim Shu — a kind of fruit, like the blackberry, Wan — a bay. A seaside village where this fruit is grown.

Nim Shu Wan (391994) (Austin Coates)

Population 42. Surnames: Koon, Tsang, Lam, Chow Tse. This place was once a fishing village and small market larger and more prosperous than Ping Chau, 80-year-old residents of which can remember clearly the prosperity of Nim Shu Wan. It is difficult to see how Nim Shu Wan, as exposed as it is, could have remained prosperous; its decline and eventual complete extinction are locally blamed on bad fung shui. What is known is that around 1890 it was more important than Ping Chau, but that after the lease of the New Territories Ping Chau steadily grew and Nim Shu Wan declined till by the end of the Second World War it was totally deserted. Only a ruined temple and some remains of houses suggest that the bay has once been inhabited. In 1947 Tse Chung, the new Village Representative (the first at Nim Shu Wan), was dismissed as redundant staff from the Great China Match Factory at Ping Chau. He had worked there for 3 months, before which he tried to earn a living farming in Yuen Long. Left without a job, he crossed to Nim Shu Wan and began to cultivate seashore land. He is of mixed Cantonese-Hakka descent, and others of his kind soon joined him. Remarkable progress has been made in reclaiming most unpromising soil, and there is an excellent community spirit in the place. The Director of Agriculture is being asked to assist in various ways; and this office will arrange for them all to receive permits for the land they are cultivating. This is the brightest refugee spot in the Southern District.

Nim Shu Wan (*Gazetteer* p. 86) (James Hayes)

A settlement of Hakka newcomers who farm vegetable crops for sale, 90% of them to the wholesale market at West Point, Hong Kong Island. Main crops are cabbage in winter and string beans in summer, with melons and spinach. They buy their rice from Peng Chau. Pig-breeding is also carried on, currently with 30 sows and 40 porkers.

According to local people, Nim Shue Wan had been uninhabited for 20–30 years before the newcomers revived the settlement in the late 1940s. At an earlier time, it was a prosperous place, with a Tin Hau temple built by a fisherman named Lam [the site said to be registered in the name of a Lam To-ho). Families of Liu lived there at that time, related to the long-settled Liu's of Peng Chau.

Tai Tsun and Yi Tsun (S. H. Peplow)

Tai and Yi Tsun. Main and second village. These two villages are on the east coast of Lantao, and very close to each other. The name Tai being given to the larger or most important.

Tai Pak (384001) (Austin Coates)

Population 66. Surnames: Cheung, Leung, Ng, Lam, Wong (king 王), Kan, all of whom are Hakka; and Mui and Ng (man-five 伍), who are Cantonese. The Cheungs were the first arrivals, coming from Shataukok. A man of Cheung of 56 is of the 5th generation, giving a foundation date c. 1825.

This village is fairly poor, and receives little in remittances. Apart from rice cultivation, there is some fishing, from boats and from stake-nets, and pig-keeping. They sell wood and grass to Ping Chau, to the lime kiln, and buy pig food from there. Few vegetables. The sea in front of the village is very shallow, and transport is always likely to be a problem. In 1953 a large-scale forest project was started in this area by the Lantao Development Company, of which Mr G. O. Jones, who has a record of honourable failure as a candidate for the Urban Council and as a jockey, is the manager. The Company intends to plant Australian hoop pine over a very large area in the hinterland of Tai Pak; their present forestry lot extends right over the hills behind the village and embraces the upland hamlet of Hang Tso. After some early months of anxiety, it looks as though the project may be successful, and the Company has already improved a large tract of what was more or less wasteland near the sea, where they now have their nursery, and is building a jeep track to the upper levels of their forestry lot. Hoop pine is slow-growing in the early stages, and it is not expected that there will be any appreciable sales of timber before 1963. Mr Jones is employing several men from Tai Pak, and several from Nim Shu Wan. He has found the latter more hard-working, which is only another small example of the indolence which is at the core of so many New Territories problems. He is also employing a few former patients from Hayling Island, who are non-contagious and go back to the island for regular medical checks.

Hitherto there has been an illiteracy problem in Tai Pak, some children went to school at Ping Chau, but it is a very awkward journey, particularly in summer. After two years of persuasion the District Office has finally managed to get the villages of Tai Pak and I Pak to agree to have a joint school, each village being too small for a school of its own. This school, which will be built in a few months' time, will be situated on the steep hill between the two villages, and the children of both will have to climb about an equal distance.

Tai Pak (*Gazetteer* p. 85) (James Hayes)

Population of 90 persons. A small village, but of mixed settlement. There are five families of Leung, here for five generations, from Po Kat in San On County; three of Lam from nearby Hei Ling Chau; one family of Cheung, here for five generations from Kwai Chung in Wai Chow; two families of Wong here for 20–30 years from nearby Hung Shui, above Mui Wo, where settled for several generations, also from Po Kat; and one family of Kan, origin and length of settlement not stated (perhaps absent on the day of my visit). All are Hakkas, speaking the Hakka language.

Twenty children attend the new school shared between this village and Yi Pak, and therefore built halfway between the two settlements.

All told, there are some 30 taochung of rice fields, and they say they eat their own rice (instead of the then common practice of exchanging it for a greater quantity of imported rice), buying extra at Peng Chau at need. Some vegetables are also grown, sold to Peng Chau or Hong Kong Island. They have 90 pigs, including ten sows. There is very little fishing here, owing, they say, to the low coastline and the very shallow bay.

They asked for materials from our Local Public Works Vote, to build a concrete path and small bridges to the new school, so as to make it easier for children in bad weather. Also to provide a piped supply of hill water for drinking.

I Pak *(Yi Pak)* (385408) (Austin Coates)

Population 77. Surnames: Leung, Chan, Ngai, Ng, Lee, Lau, Sun. Hakka. Leung is the founder of the village, followed 6 years later by

Chan. Leung came from Shumchun, and a man of 39 is of the 4th generation, giving a date of arrival c. 1865. Most of the people are illiterate, and will benefit greatly from the joint school being built this year for Tai Pak and I Pak. The means of livelihood are as at Tai Pak. I Pak owns 3 fishing boats without masts, 2 stake-nets, and 2 sampans. Some of the womenfolk are employed in Ping Chau at the match factory. The sea in front of I Pak is very shallow and communications are awkward. It is about 1½ hours by sampan to Ping Chau.

Yi Pak (*Gazetteer* p. 86) (James Hayes)

There are 97 people here. Also Hakkas, and of mixed settlement, with five families of Chan from So Kun Wat in the mainland New Territories, here for five generations; one family of Lee, also from So Kun Wat and here for five generations; and one of Ngai, for four generations, from Po Kat. In the case of the Lee family, it is specifically stated that they came here for the stake-net fishing, there being insufficient at So Kun Wat, a large village. Probably also applies to the Chans. Besides these, there are also two families of newcomers, Tsangs removed here from Hei Ling Chau.

They have 60 taochung of rice fields, including some at adjacent Sam Pak, and 20 taochung of vegetable fields, sending some produce for sale to Hong Kong Island. They also say they eat their own rice. Currently, there are 30 pigs, with two sows. They fish with nets (stake-nets or from boats?) in season.

Fourteen or more children go to the new school, shared with the Tai Pak folk.

Their Local Public Works needs are to bring piped water from the hill for a cleaner drinking supply, and to aid vegetable cultivation.

There were two small hamlets beyond Yi Pak, named Sam Pak and Sze Pak (*Gazetteer* p. 86) but I have no details.

I recall that grass-growing was an extra source of livelihood in the Tai Pak to Sze Pak area.

When on the District Office launch, I used to see village rowing boats laden with grass, on their way to Cheung Chau, where its many boatyards needed a supply of fuel for the regular breaming of the wooden hulls of the larger fishing craft.

Sam Pak (Austin Coates)

Fields owned by I Pak people. No houses.

Sei Pak (*Sze Pak*) (429031) (Austin Coates)

Sometimes called Pak Tau Ku, population 14. The hamlet, which is on the coast, consists of 3 families of Hakka (surnames: Wong, Ng) who were cultivating here before the war, left during the Japanese occupation, and returned after the war. They have since been joined by one Cantonese family of Tsang, who moved over from Ping Chau. There are 6 taochungs of rice land. The principal source of livelihood is grass-cutting.

Lantao South Coast East of Mui Wo (James Hayes)

Villages on adjacent parts of Lantao Island coming under the auspices of the Peng Chau Rural Committee include Nim Shu Wan, Tai Pak, and Yi Pak.

There were two small hamlets beyond Yi Pak, named Sam Pak and Sze Pak (*Gazetteer* p. 86) but I have no details.

Further along the coast, towards Kap Shui Mun and Ma Wan, are the other hamlets of Wan Tuk, Mong Tung Hang, and Ngong Suen Au, all located at Chuk Ko Wan (Penny's Bay), and others at Fa Peng and Pa Tau Kwu. Altogether, 11 families with 34 persons live in these places, which are listed in *Gazetteer*, p. 87. There are eleven taochung of rice fields, and some vegetable land, with, at present, ten cows, six sows, and varying numbers of porkers. These people all shop on Peng Chau, where many of the men also work. This information was obtained from the Rural Committee. Wan Tuk, I was told, had been burned by Japanese troops during the wartime occupation.

NORTHEAST LANTAO

Yam O Tuk *(Yan O Tuk)* (401041) (Austin Coates)

Population 7, consisting of a refugee couple (surname Au) who came here from Punyu District in 1951. With 2 fokis they run a small farm on the north side of the small bay. 1 sailing boat. At the south end of the bay are 3 grass-cutters. The fields in this part are cultivated by Luk Keng people.

Luk Keng (395042) (Austin Coates)

Population 14. Surnames: Wong, Cheung. Cantonese. Refugees from Shiu Hing District. Vegetable growers. 4 small boats, used for fishing and for transporting vegetables to Ma Wan. They take their own vegetables right into Hong Kong by ferry from Ma Wan, to conduct their own sales.

Tung Yat Hang *(Tung Yip Hang)* (383034) (Austin Coates)

A settlement of stake-net operators from Tai Lam Chung. 5 persons at present. In the area there are other similar settlements, consisting of not more than half a dozen people all from Tai Lam Chung.

Taiwan *(Tso Wan)* (430046) (Austin Coates)

Population 11. Surnames: Yiu, Cheung, Wong, Hung. They moved here from Ma Wan in 1952. Hakka. Grass-cutters. 1 sampan, used for transporting grass for sale at Shamshuipo.

Fa Ping *(Fa Peng)* (S. H. Peplow)

Fa Ping. Fa Ping — a vase. The village is so called from the fact that the surrounding hills are shaped like a vase.

Fa Ping *(Fa Peng)* (430042) (Austin Coates)

Population Surnames: Cheung, Tsang, Lee, Lam. Grass-cutting and poultry.

Ching Chau (414053) (Austin Coates)

(The editor has not been able to locate this village on the map.)

A stake-net settlement operated for over 20 years by villagers from Tai Lan Chung, in Yuen Long District. At present there are 28 people here. 9 sampans. Fish collectors come down this part of the Lantao coast daily, to take fish to Ma Wan, for transport by ferry to Hong Kong.

This village was probably on the site of what is now the Sunny Bay MTR station.

Ngong Shuen O *(Ngong Shuen Au)* (Upper Boat Dock) (417032) (Austin Coates)

Population 11. A refugee settlement consisting of one family from Toi Shan District. Cantonese. They came to the New Territories in 1950, settled first at Tai Pak, then moved here. 16 taochungs of rice land, most of it hilly, grass-cutting. 1 good medium-sized sailing boat, single masted, used for transporting grass to breaming yards at Taikoktsui.

Chokkuwan *(Chok Ko Wan Tsui)* (412037) (Austin Coates)

Population 8. A private farm owned by a Hong Kong businessman and maintained principally to feed his refugee clansmen. Dried fruit.

Ta Pang Po (S. H. Peplow)

Ta Pang Po. Ta Pang — to make mat-sheds, Po — a mart. A place where the villagers are experts at making mat-sheds.

Notes

1. See *Friends and Teachers Hong Kong and Its People 1953–87* (Hong Kong: Hong Kong University Press, 1996), p. 53. Chapter 2 recounts my/our experiences at Shek Pik between 1957–60.
2. Like a few others on related matters, they only expose my ignorance! However, when the devil drives, needs must, and I am happy to report that, ultimately, we were able to negotiate a cash settlement and achieve voluntary surrender of all village holdings.
3. See the chapter on the settlements in the Shek Pik Valley in *The Hong Kong Region 1850–1911: Institutions and Leadership in Town and Countryside* (Hamden: Archon Books, 1977).
4. There were 29 persons here at the 1911 Colony Census, 18 males and 11 females.
5. See also, and generally, *Hong Kong Region*, pp. 27–30.
6. This village had greatly declined in numbers in the past fifty years. At the 1911 Colony census, the recorded population had been 277.
7. As at Shek Pik, the villages of Shui Hau and Tong Fuk had larger populations at the 1911 Colony Census: 214 at Shui Hau, and 198 at Tong Fuk. Recurrent epidemics were again the cause, not out-migration or men away.
8. See my *Rural Communities of Hong Kong: Studies and Themes* (Hong Kong: Oxford University Press, 1983), pp. 12, 187–191.
9. Pui O is the centre of the sub-district and the seat (from 1958) of the South Lantao Rural Committee. See the chapter on Pui O in *Hong Kong Region*.
10. See *Hong Kong Region*, table at p. 131.

11. In Lo Wai and San Tsuen, the populations are lower now than those recorded at the 1911 Colony Census, when they were 165 and 132 respectively. The population at Lo Uk was slightly less than now, at 37. No population figures were given for Ham Tin and some other villages of the group, probably omitted to an oversight.

Tung Chung, c. 1960. Chap Lap Kok can be seen in the distance. Courtesy of Tim Ko.

Tung Chung Fort, 1950s. Courtesy of Tim Ko.

Po Chu Tam (left) and Hau Wong Temple in Tai O, 1960s. Photograph by Chan Chik.

Stilted huts in Tai O, 1956. Photograph by Chan Chik.

The famous General Rock (將軍石) in Tai O, 1953. Photograph by Chan Chik.

Ngong Ping, 1950s. The arch leads to Po Lin Monastery. Later the monastery was greatly expanded and rebuilt. Photograph by Chan Chik.

Terraced fields in Shek Pik, c. 1958. Photograph by Chan Chik.

The tranquil Ngau Ku Long, 1992. The fields in front of the village had long been abandoned when the photograph was taken. In a few years' time, the whole area would see tremendous changes with the construction and completion of the North Lantau Highway. Photograph by Tim Ko.

Fan Lau Fort, 1990. Photograph by Tim Ko.

Landing pier at Cheung Chau, 1930s. Courtesy of Tim Ko.

Cheung Chau, c. 1950. Courtesy of Tim Ko.

Peng Chau shortly after the 1941–45 war. Courtesy of Tim Ko.

Ma Wan, c. 1960. Photograph by Chan Chik.

Clear Water Bay (Hang Hau) Peninsula, 1935. Photograph from the collection of the late Mrs Joan Fuller, now in the possession of Jason Wordie.

Terraced fields in Clear Water Bay (Hang Hau) Peninsula, 1935. Photograph from the collection of the late Mrs Joan Fuller, now in the possession of Jason Wordie.

Sai Kung Market, 1950s. The two-storey building is a Protestant seminary. Courtesy of Tim Ko.

4
Lamma Island

LAMMA ISLAND (Paul Tsui, written after he retired)

Off the coast to the southwest of Aberdeen on Hong Kong Island was Lamma Island. There the late Fr Finn, a Jesuit historian, first discovered and dug up some prehistoric relics some years before the Second World War. Ever since then, the island began to attract visitors, not a few of them were Kwai Los. Because of its geographical proximity to Hong Kong Island, it also attracted a lot of picnickers, particularly those on private pleasure craft.

Topographically Lamma Island can be conveniently divided into two parts, North Lamma and South Lamma. Easy access to the latter can be by sampan from Aberdeen; but access to the north would normally be by junk from Kennedy Town or West Point. As there is not that much arable land on the island, the inhabitants have been more urbanised than those on the mainland of the New Territories. Socially and economically they are closely tied to the people at Aberdeen and Apleichau.

LAMMA ISLAND (Austin Coates)

Lamma Island is overwhelmingly Cantonese. There are a few Hakka at Yung Shu Wan, and a handful of refugees at Sokkuwan.

Although there are some poor villages (Lo Tik Wan and Luk Chau) and a number which have been hit hard by the restrictions imposed on trade under the United Nations Resolution of May 1951 (Tung O and Yung Shu Ha) this island can safely be called the most prosperous part of the Southern District. In such heavily populated villages as Tai Wan To, all resources have to be used to maintain a reasonably standard of living, and nowhere can life be called comfortable by urban standards; but the various villages of the Chow clan (all near Yung Shu Wan) and other villages such as Pak Kok Old Village, Tai Ping and Mau Tat enjoy an almost absurd state of well-being, in which local thrift and activity are rendered almost superfluous by a flow of remittances which to people in other parts of the district must make life on Lamma seem a paradise.

The population is about 1,640, of whom 800 are fisherfolk permanently based in Sokkuwan. Of the land population, however, at least 150 live and work in Hong Kong, or abroad, and many more have connections of one sort or another with Hong Kong. There are many Western-style houses, some of them set imposingly between pigsties and threshing floors. Remittances also account for the (at first sight) mysterious existence of Yung Shu Wan, an evidently flourishing place without any apparent sources of livelihood, other than as the port for a few very small villages. Many of those abroad are seamen, and many of the older men are retired seamen. A rich nautical English is spoken, its use stimulated by alcohol. Every other kitchen on Lamma tends to be a galley, and when going upstairs one goes above. On the various occasions when I have held morning meetings in Yung Shu Wan, sometimes quite early, at least one member of the Rural Committee smelt of Chinese liquor. It was of course on Lamma that the incident took place in which a villager, complaining of the overbearing manner of a certain police officer, told me that he did not mind the officer stopping his illegal stone quarrying, but he did object to being called a bastard.

Although the earliest of the island's present inhabitants evidently did not arrive until about 1615, human activity on Lamma is of great antiquity, and there is an archaeological site at Hung Sing Yeh which has aroused justifiable interest among students of Far Eastern pre-history.

Lamma is close to Hong Kong, is closely linked economically with Victoria and Aberdeen, and has risen to prosperity as Hong Kong grew during the last century. Yet the Government of Hong Kong has done very little on this island. There are schools at Yung Shu Wan and Lo So Shing, which between them serve the children of the entire island, but there is no clinic, no post office, no midwife, no electricity, no roads, no ferry. In spite of the prosperity of some parts of the island, the Government's record is one of neglect.

The main impediment to Government assistance is that geographically the island is so divided into a northern sector (centred on Yung Shu Wan) and a southern sector (centred round Lo So Shing and Sokkuwan) that the place is virtually two islands. The walk from Lo So Shing to Yung Shu Wan takes about 1½ hours carrying a load, but the section between Hung Sing Yeh and Lo So Shing (the southern stretch) is steep and exposed, and due to the nature of the terrain (rock covered with only the thinnest patches of grass, here and there) it would be very difficult for the villagers to construct a good path. This means that every development at Yung Shu Wan has to be duplicated at Lo So Shing, and vice versa, if one sector is not to be developed to the neglect of the other.

With its small population it would be out of the question for Lamma to be provided with two clinics, two post offices, two ferry piers, etc. The prerequisite to any real development therefore is a road from Yung Shu Wan to Sokkuwan, and details of this are being prepared for full consideration. The first stretch, from Yung Shu Wan to Hung Sing Yeh, will need minor resumptions, and represents few difficulties. The second stretch is harder and more expensive.

If this road can be built in the next few years, it would be possible to concentrate development at Yung Shu Wan. There is already a scheme afoot among the local people to put up money for a small health centre, and this should be encouraged. The Hongkong and Yaumati Ferry Co. did, for a short time in 1946, operate a service between Yung Shu Wan, Cheung Chau and Hong Kong, but it closed down as a loss. If a pier and a road were provided, however, there is no doubt that a regular ferry service would quickly begin to pay. Lamma is an island of great beauty, with adequate water supplies, and the northern sector would quickly attract people from Hong Kong to build houses on the same lines as at Cheung Chau, from which a number of people go in to Hong Kong to work each day.

Lamma's main needs are therefore a road, a pier at Yung Shu Wan, a post office, a small health centre, and a resident nurse/midwife.

Medical services are unsatisfactory at present. A doctor from St John Hospital, Cheung Chau, visits Yung Shu Wan and Lo So Shing once a week, and at Yung Shu Wan there is a small missionary clinic. There are no proper midwife services.

LAMMA ISLAND (Paul Tsui)

Except for the village of Yung Shue Wan itself which was built at the sea beach, all the other 6 villages in northern Lamma stand either at the foot of or on sites slightly higher along the skirt of the low hills. Yung Shue Wan itself is a fishing village but the others are mainly agricultural. Being close to the sea, most of the inhabitants are familiar with the livelihood of coastal fishermen. Not a few amongst them have big square trap nets of their own, but not every family can afford the capital to provide one for themselves.

The houses on this northern part of Lamma are built in a pattern typical of that found in various parts of South China (i.e., close and attaching to one another, sharing common walls with their neighbours in flights of 10 to 15 per line). Streets, like all other Chinese villages found in the New Territories, are narrow, but in 9 cases out of 10 each house has its own small courtyard of 10 to 15 sq. ft. fronting the main door. These small courtyards suggest immediately their use for drying farm crops as well as sea products. But the significance here is the existence of 1 or 2 dwarfish structures built on the side of every one of these courtyards. Such dwarfish structures are being used either as pens for domestic birds or as pigsties; but younger chicks could be seen being kept indoors in the living room, which also serves as a dining room as well as a parlour. The rear section of the parlour is partitioned into a small bed chamber with one or two double-beds of the usual plain board type; but children of school age would normally be made to sleep in the cockloft immediately above the main sleeping chamber. Not all such sleep chambers have windows of their own, but the chamber door (usually on the right side of the partition) leading to and from the parlour provides the necessary lighting and ventilation. The partition wall may, in some cases, be cut into vertical strips of openings serving as additional channels for light and air.

In every village two or three two-storied houses can be found. Their occupiers are usually the better-off individuals of the village who may either have made their money as seamen on board an ocean liner or, in some cases, having been fairly successful in business in town. Not a few in every village here are still at sea, but many more have spent a few years working as stokers on board Blue Funnel liners. There are usually one or two families in each village who still depend on the income derived from their big square trap nets on the island coast, but all the women are kept very busy at their work in the fields.

The floors, in many cases, are well surfaced with cement, but more are of hardened earth. Sweepings are carried out more than once a day. But young chicks do not seem to show much respect for the housewife's increasing efforts by their continuous casual droppings. Furniture in the parlour consists of one or two old chairs, two or three worn-out wooden benches usually occupied with odds and ends only to be removed when required for use at meal times, or for distinguished visitors.

Casual calls by neighbours may take the form of shouting in from the streets outside with voices carried over the distance of the courtyards; seldom would such neighbours remain in the house for a period longer than necessary. Conversation whilst standing in the parlour is not looked upon as inhospitable. Boiled water or tea may be obtainable by self-help, but it is not considered impolite to miss out on this part entirely except in the case of visits by distinguished guests. Arrivals of neighbours are not marked by any initial greetings, and their departure is not marked by a goodbye either. A complete silence following a battery of words and phrases would normally indicate the end of such business calls, though such silence may be followed by indifferent remarks of assurance.

Villagers in northern Lamma tell about the DDT sprays which took place in early 1946, with great satisfaction. They claim that the two sprays which the medical authorities did in 1946 have reduced considerably the menace of mosquitoes. They even claim that their beds are now free of bugs and fleas. When I enquired about the visits of a Medical Officer, the reply was "not known". Malaria is said to be very rare since the sprays in 1946 but villagers do not seem to know how many different types of malaria there are. A fever without shiver, for instance, is not looked upon as malaria, and they are usually cured by a dose of herb coupled with a hot bath. Tobacco and tree leaves are still the chief means used to stop bleeding. Not more than two households keep a stock of cotton wool and iodine for first aid. One or two families keep a small stock of Chinese patent medicines for emergency needs, but that is available for anyone who may need them in an emergency.

Pregnant women work up to the time close to delivery, but the tendency has been to send expectant mothers to maternity clinics in Aberdeen or Hong Kong, wherever possible. This practice has become popular not as much through the administering of medical science as the attraction of the birth certificates. The elderly women who used to attend to the birth of their own relatives claim no knowledge of cleanliness; unwanted or rusty scissors were used for the cutting of the infant cords. The reason for using the unwanted old dirty scissors was mainly superstition as all such instruments used in connection with childbirth would be thrown away soon after they have been used. The idea that cleanliness is essential to the safety of childbirth was something quite new to the fairly enlightened village elder I spoke to.

The moral condition of the womenfolk is said to be very good. Not a single case of adultery was admitted to have happened in any of the villages. The men, however, who work as seamen elsewhere, are expected to have indulged in prostitutes, but few cases of venereal disease have been known or detected. Only two of a few dozen veteran seamen have been known to have suffered seriously as a result of their sexual indulgence. No special shame is attached to men who contracted VD, though it is considered an economic misfortune to have young men so infected, as it costs the family a great deal of money to have the disease cured.

Two elderly men in this community practise as part-time herbalists, but neither of them has real training. One of them just happened to have read a few books more than the other and acquired his herbalistic knowledge through his own personal experience. The other simply inherited his knowledge from his deceased uncle. Any case of serious illness will have to seek attention in town where the family has the finance to do it, or else the patient will have to be left alone to cure himself or die through the natural course of events. But resistance against ordinary diseases appears high; fatal cases of youthful patients are not many.

Annual expenditure for a household includes the following items of essentials for a family of 5 persons:

Rice	300 catties per person	1,500 catties
Oil	6 catties per person	30 catties
Salt	12 catties per person	60 catties
Sugar	12 catties per person	60 catties
Meat (mainly pork)	6 catties per person	30 catties
Fish (salt)	10 catties per person	50 catties
Fish (fresh)	25 catties per person	125 catties
Veg (preserved)	10 catties per person	50 catties
Veg (green)	36 catties per person	180 catties
Sweet Potatoes	150 catties per person	750 catties
Taro	100 catties per person	500 catties
Chickens or Ducks	4	20
Fuel (grass)		18,000
Firewood	6	3,000
Matches	12 packs	60
Wine (alcohol)		4

YUNG SHU WAN GROUP OF VILLAGES

Yung Shu Wan (S. H. Peplow)

Yung Shu Wan. Yung Shu — banyan tree, Wan — a bay. A seaside village where these trees are grown.

Yung Shu Wan (496915) (Austin Coates)

Population approximately 315, a multi-clan village founded by the Sum and Fong clans, who are now in the 5th generation. The place is thus younger than the small surrounding villages, which are the homes of the island's two principal and oldest clans, the Chow and the Chan. Yung Shu Wan grew to become a small market village and port, as the prosperity of the main clans increased, and as Hong Kong grew in importance. The population includes a land-based fishing group of about 80, many of whom are Hoklo. Other surnames are Ng, Yiu, Tsang, Cheung and Chung. There are about 25 Hakka, surnames Lee and Ng. Apart from the Sum and Fong families, most of the inhabitants have settled here within the last 50 years.

There are several tea-houses and shops, and the place is obviously prosperous. Miss Robertson, the missionary nurse who conducts a small clinic in a house built by the Chinese Imperial Maritime Customs, stated that during her years there she had only come across one family that was actually short of food, although there were a number of cases where the diet was not sufficiently varied. The principal source of livelihood is acting as a market and port for the northern sector of the island. Many of the well-to-do are retired seamen with sons at sea, sending remittances.

The Hoklo and other fishing people own about 80 sampans, which are used for net fishing in the waters adjacent to the island. As they have not enough men to keep all the boats in operation they employ land-based Cantonese, on the basis of an equal share in the profits from each catch. Each sampan is operated by 3 men, or occasionally by 2. The Hoklo are considered to be better off than many of the villagers, but in spite of this they seldom live in other than grass huts near the shore. Fish is sold through the Government Wholesale Fish Marketing Organisation in Aberdeen. Vegetables, grown in fairly large quantities in the smaller villages, are sold either at Aberdeen or at West Point. There are three passenger-and-cargo junks, one of which operates a regular service between Yung Shu Wan and Aberdeen. Requests were for two dry latrines, one at each end of the village, with 3 male seats and 2 female. Water pipes are also needed to bring the water supply from a small reservoir just south of the village, nearer to the houses.

The doctor from St John Hospital, Cheung Chau, visits once a week on Thursdays. There are no proper postal facilities, and it is proposed to approach the Postmaster General with a request for a postal agent to be established here, mail to be delivered by the ferry. There are occasional troubles with the Marine Police when the ferry junks are overloaded, and this is a source of complaint throughout the island, voiced equally by owners and patrons. The harbour is shallow, but it would be possible to construct a pier here. A small illegal pier exists already, for use by sampans.

Yung Shue Wan (*Gazetteer* p. 99) (James Hayes)

Has a population of 560 people, with many family names. Those of Ng, Lai, Fong and Lung were particularly mentioned, with a settlement length of 8–10 generations, around 200 years. They are predominantly Punti. This figure does not include some 280 Hoklos, living on some 100 boats, with only 15 persons on land. The Hoklos come from places like Wai Chow, Kwong Hoi, and Hoi Luk Fung in Chinese territory.[1]

Ten men from the village are seamen. There used to be more seamen from the village before the war.

There is very little paddy land.[2] Most of the land associated with the village is under vegetable cultivation, and in villagers' ownership, with very little of it rented. Consequently, there are few of the brown cattle used to plough and harrow irrigated fields across the District.[3] There are 20 to 30 sows, but no pig cooperative society.

It is felt that the ferry pier should be extended to form a typhoon shelter. They think that around 500 boats would use it, whereas at present, in worsening weather conditions, local craft must disperse to various other shelters.

Other topics raised include the school, water supply to the village, and a possible reservoir.

The notebook records that Mr Chow Sau, a leading villager of Yung Shu Long, with a house at Yung Shu Wan, says he has no covering land certificate for the latter, and would like us now to regularize its status.

Shan Po (497914) (Austin Coates)

Population 38. Surnames: Chow and Tsang. This hamlet consists of two rows of houses, one row for each clan. It is so close to Yung Shu

Wan as to be almost indistinguishable from it. The Chow family settles here as an offshoot of the clan's village of Ko Long. It has 6 families, a lad of 20 being of the 6th generation at Sha Po. The Tsang family settled a little later; it has 4 families, a man of 39 being the 4th generation. The Chows own their own fields, but the Tsangs, later arrivals, rent some of theirs from the Chows. Remittances are a principal source of income; 7 men are working abroad or in Hong Kong, and the number was formerly higher. Younger men work in Hoklo boats in the manner described above. Pigs, poultry and vegetables are sold in Hong Kong. A T-extension to the water pipe for Yung Shu Wan could provide water for this village.

Shan Po (*Gazetteer* p. 100) (James Hayes)

Said to be 49 persons in the village, with six families of Chow and four of Tsang. The Village Representative is a Tsang of the fifth generation. The Chows claim 14 generations, coming from Po Kat in the Po On District.[4]

The Chows have rice fields, but not the Tsangs. The former have 20 taochung of paddy, and the latter five taochung of their own land but it is not irrigated.[5]

There are 12 or more cows in the village, and six sows.

Twelve men are at sea, but very few work in Hong Kong or Kowloon. All children of 6 and over attend school.

They say they do not have enough water to drink. They have a well, and would like 60 bags of cement immediately from our Local Public Works funds to construct a footpath to the well.[6]

Ko Long (S. H. Peplow)

Ko Long. Ko — high, Long — waves. A village near the shore where the waves are high.

Ko Long (498913) (Austin Coates)

Population 16. Surname: Chow (2 families), with a family of Ho, who arrived in 1948. This is one of the original 4 Chow villages. Principal source of income: 4 men are working outside, 3 as seamen, 1 in a shop in Hong Kong. A well needs enlarging, and a 100-yard path connecting the village with Yung Shu Wan. Both can be done in the autumn as Local Public Works jobs. Vegetables, rice, pigs and poultry.

Ko Long (*Gazetteer* p. 100) (James Hayes)

27 people in five families. Two are Chows (indigenous), and one each of Ho (from Nam Hoi, near Canton), Wong, and Leung (from San Hing, Kwangtung). The Hos have been here for 13 years and the Leungs for twelve.

The Chow families have one taochung each, of paddy and vegetable land. No cows, four sows.

One child of school age at school, three do not.

Several men at sea and one in the United States, presumably all Chows.

They would like cement to repair a well and surface a small path, and say they can do the work as soon as they get the materials.

Tai Yuen (499915) (Austin Coates)

Population 20. Surname: Chow. One of the original 4 Chow villages. This village resembles something out of Shun Tak district, and is evidently in the process of being converted from Chinese to European architecture. A comparatively grandiose modern house has just been completed by the local big noise, Chow Sing, who has made a good deal of money by running a seamen's boarding house in Hong Kong. Of the 20 inhabitants, 12 are working in Hong Kong, but, true to the best Chinese tradition, there is no sign of neglect of the home village. It would be pure philanthropy to let this place have anything free, and they were wise enough to ask for nothing.

Tai Yuen (included with Wang Long) (*Gazetteer* p. 100) (James Hayes)

All Chows. There are six families here, with 26 people, including the present chairman of the North Lamma Rural Committee, Mr Chow Shing. He is a small, well-dressed man, always in a European suit, with a modern Western-style house.

They say they have been here for 10 or more generations, and descend from a younger brother of the first Chow settler on the island. Perhaps due to Mr Chow's money and influence, all save one of the children of school age study in Hong Kong schools, and not locally.

There are 10–20 taochung of paddy fields.

Wang Long (500913) (Austin Coates)

Population 46. Senior village of the Chow clan. A man of 39 is of the 13th generation. The Chow clan is native to Po On district, of which the New Territories were once part. The Lamma Island Chows are an offshoot of the clan whose place of residence was near Aberdeen, and one of whose members is the Hon. Sir Shouson Chow. The clan moved to Lamma probably between 1615 and 1640. Four brothers originally came together, setting themselves up at Wang Long, but soon after they split into four villages, Wang Long, Ko Long, Tai Yuen and Yung Shu Long. An offshoot of Ko Long founded Sha Po. Remittances are a principal source of income, 8 men being seamen, and the number has been higher. Vegetables and rice, pigs and poultry.

Wang Long (*Gazetteer* p. 100) (James Hayes)

Gazetteer gives a population of 70, including Tai Yuen, above.

Thirteen families, all named Chow. The Village Representative is Chow Li-fuk.

They are of the 14th generation to date, from Wai Chow, via Little Hong Kong village on Hong Kong Island. They claim Sir Shouson Chow and Sir Tsun-nin Chow (Chau) as their relatives, probably wrongly in the case of the latter.

They have 25 taochung of paddy fields, and 15 of vegetable land.

It was mentioned that they descend from the elder of two brothers who divided their father's land equally "a long time ago". This was clarified as being six or seven generations before.[7]

They have 15 cows and 10 sows. Ten children of school age go to school, and another ten do not.

Ten men are at sea, but there is no-one working in Hong Kong or Kowloon. Some 4–6 men have no employment.

Yung Shu Long (*Yung Shue Long*) (501917) (Austin Coates)

Population 20, of whom every able-bodied man works either abroad or in some other part of the colony. The head of the village is Chow Li Ping, several times chairman of the Cheung Chau Residents' Association. One of the 4 original Chow villages. In spite of it all, the women do well with their rice and vegetables. Foreign influence is noticeable in good cowsheds and pigsties kept well apart from the houses. Chow Li Ping has recently built a European-style house here.

Yung Shue Long (James Hayes)

All belong to the Chow clan. Mr Chow Li-peng, Chairman of the Cheung Chau Rural Committee, comes from this village, and his brother is the current Village Representative. However, the Village Representative was not present for my visit, and it is not clear to me for how many generations they have been here.

They own all their own paddy fields, and have little vegetable land. They have 8 cows and 8 sows.

They have one well, and say they have enough potable water. They have already applied for some Local Public Works materials [for something unfortunately not recorded] and say they will start work in the autumn.

Only two children go to school, the others are said to be too young.

Four men are seamen on ocean-going ships, but no men work in Hong Kong or Kowloon.

Tai Ping (501920) (Austin Coates)

Population 122, a multi-clan village 15 minutes' walk from Yung Shu Wan by a road which the villagers would like to improve if this office will provide cement. A Land Bailiff will inspect during the summer. Surnames: Ng, Wan, Fong, Cheung, Tsui, Lee, Luk, Kee, Tsang, Chu, Chan. The Ng family are the oldest in the neighbourhood followed by the Wan and Fong families. Apart from them, no family has adult men in more than the 3rd generation.

This is evidently a village that rose to prosperity as Hong Kong grew in importance. A member of the Ng family, Ng Yung Kan, runs the ferry junk from Yung Shu Wan to Aberdeen, and the villagers sell vegetables in Hong Kong. They are a local centre of the pig industry, and sell piglets to other villages even as far away as Luk Chau. In general the people own their fields, but some rent from the Ng family, who, being the earliest arrivals, own most of the best land. The village is comparatively high up and water is a problem.

Tai Peng (*Gazetteer* p. 99) (James Hayes)

A mixed settlement of 154 persons. There are 6 families each of Ng and Fong, 4 of Wan, 2 each of Cheung and Tsang, with single families of Chu, Chan, Tsui and Lam. The Village Representative is of the 3rd generation, and most are of the 4th generation in the village.

Village people own 4–5 taochung of paddy fields, and farm a lot of vegetable fields (six acres was mentioned), either their own or Crown land on permit from the District Office.

They say that practically all their children are at school, around 30 of them, with very few of school age not attending. They go to the school at Yung Shu Wan, which was reported to have 197 children (probably operating in two sessions). There was a school at Tai Peng pre-war, but not post-war.

Seventeen men are at sea, about the same as in pre-war days, they said.

There are 14 sows, and probably 100 pigs all told. All the pigs are sold to Kennedy Town (Sai Wan), but they are bought here, probably by the wholesale dealers.

They grow enough vegetables for their own use, and at need buy in Hong Kong, not in the Yung Shu Wan shops.

They mentioned land suitable for growing the black sugar cane ("hak chea").

They wanted chloride of lime for mosquito spraying.

A request was made for 150 bags of cement from the Local Public Works vote to surface the footpath from the village to Yung Shu Wan, which is heavily used.

Pak Kok Old Village (498927) (Austin Coates)

Population 120. Surname: Chan. 25 minutes' walk from Yung Shu Wan, uphill to Taiping, and down again. There is a small pier built by Local Public Works some years ago, and used by ferry junks delivering supplies. 20 minutes by motor launch from Aberdeen. Only about 85 people actually live in the village; the rest work abroad or in Hong Kong. Remittances are the principal source of income. The Village Representative is aged 71. There is such confidence in remittances here that the villagers are somewhat indifferent to local problems. In some fields agricultural labourers are employed, and there are in addition 4 strangers growing vegetables, 3 of them Hoklo. All available land is cultivated, rice on the lower fields, vegetables on sloping land. Pig-breeding is a minor industry.

Pak Kok Old Village (*Gazetteer* p. 99) (James Hayes)

140 persons in 26 families. Nineteen families are Chans, plus one each of Tsui, Leung, Lam, Yau, Lei, Tam and Wong.

The Chans are of the same generation as at Tai Wan, and say they are from Po On.

The other families are post-war immigrants from Hoi Luk Fung in northeast Kwangtung.

The Chans have 60 taochung of rice fields, and 50 of vegetable land. There are 45 cows, and 15 sows.

Ten Chans are at sea, with 30 unemployed. Several men work in Hong Kong or Kowloon.

26 children are at school locally, with several more studying in Hong Kong. The proportion of children of school age at school/not at school is roughly 50/50. The school is about 15–20 minutes' walk from the village, for an adult.

They would like to repair and extend a footpath with Local Public Works materials. Assistant Inspector of Works Chan will visit to inspect and discuss with them.

There is a large well in the village which supplies their drinking water.

They wish to build a shelter at the ferry pier. It is visited three times daily by a ferry. On average, about ten persons use it each day, to take vegetables to Sai Wan on Hong Kong Island for sale.

There was a request for specialist medical assistance here, for a Chan Kwun-lin, and I had jotted down, "to see the doctor at the local clinic in Yung Shu Wan first".

Pak Kok New Village (499928) (Austin Coates)

Population 56. Surname: Chow. This village is not quite so well-off as the others, and more ingenuity is required to keep up the high standards of the island. There are 9 able-bodied men, of whom 3 grow vegetables, 3 operate a fishing sampan, 2 work in a textile factory in Kowloon, and 1 works in a shop in Hong Kong. The fishing area used by this village is on the southwest coast of Hong Kong Island, just below the Queen Mary Hospital. The type of fishing is by nets attached to coast rocks. Middle-aged people are of the 7th generation, which gives 1775 as the probable date of the foundation of the new village. The Village Representative is 68. Both the Pak Kok villages draw water from the same spring. The new village asked for a small reservoir, but as this may cause friction between the villages the Chairman of the North Lamma Rural Committee must be consulted first, before any cement is issued.

Pak Kok New Village (*Gazetteer* p. 99) (James Hayes)

Despite its name, the village is about the same age as the other. There are 58 persons, and nine families of Chow here, from Po On they say, as does Chairman Chow Shing.

They have 10 taochung of paddy fields, and 8 of vegetable land, with 10 cows and 7 sows.

Five men are at sea, and one is in the United States.

The potable water supply is inadequate. They want to build a well when they are able, and ask for advice and the necessary materials: also, to make an all-weather footpath.

As in the Old Village, the proportion of school-age children at school/not at school is about 50/50. Ten attend school in their case.

The notes mention Ko Po Shan, near the village. They pronounced it "Kau Po Shan".

There are another 10 persons in two families living at Pak Kok Tsui, named Fung.

Tai Wan Old Village (Walter Schofield)

One of the subjects which used to excite much feeling in the Chinese countryside was the disturbance of graves. In 1930 this occurred at Tai Wan in Lamma, on the big sandbank later excavated by Father Finn, once a leading local centre of Bronze Age culture. The sand diggers had cut away so much sand that coffins buried 2 feet deep in the bank were sticking out, and their contents could be seen. I at once ordered digging to stop till the coffins could be properly disposed of. Enquiries in the village showed that the villagers were not interested; so it was clear no local cemetery had been violated, and the persons buried had most likely been boat people. I believe the sand contractors got the Tung Wah Hospital authorities to remove the coffins: certainly there was no trouble with any local people. The high level and good preservation of these coffins showed that their burial took place long after the Bronze Age. One troublesome class of case was the "fung shui" difficulty caused by digging a new grave on a hill ridge not far above an older one. If the family owning the latter lost a child or two by smallpox or other complaint, they would conclude that their ancestor was displeased with them for letting a deceased stranger "ride" his grave, and so hinder the good influences of the site reaching him. Such cases might have to be settled by removal of the later grave, or by some compensation to the aggrieved family.

Tai Wan To Old Village (502914) (Austin Coates)

Population 85, of whom 15 work either abroad or in Hong Kong. Surname: Chan. The two Tai Wan To villages, and Lo So Shing, are the principal Chan villages. The Chan clan settled on Lamma in the 50 years after the Chow's arrival. The probable date of their settlement is c. 1660.

Tai Wan Old Village (James Hayes)

Eighteen families of 104 persons live in the village. All are named Chan, and claim to have been here for 14 generations.

They have enough drinking water.

They have 60 taochung of rice fields. They reported having 12 cows and 17 sows, but currently with few other pigs.

13 children of 8 years of age and above go to school, but it seems that a number may not, which includes those children aged 6–8.

A widow was reported as having been overlooked in a recent survey for a "widows assistance scheme", funded by the Kadoorie Brothers, through the District Offices and the Agricultural Department.[8] This widow, Chan Cheung Tai, has a son who is still at school, but already has a wife.

The Village Representative mentioned the need for repair of a drainage ditch in the Old Village. This had been reported 2–3 months ago, and I said I would ask Mr Chan, our Assistant Inspector of Works, to visit soon.

Tai Wan To New Village (501913) (Austin Coates)

Population 74, of whom 5 work either abroad or in Hong Kong. Surname: Chan. Chan Yau, a prominent member of the North Lamma Rural Committee, lives here. The New Village, as can be seen from the population and employment figures, is slightly less well-off than the old village, and their thrift (and possibly their needs, which are evidently greater than the Chow villages) can be seen in the intensive way they cultivate their fields. All available land is used for rice and vegetables. From both villages there are 12 able-bodied men who in the ordinary way by now would be seamen. At present they are fishing and farming. If employment could be found for 3 of them it would lighten the burden.

Water supplies are fairly good, and irrigated fields were being planted out with rice in spite of the drought. There is a good site for a large dam in the hills above the village, but the work is greater than what the local people are capable of carrying out. The Irrigation Engineer will have to inspect this and advise. It might be a suitable scheme for Colonial Development and Welfare.

The new village owns 3 sampans, used for fishing, and most householders keep pigs. There are about 150 taochungs of cultivated land, chiefly for vegetables, and mainly looked after by women. The under-employed figure is about 3, who need inclusion in overall arrangements for employment.

One of the recurring problems at Tai Wan To is the determination of the Administration of the Sand Monopoly to take sand from their beach. Many of the beaches in the New Territories have sand that is either too thick or too thin for construction work, but Tai Wan To's sand is apparently perfect. Some years ago the District Commissioner gave a guarantee to the villagers that their beach would not be touched. The Sand Monopoly then resorted to dredging sand from the bay. During 1953 they kept some distance away from Tai Wan To, but in 1954, with my approval and after the Chief Engineer, Port Works, had given an opinion that it was most unlikely to cause any harm to the sand beach protecting the rice fields behind it, the dredgers came to take sand from within 50 yards offshore. In fact, they never took any sand. In 1953 the villagers had already taken the matter into their own hands, and had been severely warned for doing so by the Police, when they arrested the sand contractor and only allowed him to operate under armed escort from the village. In 1954 they set upon the new contractor, beat him up badly, and left him for several hours locked up in a temple. The Police came within 6 hours, and subsequently at the Kowloon Magistracy four of the villagers were sent to prison for several months. Chan Yau later appealed to Mr Brook Bernacchi for help, and the sentences were reduced on hardship grounds. An exhaustive investigation has now been made by the Port Works Office, and I see from it that they understood the problem as little as I did when I first came in contact with it. By all appearances it does not seem possible for offshore dredging to have an effect on the beach, and the anxiety of the villagers seems to be groundless.

So on this visit I took the opportunity to dawdle at the village, talking of this and that with Chan Yau, and at last another fact emerged. He claimed that the sea had been gradually encroaching here over the last 30 years and more. The beaches off Tai Wan To and Hung

Sing Yeh were formerly one beach, but have now shrunk to two small beaches. This may be the key to everything, and, having lived for 20 years in England beside an encroaching sea, and seen the scientific and elaborate arrangements made to hold it back, all of which failed. I can say that if Chan Yau is telling the truth, there is nothing else for it but to abandon all further thought of sand lifting here, either from the shore or by offshore dredging. I have contacted the Director of Marine in an attempt to find nautical maps showing the beaches of Lamma prior to 1920. Damage was done to fields in the typhoons of last summer, and compensation of rice for paddy lost is being given.

Tai Wan New Village (*Gazetteer* p. 101) (James Hayes)

59 persons in eleven families, all surnamed Chan. Have been at Tai Wan for 14 generations to date, but 37 years ago [circa 1920] many families moved here to establish the New Village, because the site of the Old Village was, they felt, too low and wet. An ancestral hall was built here at this time.

They farm 60 taochung of paddy, and 40 taochung of vegetable land, all of which are their own fields.

Five or six men are unemployed, but were formerly at sea. Only two men are still at sea.

Some 3–4 men work in Hong Kong and Kowloon.

There are 12 children aged six plus. All go to school. There are a further 5–6 aged between 6–8, and these "have no school to go to". [I am not sure now quite what this means.]

Both the Tai Wan villages would like to apply for an orchard from the Kadoorie Agricultural Aid Association, administered by the Agriculture and Fisheries Department.[9]

I also noted that "the two Village Representatives are a great contrast", but without specifying why. Probably because one, Mr Chan Yau, who is also Vice-Chairman of the Rural Committee, wears a Western suit or sports jacket, whereas the other is old-fashioned in all respects.

Hung Sing Yeh (507907) (Austin Coates)

The small bay next to Tai Wan To, with a small area of cultivable land behind it. This area has been considerably enlarged since 1950, and 3 refugee families from China (all Cantonese) are working the fields and keeping pigs. One of them appears to be a former official, and it was difficult to obtain full information without wasting a lot of time sitting down and establishing confidence.

My estimate of population is 12, and the fields are tilled on some sort of joint basis with Tai Wan To people, who work with the refugees and appear to be friendly towards them. Hung Sing Yeh is the site of many of the most important archaeological finds before the war; Father Finn and Stephen Balfour did a good deal of work here. It is clear that Hung Sing Yeh was a prehistoric seasonal trading station, with contacts northwards to the Yangtse valley and southward to Indonesia. The finds correspond with those discovered at Shek Pik, Lantao Island.

Lo Tik Wan (S. H. Peplow)

Lo Tik Wan. Lo Tik — a water reed, a kind of artemisia, Wan — a bay. A village near a bay where these reeds are plentiful.

Lo Tik Wan (511915 and 513914) (Austin Coates)

Population 60. Surname: Ng. On maps this village is incorrectly described as Long Ming Wan. This place is considerably poorer than any other North Lamma village. Although it is only 20 minutes by motor launch from Aberdeen, it is not on the route of the ferry junk and owns nothing faster than a rowed sampan. It is geographically removed from the other villages by hills crossed by a somewhat awkward path. Carrying a load it is about 50 minutes' walk from Yung Shu Wan. The Ng clan came from Nantao, and a man of 42 is of the 6th generation at Lo Tik Wan, giving c. 1815 as the date of foundation.

This is the only village in the group to receive no remittances from abroad or from Hong Kong. Employment on ships depends largely on the introduction of relatives, and the Ng clan has no entree. 40 are permanent residents of the village, the other 20 being fishermen.

The village owns 8 sampans used for inshore and rock fishing off Stanley Peninsula. Catches are sold at Stanley and Aberdeen. During the fishing season in these waters the villagers are absent from Lo Tik Wan for 6 months. At least one sampan, however, remains in the village for transport between the village and Aberdeen and Yung Shu Wan. 2 of the fishermen have been employed as beach coolies at Stanley.

There are a few hill fields, but the main industry, apart from fishing, is pig-keeping. The village is gradually turning over to breeding, and more sows are still needed. This is a case that merits consideration of a free grant of piglets by Kadoorie Agricultural Aid Association and this will be taken up. This is priority village for outside employment. Children go to school at Yung Shu Wan, and pigs are sold there.

Lo Tik Wan (James Hayes)

33 persons here by the shore, with another fifty in the old village up and over the hill. Styled Lower and Upper Villages respectively.

All belong to the Ng clan. The Village Representative is Mr Ng Shue-hing.

Ten children from both settlements walk to school in Yung Shu Wan. The journey takes 30 minutes.

I had wondered whether a school was needed for Luk Chau–Lo Tik Wan, with connecting concrete footpaths.

Because of their dependence on Aberdeen, rather than Yung Shu Wan, and the fact that they use rowing sampans to get there and back, weather is important here, on the east coast of North Lamma. They say that an east wind brings bad weather conditions, and extra problems.

They have 5–6 taochung of rice fields. Their supplementary rice is bought from Aberdeen, not from the Yung Shu Wan shops: presumably because it's cheaper than at the island shops, where the shopkeepers have to add in their own transportation costs.

They have also 8–9 taochung of vegetable fields. Like the Luk Chau people, they only grow sweet potato, and buy their vegetables from Aberdeen.

They go often to Aberdeen, sometimes more than once a day. But they have to row their sampans there, as they have no motor junk or kaifong ferry service.

Both they and the Luk Chau people mentioned that they needed and wanted to buy a motorized junk. They will collect money for this purpose, and will write to me to help with getting a licence for it from the Marine Department. It seems they prefer to ask the District Office for help whenever something new is required, since they are unfamiliar with the Marine Department staff and their regulations.

LO SO SHING GROUP OF VILLAGES

Lo So Shing (S. H. Peplow)

Lo So Shing. Lo — reeds or rushes, So — a beard, Shing — a city. A large village where these reeds are grown and used by the inhabitants for making mats, raincoats, etc.

Lo So Shing (512891) (Austin Coates)

Population 140, of whom 8 work in Hong Kong and 6 are seamen abroad. Surname: Chan. The senior Chan village in South Lamma. A man of 36 is of the 8th generation, giving c. 1770 as the date of settlement. Remittances are an important source of income. The principal village occupations are grass-cutting (the grass being sold to Aberdeen for breaming), fishing (the village owns 6 small boats) on the west side of the island, Lo So Shing has a beach on both east and west coast, and pig-keeping. Like most villages, this one is going over from keeping to breeding. Rice is the chief crop, with sweet potatoes as the only vegetable grown in any quantity. The Lo So Shing valley is rich in trees. The only school in South Lamma is situated here, subsidised by the Government.

Lo So Shing (*Gazetteer* p. 101) (James Hayes)

The population here is said to be around 150.

There is a school here, supposed to serve this place and Sok Kwu Wan, but whilst 40–50 children attend from this village, only 20 come from Sok Kwu Wan, with its larger population. It is likely that some 30–40 more places are needed, probably through providing an extra classroom. I asked the teacher to write in, giving the figures, and the Village Representatives to sign in support.

Many men are working away from home. They include 13 men at sea, but I was told that this is more difficult than before, owing to the higher physical standards now being required by the shipping companies. One man works in the Dairy Farm.

There are more than ten men under 25 who are currently without employment. Among these, three worked as labourers on the airport extension, but owing to reductions they have only had 7 months' work this year (1957) before being laid off. Another four men had returned recently from Nauru. Some of them are married.

Sok Kwu Wan (Eric Hamilton)

On two occasions at Ha Mei Wan *(west of Sok Kwu Wan)* in summer the surf was too bad for the boat so I swam ashore with my papers tied onto my head. The second time I found to my horror that most of the sand there had been sucked out to the sea and I had a hell of a time. But there was a lot of cultivated land round about there and the only alternative was a long steep walk over the hills from Picnic Bay. It was a great pity as it was a lovely half moon of sandy beach and popular with turtles. Capturing turtles is great fun but it meant being up all night. The Hong Kong Hotel always took them, polished the shell and gave you a quart of turtle soup and big turtle steaks. One kept the eggs of course, but believe me the two latter items are much overrated.

Sok Ku Wan (S. H. Peplow)

Sok Ku Wan. Sok — to make or fasten, Ku — fishing nets, Wan — a bay. A village near the bay where fishing nets are made.

Sok Kwu Wan (519890) (Austin Coates)

Population 125, of whom 10 are working in small jobs in Hong Kong. There is a Tanka population of approximately 80, who make this their permanent base. The latter fish and live in small single-masted boats; land-based people also go fishing, but in sampans and only in waters immediately adjacent to the village (Picnic Bay). The village owns 3 junks, 1 of which, motorised, provides a regular ferry service between Sokkuwan, Mau Tat Wan and Aberdeen. Passengers and goods from Lo So Shing join the ferry at Sokkuwan; those from Mau Tat, Yung Shu Ha and Tung O join at Mau Tat Wan.

Sokkuwan is developing and expanding rapidly, and it is clear that, at the present rate, it will soon replace Lo So Shing as the "capital" of South Lamma. The same process is happening here as took place earlier in North Lamma; a market and port town is developing as an outlet and feeder for the less adaptable clan villages. The earliest known family at Sokkuwan was Im. They were murdered by being towed out to sea in a boat, which was then sunk. They were succeeded at Sokkuwan by a branch of the same clan, which appears to have arrived shortly after the lease of the New Territories. In 1920 there were two clans, Im and Law. Between then and the Second World War a few families came, and since 1945 the population and its activities have risen quickly. Surnames: Im, Ho, Wu, Chow, Tiu, Tsang, Chan, Chung, Lam, Lai, Wong, Yeung, Tong, Ko, Tang, Fong, So, Cheung.

In the last few weeks some refugees, hitherto living at Mongkok, have set up stone and corrugated iron huts; the principal earners in these families live in Kowloon and work as hawkers, etc., but the reactions of the women are initially favourable to Sokkuwan, and it seems likely that more will follow. Marginal land is being intensively cultivated for vegetables, sent by the junk ferry to Aberdeen, and there is a flourishing pig industry. The pigs are sold through laans at Aberdeen, and not at Kennedy Town. The reason for this is that pigs are stuffed with food before being sent for sale, and the sooner they can be sold after this last meal, the heavier they weigh and the more money they fetch. Aberdeen being nearer than Kennedy Town, the prices are always better there. The two most prominent families are now Im and Wu. The former are vegetable growers, the latter fish dealers, the owner of the ferry, and manager of the breaming facilities, which are good and cheaper than in most parts of the district.

Projected local public works for this autumn include a pier, a well, a fairly large bridge on the main track between the village and the school near Lo So Shing, and a stepping-stone bridge. There is a fair demand for employment, and several of the villagers are skilled labourers who can be put on the special arrangements for such people with Messrs Gibb, Livingstone for work in the Seria oilfields.

Sok Kwu Wan (*Gazetteer* p. 101) (James Hayes)

Population reported to be around 200, compared with 100 before the war. Post-war, some families came over from Hong Kong, and the men have continued to work there.

There are no rice fields here, but eight families grow vegetables. Others get income from selling firewood and grass, mainly to boat people.

Many men work outside. The list I was given comprised two in a textile factory at Sham Tseng, two in the War Department, one in the Fire Services department, one in the Royal Army Service Corps, one lorry driver, two apprentice carpenters, and a cook boy with the Dairy Farm. Another man works in Mong Kok. Two families provide services for the local boat people. The very large wooden vats they used for dyeing nets and other paraphernalia could be seen under trees near the shore.

There is a request for a proper pier. They explained that the village is a place from which the people of six villages go to Hong Kong. Local Public Works wants include materials for a bridge at a point on the footpath along the shore to the next village, Lo So Shing. There was also a request for DDT to help reduce mosquitoes. Apparently, a sanitary coolie is supposed to make a regular visit, but he does not always come.

Some of these requests, the pier, and the bridge over a stream on the footpath leading to Lo So Shing school, had been raised when I had visited the Rural Committee a few weeks before. Employment, for the younger men in particular, had been another talking point. The need for an all-weather footpath linking the southern and northern parts of the island was another, together with the request for a police post. Owing to early post-war insecurity, the police had supplied rifles to village guards. The weapons had been withdrawn, for repair they thought, two months before, and they wanted them back. They had also wanted rockets which they could fire off to summon help, in case of attack.

The chairman, Mr Chan Pak-yip, a local-born Hong Kong businessman, and a Village Representative, was most voluble on the needs of his sub-district, and clearly thinks that the District Administration (and the government in general) should be doing more for his people.

Mau Tat Tsun (S. H. Peplow)

Mau Tat Tsun. Man — a white flowered rush, a kind of grass, Tat — a mat awning over a boat, Tsun — a village. A place where these rushes are grown, and made into mats for the above purpose.

Mau Tat (536892 and 536895) (Austin Coates)

Population 135, of whom 4 are working abroad and 12 are employed as shop fokis in Hong Kong. The village is in two sections: Mau Tat Wan, on the coast of Picnic Bay, and Mau Tat Tsuan, higher up inland, the agricultural section of the village. Surnames: Chow, Chan, Ip. Remittances form a large part of the general income, and the village is clearly prosperous. It possesses excellent rice fields, which even in this year of exceptional drought are saturated with water. Vegetable cultivation on a large scale is precluded by the large amount of water reaching the main fields. Pig-breeding has recently replaced pig-keeping, and the village's 8 remaining able-bodied men are all employed on sampans which go to Beaufort Island (Lo Chau, one of the islands in the Po Toi group) for fishing and collecting seaweed as pig food. The youths would, however, prefer to have jobs in town. This would provide enough money for the village to give up collecting seaweed and purchase pig feed directly from Aberdeen. They own 6 sampans.

The principal request was for employment, but it is difficult to give Mau Tat any priority.

The village has been fined twice for overloading its sampans plying between here and Aberdeen; the second fine was apparently $100, which seems high for an offence of this kind. The general question of transport within the terms of the Merchant Shipping Ordinance is being brought to the attention of the District Commissioner, as a prelude to possible discussions with the Director of Marine and the Commissioner of Police.

A mining company prospecting in the hills near Mau Tat has left the landscape badly scarred, and the villagers request that the miners be made to tidy the place up has been referred to the Superintendent of Mines.

Yung Shu Ha (531884) (Austin Coates)

Population 68, of whom 3 are working abroad. Surname: Chan. A man of the 6th generation here is 69, which dates the settlement at c. 1785. A village somewhat poorer than the rest, situated on the rock bay of Tung O. The bay is only navigable by small craft in calm seas, and the most reliable connection with Aberdeen is on foot at Mau Tat Wan, carrying a load, 40 minutes.

The village formerly had more men working abroad, but they are self-reliant and uncomplaining. The principal source of income is from grass-cutting, the grass being sold at Aberdeen for breaming. There are a few fields, and several recently opened marginal fields for dry cultivation. Vegetables cannot be grown because of the uncertainty of transport to Aberdeen, due to high seas. The walk to Mau Tat plus the cost of the junk ferry is not considered feasible.

The village owns 5 small boats, used for fishing in the waters immediately adjacent. A few families keep pigs, fodder for which is purchased at Aberdeen. There are 12 able-bodied men under 40, and the economic position would be improved if outside employment could be found for 3. Chief request was for the improvement of the track to Mau Tat.

Tung O (*Tong O*) (528879) (Austin Coates)

Population 120, of whom 20 work elsewhere. Of these latter the ones sending remittances include 4 sailors and 2 shop fokis in Wongneichong. Surnames: Chan, Chow, Ng, Cheung. This village in prosperous times has not a single able-bodied man who is not working as a seaman abroad, and has been hard hit by the trade slump. The problem now is not so much the returned seamen as the young men who have reached the age of 21 and who would, in the ordinary way, have found employment by now. I was asked to give assistance in finding jobs as seamen, but explained that if the applicant cannot pay the $100 demanded by the Seamen's Union, there is nothing I can do. The Union will accept $50 provided the applicant's father is a friend of one of the office-bearers, but refugees in Hong Kong have to pay as much as $300 to get a ticket.

Grass-cutting is, apart from remittances, the principal source of income, with fishing as second. The village has 5 small boats and fishes in adjacent waters. They also keep pigs and chickens, but only occasionally sell these outside. A seaman from this village remits $1,000 per year. There are 13 able-bodied men engaged in wood cutting and fishing. If 5 could get employment, the economic position would be eased. An urgent request for cement for a new well was added to the Kadoorie Agricultural Aid Association list and has been dealt with. Water supplies are unsatisfactory. The village has never really taken its fields seriously, and there is no irrigation system worthy of the name. Sand at Sum Wan, on the south side of the village, is suitable for building purposes, and I have obtained the Village Representative's agreement to limit sand lifting by the Sand Monopoly.

Mo Tat, Mo Tat Wan, Yung Shu Ha and Tung O (*Gazetteer* p. 101) (James Hayes)

The notes for these visits are missing, though I have a record of the inscription of the bell in the temple at Tung O (dated 1828).

Gazetteer gives populations of 130, 40, 70, and 100 respectively for these villages. The predominant surnames in each are, again respectively, Chau, Chan, Chau and Chau.

Luk Chau (518906) (Austin Coates)

Population 65. Surname: Yiu. A poorer village situated near Lo Tik Wan. Occupations: grass-cutting, fishing, pig-breeding. A few fields. One rich man from this village was the late Yiu Fu Kee, but apart from building himself a Western-style house it is not evident that the rest of his clan have benefited in any way from his success. The village sells its grass at Aberdeen chiefly for breaming, and catches fish off Stanley and Beaufort Island. They also, in the appropriate season (winter), collect seaweed from the latter place as pig food. The people complained of Police interference with sampans, of which there are 4. Requests were received for a new well and 2 small roads. The village is perched high up on rocks, and the path to it from the shore is most awkward and uncared for.

Luk Chau (*Gazetteer* p. 100) (James Hayes)

80 persons, all of the Yiu lineage. The Village Representative is Yiu Yau-kuen.

Have only 7–8 taochung of rice fields. Don't grow vegetables, only sweet potatoes.

They have 23 sows, and currently 210 pigs altogether. Pigs sell at $160 per picul at present.

All pigs are sold at Sai Wan (Kennedy Town). They are taken first to Aberdeen by sampan (no motor junk is available) and then to Kennedy Town by lorry.

It transpired that none of their pigs are vaccinated, and I will pass a request to Agriculture and Fisheries Department on the subject.

They kept some buffaloes pre-war, but these (they claim) were taken by Japanese soldiers, and they have never kept any thereafter.

There is no village forestry lot, but there are a few private orchards.

There are 20 children of school age, of whom 14 go to school at Sok Kwu Wan. However, it is a difficult journey on foot, which takes 40 minutes, and the younger children are not sent.

Two men are at sea.

Pre-war, one of the Yius was a big contractor in Yaumati, Kowloon. The firm was called "Yiu Fu Kei". He and a brother managed the business. He has since died.

Ha Mei (*Ha Mi*) (508874) (Austin Coates)

On the southwest coast of the island is marked on maps as having buildings, but none exists any more. The waters off it are a good fishing area, used by boats from Lo So Shing and Yung Shu Wan, and villagers from various parts of the southern sector bring the cattle there for grazing. No inhabitants.

Notes

1. Most Hoklo speakers come from the seaboard counties of northeast Kwangtung. Their language has little in common with Cantonese or Hakka.
2. Paddy means "rice in the straw or in the husk" (OED) and was in common use for rice fields in the New Territories. I have used "paddy fields" interchangeably with "rice fields" herein. From the Malay *padi*.

3. Save in a few places where soils are heavy — as in certain parts of Lantao Island — which require the use of the much larger and stronger water buffalo.
4. From 1914 this became the new name for what was left of the old San On County, to which Hong Kong Island, Kowloon and the New Territories had all belonged. The usual romanization for Chow is "Chau", but this lineage uses the former.
5. This is the local measure of land. For an explanation, see my *The Hong Kong Region: Institutions and Leadership in Town and Countryside 1850–1911* (Hamden: Archon Books, 1977), pp. 202–203. Put very simply, it was the area that could be covered by rice seed from a wooden container (*dau*) of a certain size and content, and filled level with the sides.
6. Local Public Works means "Local Public Works", undertaken by villagers with materials supplied from funds provided by the District Office.
7. Does this imply that they first settled at Wang Long at that point in time, or a generation before? And could the discrepancy over length of settlement at the associated hamlet of Tai Yuen, see above, be explained in part by the removal there of the younger brother and his descendants following the division?
8. The Kadoorie Brothers, Lawrence and Horace, were wealthy businessmen and philanthropists who financed the Kadoorie Agricultural Aid Association which provided much assistance to native and immigrant farmers and livestock breeders through the agency of the District Administration and the Agricultural, Forestry and Fisheries Department.
9. There is an excellent quote on the introduction of this scheme, at para. 38 of DCNT's printed *Annual Departmental Report* for 1955–56.

5
The Lesser Islands

THE LESSER ISLANDS (Austin Coates)

Lesser in size, not in importance, for the Lesser Islands include two of the District's largest centres of population, Cheung Chau, with a population of 18,500, and Ping Chau, the population of which varies between 3,500 and 3,800, depending on the number of workers being employed by the Great China Match Factory, which is situated there. Tsing I has a population growing rapidly, and may be expected to become, in the course of time, an offshoot of Tsun Wan's industrial zone.

Four of the islands are important as markets and noted for their industries: Cheung Chau (soy sauce, preserved fruit, shipbuilding and repairing, salt fish, leather), Ma Wan (shrimp paste), and Tsing I (lime kilns). Ni Ku Chau (Nun Island) was in 1952 granted by the Governor to the Mission to Lepers and changed its name to Hei Ling Chau (Island of Happy Healing). In addition, there are the following which are uninhabited, though visited occasionally and used for net-drying, grass-cutting, or as concealed storage places in connection with smuggling: Shek Wu Chau, situated between Cheung Chau and the Soko Islands, and where there were formerly several acres of terraced rice land which was abandoned about 30 years ago; Beaufort Island and Sam Kong, in the Po Toi group, southeast of Hong Kong Island, Kau I and Shiu Kau I, between Hong Kong and Ping Chau; and a few others which are really no more than rocks in the sea. Waglan, easternmost of the Po Toi group, has an important lighthouse on it, and does not lie within the area under the District Administration.

The earliest historical traces of permanent occupation are on Tung Lung, in the main ocean-going approach to the port of Victoria. On this island was the original site of the Tai Miu, or Great Temple, which since Southern Sung times has been situated in Joss House Bay (Tai Miu Wan). The existing traces of settlement are at present being studied, but it certainly seems that Tung Lung was occupied or regularly visited in the Sung dynasty, if not considerably earlier. Tsing I was probably evacuated in 1663; the present inhabitants began to arrive in 1670, promptly on the cancellation of the imperial order of withdrawal of coastal populations. Ping Chau's present inhabitants probably began to settle there at the beginning of the eighteenth century, and Cheung Chau a generation or so later. Prior to settlement, Cheung Chau was evidently a place of pirates.

The population of the islands is mainly Cantonese and Hoklo. Various possible explanations can be put forward to suggest why the Hakka, a persecuted people in search of land, did not occupy the islands, except where, as on Tsing I and Ma Wan, they were situated very near the continental coast. Of these the most important seems to be the danger of pirates. In both Chinese and European tradition the whole region was notorious for its pirates. Sea banditry in fact continued to be prevalent up to the time of the establishment of the Central People's Government, which is the first Chinese government in history to exercise effective control over the waters at the mouth of the Pearl River. There are indications that certainly Cheung Chau, and possibly other islands as well, were originally settled by pirates, who gradually became domiciled and respectable. The Cantonese populations in most cases came later, when the islands were a little safer, and they were primarily attracted by the curious relationship between land and sea people which is a feature of rustic life on the coast of South China.

The Tanka, or egg-families, the principal boat people of the China coast, were an ostracized, almost an outlawed, section of the community under the Chinese Empire; and to this day they remain the most backward section of Hong Kong's population. Because they were not allowed to live ashore or have schools, or in any way participate in civilized life, they were simple and easy to cheat. Like fishermen the world over, they were superstitious, and playing on their superstitions and stupidity, petty Cantonese merchants have always been at hand, or have gone out of their way, to batten and exploit the sea people. By buying up fishing catches and acting as agents for sales and transport inland, by issuing loans and by providing small shops, eating houses and opium saloons, they have provided required services, and to minimise the danger of fishermen taking loans and never returning to the same anchorage again, they have been at great pains to maintain temples and organise annual festivals, which few fishermen would risk the ill luck of missing.

Near the temple is usually the breaming beach, where the land population keep vats for boiling nets, and sell fresh water and breaming grass at exorbitant prices. In return, it is often the case that fishermen are friendly towards land people, looking to them for assistance in maintaining temples and organising festivals, and blaming them for bad luck if the festivals are not sufficiently splendid. This curious reaction is similar to that of a dog which will sometimes be obediently attached to a cruel master who beats it. It cannot be said today that the Cantonese are still the cruel masters, though the Chinese Empire once was. They are however the little cheats, and there are few social relationships more depressing and philosophically interesting than that between the land and the sea people in Hong Kong. It is here that the Cooperative and Marketing Department is providing its most significant service, in the provision of schools for fishermen's children. Formerly there have been such schools, organised by the land people who are the fishermen's chief exploiters, and they have not resulted in any degree of emancipation. It is a matter of the greatest social interest to watch how the present educational policy of the Department, if continued, will gradually change the manners and outlook of the floating population.

Most of the islands are rocky and hilly, and there is very little rice land on any of them. Vegetables are grown on Cheung Chau and Ping Chau, for export to Hong Kong, but only a small percentage of the population is engaged in agriculture.

PING CHAU (*Peng Chau*) (Walter Schofield)

The sea near Ping Chau was officially divided into two Marine Lots. Not long after, with constant raking of the sea bed for raw material, growing pollution of water from rubbish dumping by the Sanitary Department and increasing sewerage from Hong Kong by increase of the water carriage system, the lime kiln industry declined for lack of coral to burn: complaints were made about this to me at one time. In Ping Chau this industry employed numbers of Hoklo lime burners and in 1925 they staged a clan fight which cost several men their lives. There was no police station on the island, so investigations were delayed and no evidence of murder could be got: so after taking a lot of evidence in my "court" in the Hong Kong office, I simply bound everybody over, which at any rate gave a period of peace to Ping Chau. It must not be thought that the decline of lime burning ruined Ping Chau, for the islanders had thoughtfully provided themselves with a lucrative light industry in the shape of six or seven flourishing gambling houses, which naturally emptied whenever a District Officer's or Water Police launch appeared.

Communications with the outside world were then pretty elementary. A junk left Ping Chau about 8 a.m. for Hong Kong and returned to the island in the evening; no more encouraging to anyone wishing to "come to sunny Ping Chau" than the clouds of smoke and lime dust that rose perpetually from the kilns. Another industry for which Ping Chau and the other western islands were well adapted was distilling, as their inaccessibility was a great assistance to undertakings wishing to short-circuit the revenue regulations.

Yet another industry flourished at one time in this group of islands. The small islet of Kau Yi Tsai between Ping Chau and Kau Yi Chau, has a cleft in its granite cliffs which opens inwards into a cave of some size. About 1922 this was the scene of the greatest opium seizure in the Colony's history. Up until then eight and a half tons of Persian opium came from the cave, and the crew of the sampan guarding it were put up for banishment. Only the banishees appeared before me, as I was then in the Secretariat for Chinese Affairs, but what became of them I cannot remember.

PING CHAU ISLAND (Map Reference 4398) (Paul Tsui)

Ping Chau Island is situated approximately 5½ miles due west of Sulphur Channel. It is a small island less than ¾ mile at its longest stretch, with a shallow bay deeply indented at the northeastern side.

The original settlers are known to have come from Po On, but they were a mixed group. The families established with the longest histories are believed to the Wongs, who are related to the Wong Wai Tsak Tong of Cheung Chau. Another well-established family is surnamed Chan. There are probably some more others, but the family feeling is not very strongly felt here.

It is believed that the original settlers were attracted here by the fishing potential. A few dozens of small craft of the Koo Chew type still base their activities here. They had a tradition of shrimp fishing, as still prevails on Ma Wan, but the shrimp industry appears to have died out these days.

About 30 or 50 years ago new settlers found this place suitable for the lime kiln business. Coral shells were plentiful in the sea bed around and proved to be easily obtainable. Over 20 lime kilns were once established in Ping Chau's most prosperous days, but most of them packed up in the early 1930s when Japanese cement was selling at a price cheaper than the cost of the production of lime in Hong Kong. At the time of the outbreak of the Second World War, hardly any one was left operating. Three re-established themselves after the war and they form the only survivors out of the once prosperous 20.

One leather tannery still stands today, which is operated by Choy Foon Kuen. He has been established here for nearly 30 years.

The Great China Factory came along in the late 30s, who bought their land from the bankrupt lime kilns. This match factory is well equipped, and backed by strong and powerful Chinese financiers. The management is in the hands of Shanghai Chinese. The registered managing director is T. L. Sung who is a cousin of the well-known T. V. Sung, once premier and finance minister of Kuomintang China.

The main temple on the island is to Tin Hau, the Goddess of Heaven, a sea god, worshipped by many coastal fisherfolk, both on land and afloat.

The Great China Match Factory with approximately 300 workers of both sexes now forms the most important feature of this community. Being outsiders they have to be very tactful in their dealings with the local people. The factory workers have to find living quarters, rented from the village, and apparently the land in the valley is in the hands of the earlier settlers. The lime kiln workers form the next most important group making up the present community. They consist mainly of people who have come from Swatow districts, speaking the Hoklo dialect. Coral digging and lime kiln operatives are essentially coastal; these people therefore are fishing folk by origin and coral digging by profession. They are mobile by habit, and are unlikely to take root anywhere. Lime kilns are operated financially by late settlers. One family, Yu Yat Hing, a native of Toi Shan, came to Ping Chau some 20 years ago but has packed up and now holds 3 to 4 houses on the island with a furniture shop in Yaumatei. Choy Foon Kuen who bosses around a bit on the Kaifong Committee owns the leather tannery. He himself lives most of the time in Hong Kong though he owns properties on the island. Chan Hong, who has been the head watchman in the village, was himself born on Ping Chau. He was a fitter by profession and worked in the Naval Dockyard for some time before the war. He was asked to take up the appointment as head watchman by people like Yu and Choy and also the Great China Factory, to act as liaison officer between the rich financier and the poor indigenous villagers.

Lately there has been exceptional growth of vegetable gardening. These vegetable gardeners obtain their water from wells. Wells in Ping Chau have been reputable for the good water they produce. Water from wells in Ping Chau is said to be better than from that found in

Mui Wo on Lantao Island. It is interesting to contrast this with Cheung Chau where from every well you dig you get nothing but saltish water. Cheung Chau people buy water for drinking which has to be brought in by boat from Lantao. Water obtained from wells on Cheung Chau may only be used for washing. It is believed that the main attraction for vegetable gardeners to come to Ping Chau is the privilege of exemption from delivery of their vegetables to the Government wholesale market.

Ping Chau has been very peaceful except in October 1945 when one of the local bad eggs, Lui Luk, who having sold all the properties left to him by his father, got in touch with the bandits in Mai King Sha (C.T.), who came along under the pretence of being guerrillas, and robbed the whole village after a few days' occupation of the island. Lui Luk has since disappeared.

A few months ago the Kaifong, with the support of the Great China Match Factory, started their own electricity supply. The mechanic working in the match factory installed and put up the whole plant in no time. The Great China Match Factory also provides other amenities such as a library, weekly cinema shows, a playing field, etc., which are open to all residents on the island, but the library is not well patronised with less than 20 attendees on the best evenings. Gambling, however, attracts the workers. It is believed that at least two common gaming houses are functioning at all times. Opium divans also exist, but the number is unknown. Quite a few "secret society" type of clubs or associations exist. One or two hang Mr Mao Tse Tung's photos up in their club premises, but it is believed that they just take advantage of it for prestige rather than for political significance.

The school is being run on free basis. Pupils pay $1 per person instead of school fees. $3,000 has been promised by the Education Department lately for the extension of the school. A new concern, which intends to use the northeast bay as a salt pan, has offered to donate $1,000 towards the building of the school. So far, 60% of the Kaifong are opposed to the turning of the bay into salt pans. The arguments put up being: (a) it should be reserved as typhoon shelter, and (b) someone did apply for the same privilege some 30 years ago but was turned down on the same grounds. Those who are in business feel indifferent to this suggestion, but those who are connected with the sea rather see the bay being reserved for a typhoon shelter.

The Kaifong was known as a Kaifong until 1946 when on the advice of District Officer Southern (Capt E.?) they followed in the footsteps of Cheung Chau and restyled themselves as the Ping Chau Residents' Association. I consider that they should go back to the old name of

Kaifong as this implies the function of local government, which is what they are, whilst the Residents' Association implies a voluntary society which would need to be registered under the Society Act of 1949. Their general election will take place on 16 May 1950, on which occasion I have promised to be present.

PENG CHAU (*Gazetteer* p. 83) (James Hayes)

This island community was not included in Austin Coates' account, being apparently one of those places not yet covered at the time of the illness mentioned in his Introduction.

Gazetteer records that Peng Chau had a population of 3,870 in 1960. However, it is not clear whether this includes a Peng Chau-based boat population of long standing.[1]

The island's land population comprises Hakka families engaged in farming and fishing, and a substantial body of mainly Cantonese shopkeepers and businessmen in the main street fronting the sea.

The former include the Lui and Chung lineages, settled here for upwards of 100 years from Po Kat in neighbouring San On (Po On) County. The seven families of Chung have a ruined ancestral hall. The Luis have declined in number, and are now reduced to a single household. Both used to hold land at Nim Shue Wan, on the adjacent coast of Lantao.

The principal shopkeepers and businessmen of today, as listed here, are the Hakka Lams from Nam Leng in Po On, the Cantonese Chans from Nam Tau in Po On, the Cantonese Ngs also from Nam Tau, the Cantonese Tses from Tung Kwun, and different Wong families, both Cantonese and Hakka. All have been residents for 3–4 generations.

There are a number of industries located on the island, notably the lime kilns which have been its mainstay for decades, and probably date back to before 1898. There is also a large match factory, and a tannery. A flourishing traditional cottage industry employs a number of the inhabitants to decorate the white porcelain used in dinner services and the like which is imported from Canton. Fishing-related industries and boat-building also exist.

There are a number of localities on the island. They include, the main bay or Ching Wan, South Bay or Nam Wan, Little South Bay or Nam Wan Tsai, North Bay or Pak Wan, the hill Wai Tsai, Long Kei Pai, Tai Lung Hau, and Tiu Yue Kung.

There is apparently one of the "sounding stones" (known as "tung kwu shek" or "brass drum stone", according to my informants) on the island. They said that when the north wind blew, the stone would emit a drumming sound, lasting for three days.[2]

I turn now to the Peng Chau Rural Committee. This energetic body is, in practice, that same body of shopkeepers and businessmen, known as the Kaifong, which had managed local affairs during Peng Chau's existence as a settled community and boat anchorage. Its leaders reviewed with me their present concerns, which are many.

They are pressing for a fire station and post office, suggesting that these and a public health (sanitary) office should be combined in the one building, and built on a new reclamation, if possible. These are, in fact, already in the approved Public Works Programme estimates.

They need a water pump for the new clinic, and have also expressed concern at the poor quality of water from a well. However, the Irrigation Engineer in our HQ has inspected it, and found that the water was not brackish, and is therefore OK.

Naturally, they are urging government to provide a main supply to the island, but this cannot be given unless and until the Shek Pik Water Scheme is found feasible and comes into operation, at least five years from now.

They have received 100 bags of cement to surface the surrounds to the wells at the beginning of this year (1957) and are now more or less satisfied on this score.

The Kaifong runs the island's electricity plant, charging consumers $1 per unit. It provides free street lighting in the main streets, and also maintains the island's wells. Formerly, it employed and paid for 4–5 village guards, and also scavenging coolies before the government stepped in.

I asked about the Kaifong's financial resources. It appears that there is a monthly levy on shops. Another source of regular revenue comes from its control over boat breaming, currently let to a syndicate for $200 per month (which then charges the boat people for use of the beach areas). There is also a regular income from another source. The Kaifong owns the cargo junk which carries all the vegetables, etc., to be marketed in Hong Kong, for which its master pays $200 per month for the right to charge his clients for the service.

In regard to new Local Public Works projects, they said that the Hong Kong Junior Chamber of Commerce which is building and paying for the new clinic, has asked if we can provide a water tank there. I said I would get an estimate of cost, and look into this.

They are going to send in a letter about the need for a police post at the island. Currently, only occasional visits are made by a police launch, but the lack puts the Rural Committee in a difficult position, and means that they are often blamed for "allowing" illegal immigration. I wonder whether the level of crime on the island justifies building and staffing a police post.

We then got onto the scheduled ferry services provided by the Hongkong and Yaumati Ferry Company. If people miss the only morning ferry to Hong Kong at 07.45, can a later ferry at 10.00 be provided, in addition to the existing 12.00 service? And on return services from Hong Kong, the 17.00 ferry is the last. This is too early, and they would like one at 19.00 to convenience office workers, who might number 20 to 30 daily.

They also pointed out that extra services would encourage more people to stay on the island, and that it would assist students. They added that they had been pressing for this for a long time now.

The 10 a.m. extra ferry is the one they think is most needed. Since all ferries go to Cheung Chau (and some via Silvermine Bay) these other places would also benefit. If ferries are available, they suggest a few trial runs to test out their suggestions.

In regard to electricity, they say the supply provided from their own plant is inadequate. Having noted that the China Light and Power Company has been laying cables on Lantao, including at Nim Shue Wan opposite Peng Chau, they are hoping that it can extend its service to the island. I asked them to write in with a request, after which I would call on the General Manager to discuss.

They also brought up the unsatisfactory state of the road to South Bay, where the lime kilns are. It is used by many people, but needs raise as some parts are under water at high tide. They estimate that 600 bags might be needed to turn it into an all-weather path. The lime kiln operators are prepared to spend $2,000 on the work, and the Rural Committee will provide food for their workers and meet other casual expenses. This is a big order for cement and other materials, and I wonder if the Kadoorie Agricultural Aid Association can assist.

Happening to make another visit in connection with the opening arrangements for the new clinic, I had taken the opportunity to ask about the local festivals celebrated on the island.

The main procession/festival days are stated to be the Devil Day, the Hoklo Day, and the Tin Hau Festival.

The first of these takes place on the 29th day of the 5th lunar month. This consists of making offerings to the gods, and staging the traditional puppet opera show. It is also credited with driving away evil and harmful things. It can be delayed, but 47 years ago (about 1910) some 40 persons are reported as having died on this account. People associate it loosely with the local Tin Hau Temple, but it is not really in honour of Tin Hau.

The second is held on the 21st day of the 7th lunar month. This year it will fall on 21 August in the Western calendar. Every year, on this day, there is a procession accompanied by "to si" (Taoist priests with their long hair piled up in a knot on top of the head) and "nam mo lo" (lay Taoist priests). Carrying various old weapons kept in the Tin Hau Temple, the object is to show community strength and help drive away devils, evil spirits and plague. The chanting priests, all said to be Hoklo people, include one or two from Peng Chau, and the rest are engaged from Cheung Chau.

The kaifongs (meaning the public) subscribe to the event by providing pork, chicken and other offerings, but each subscriber can get something back at the close of the proceedings. A figure of $750 was given as the cost of holding the event in 1957, which includes the hiring of the chanting priests.

The Tin Hau Festival itself is celebrated on or around the 23rd day of the 3rd lunar month, as elsewhere in Hong Kong.[3]

NI KWU CHAU *(Hei Ling Chau)* (S. H. Peplow)

Ni Ku Chau. Ni Ki — a nun. Chau — an island. The island is so called because it is said to resemble a nun's head in shape.

Ngau Tau Tong, Ni Ku Chau Island. Ngau — as above, Tau — head, Tong — a pool. A village near a pool shaped like the head of a cow or buffalo. In most cases where these shapes or appearances are supposed to be, it is very difficult to pick them out. The Chinese will point out the different parts bit by bit, but I am afraid they possess a far more wonderful imagination than I do, for up to the present I have not been able to follow but very little of their delineations.

NEI KWU CHAU *(Hei Ling Chau)* (Walter Schofield)

I once discussed with the Chung Chau Kaifongs the question of making the ferry call at Nei Kwu Chau or Ping Chau, but they never agreed to letting the boat go there or to any other island, though a call at Nei Kwu Chau would have solved the education question there by enabling its children to attend school on Cheung Chau. I once spent a morning visiting Nei Kwu Chau to discuss the problem of their fifteen children. They had been taught by a private teacher in the ancestral temple, but he had left, and the Education Department were asked to find someone to replace him. They had replied that the children would have to go to Cheung Chau, thus raising in an acute form the problem common in villages all over England today. Few, if any teachers, would volunteer to teach fifteen children in so poor an island as Nei Kwu Chau in those days. The problem of providing one was then found insoluble, and as the only means of transport to Cheung Chau was by junk or sampan. I fear the rate of literacy on Nei Kwu Chau must have declined badly.

HEI LING CHAU (*previously Nei Kwu Chau*) (James Hayes)

Now a leprosarium, but hitherto an island occupied by 10 families of Lam and one of Tsang when they were required to remove by the government in 1952–53, to make way for the leprosarium. They had lived there for at least 5–6 generations. Curiously enough, I only discovered this when I came across them by degrees during my visits to the other villages to which they had removed themselves after being told to leave the island.[4]

Two Tsang brothers are now at Yi Pak, some Lam families went to Tai Pak, with others going to Shap Long, and to Ngau Kwu Long near there.

There had been over 200 persons on the island, living mostly in the large village of Ngau Tau, but about twenty years ago (1930s) they had begun to open a new road or track, whereupon a geomancer told them that they had cut the goose's neck, and would spoil the island's fung shui. They paid no attention, but an epidemic had killed off most of the people, leaving only 50 behind.[5]

I visited the leprosarium at least once in these early days, and had been enormously impressed by the construction and development work being undertaken by the Medical Superintendent, Dr Warren, and indeed, also by the man himself. He had been a medical missionary in China before 1949, and could turn his hand to anything, having obviously been used to having to do this in his previous mission stations on the Mainland.

CHAU KUNG TO (TUNG KOK TAU) (SUNSHINE ISLAND) (*Gazetteer* p. 84) (James Hayes)

This small island near Peng Chau has paddy fields cultivated by (or belonging to) Peng Chau people, but they stopped going there over ten years ago.

In recent years, it has been the scene of a successful agricultural resettlement scheme organized by Mr Gus Borgeest, who has renamed it "Sunshine Island".[6]

TSING YI ISLAND (S. H. Peplow)

Tsing I Island. Tsing — green, I — clothing. An island covered with green clothing, i.e., grass or vegetation.

TSING YI ISLAND (12 September 1946) (Paul Tsui)

Tsing Yi is a hilly island of approximately 4 square miles situated to the southwest of Tsuen Wan, separated by a narrow strip of water from the Mainland. The nearest way to reach the island is to hire a sampan at Sam Pak Tsin Bay (at the back of the Texaco oil storage depot) but a sampan ferry service is available at Tsuen Wan Market at any time of the day charging a moderate fee of 20¢ a head. These ferry sampans run on no definite schedule but leave whenever the sampan is full (12 to 15 passengers). The trip takes about ¾ of an hour. The Hongkong and Yaumati Ferry Co. has, however, recently resumed its pre-war service to Tsuen Wan; this ferry will call at Tsing Yi twice a day both ways. This will, of course, add a great deal of convenience to the communication situation of the island.

The population on the island at the present moment is approximately 540, but the island has provided livelihood for nearly a thousand in pre-war days when there used to be as many as 15 different kinds of factories in operation. The main bulk of inhabitants concentrate themselves in 8 villages in a valley running in from Tsing Yi Bay in the eastern corner of the island. Here we have 6 native villages and a group of factories and shops which form a sort of small market town.

The 6 villages are:

(a)	Haam Tin	咸田	22 families
(b)	San Urk	新屋	6 families
(c)	Yim Tin Kok	鹽田角	10 families
(d)	Tai Wong Ha	大王下	22 families
(e)	Chung Mei	涌尾	25 families
(f)	Lo Urk	老屋	4 families

The predominant families are the Chans（陳）and the Tangs（鄧）. There is another smaller clan. These families form the main bulk of the native population. Other odd clans are mostly outsiders who live on the island just because they are employed in the various factories or shops.

The headman of the island is Tang Hon Fung（鄧漢封）who has been a Chinese middle-class official as well as a miner in China. He used to have business connections in Siam as well as in the interior, but the four years of war seem to have crashed most of his previous business projects, and he is now struggling hard to re-establish his former business connections but finds himself handicapped by his old age.

The total area of cultivable ground on the island is approximately 150 taochung of fields (including dry cultivation as well as salt paddy). This is certainly not enough to feed a native population of near 550. But the people here were quite diligent farmers, and they made full use of their time and energy in pre-war days to raise pigs and plant pine forest or pineapples on the hillsides. I was told that almost every family used to rear from 4 to 24 pigs and all the hills were covered with pine trees in the pre-war era. The 4 years of Japanese occupation interrupted the supply of food for their pigs and they had to cut down almost every pine tree for sale as firewood.

In pre-war days there were 6 lime kilns, 1 oil factory, 1 rice noodle factory, 1 motor boat slipway, 1 paper factory, 2 kerosene refineries, 1 earthenware kiln, 1 chemical works and 1 black plastic ink (for painting plastic walls) factory. But since liberation only 4 lime kilns, a few rice noodle shops and 1 oil factory have resumed operation, but not in full swing. The rest are just being left there idle. These factories offered a good number of jobs to local villagers, mainly as coolies, which had made their livelihood a great deal better and easier.

They have two fairly good school buildings in the valley, but except for a very short period, only one school ever functioned at a time. The story I learned was that a Chan family in Chung Mei (they have two quite different branches of the Chan clan on the island) offered a piece of ground in Chung Mei sometime in 1936 for the erection of a school in Chung Mei village. He obtained a grant of $800 from the Government through the Education Department, but concealed this from the public. When the school was built, he claimed that the $800 was a special donation from his own family and he wanted to exercise complete control over the school in favour of his own clan. The rest of the population on the island learned about the government grant later, and raised hell, so the quarrel began. On the following year, the rest raised their own funds, and built another school near the sea front. When the new school was built only the Chung Mei branch of the Chan clan (about 8 children) remained in the old school; the rest all started going to the new school. When the Education Department compared the number of attendants of both the schools, the annual subsidy for the old one was cancelled and the new one received it instead. The old one functioned for another year and then closed down. The old school building still stands intact, but is being left idle and the owner of that particular piece of ground never arranged a change of name in the land register at the Land Office. At the present moment, only the school near the sea front is functioning with two teachers, and 60 children, receiving a government subsidy of $50 a month.

Though the average health of the villagers on the island cannot be considered the worst we have seen in the New Territories, it is by no means good. Children with swollen stomachs caused by enlarged spleens are very common, which indicates prevalent malaria. Under-nutrition face complexions are also very common. It seems that no doctor or health officer has ever visited the place since liberation except one or two visits of District Officer patrols with their medical orderlies. Minor sicknesses are being cured by native methods and serious cases may from time to time be sent to Hong Kong. Few of the population make use of the twice-weekly sick parade at Tsuen Wan Market, yet the village appears to be fairly clean when compared with many others.

On the whole, the island is a very pleasant spot. It has beautiful scenery. The valley is enclosed by a low range of hills which hide its beauty from the distance. On walking up further into the valley, it gives you an impression that you are visiting a much larger place than you expect it to be. A few bungalows, owned by some urban rich, stand dominatingly on hill-tops, creating a very comforting atmosphere. One village at least lies hidden behind a grove of trees right inside the end of the valley giving you a mystifying interest of exploration.

We are told that many of the paddy fields they have have not been registered. On questioning one of the villagers we are assured that they are quite anxious to get someone to sort the matter out for them.

There is one more small valley on the southwestern side of the island, known as South Bay, where they have a small acreage of paddy fields. An unfortunate mishap took place on 25 August 1946 when a party of English soldiers believed to be from Stonecutters Island, landed there with some explosives and blew them up causing a certain amount of damage to the crops as well as to the paddy fields. The owners are now very anxious to get compensation for the damage. It takes about an hour and a quarter's walk over the hills to get there from the main valley.

The island normally comes under the water police for security and protection, but since the liberation the inhabitants have been dealing directly with the police station at Tsuen Wan. They have not been included in the recent issue of arms by the police in New Territories but Mr Wilson has informed me that he will put this in mind. The island was raided by a gang of armed robbers a few months ago, and they are now quite anxious to get arms for their own protection. I consider that a few rockets will do very well on top of an issue of 4 to 6 rifles to the group of villages in the valley.

TSING YI ISLAND (20 February 1951) (Paul Tsui)

Present population: about 800. I estimated it to be over 1,000 last year. The reduction is due to the closing on account of no business of 2 brick kilns, one owned by Chan Kai, the other by Kong Yui Ching. The closing of brick kilns caused unemployment of over 200, all outsiders, now gone. Chan Kai's kiln has resumed partial operation.

Four lime kilns are now operating, all owned and operated by outsiders. The maximum output is 13,000 bricks per month.

Shing Hing	2 kilns
Yuen Lee	7 kilns
Sun Shing Lee	6 kilns
Lam Si Hop	6 kilns
Tai Shan	closed down

1 charcoal kiln and another one coming (?)

Fields. Over 30 taochung of paddy fields plus 5 taochung under construction. About 2 taochung under vegetable cultivation.

$1.50 per trip, sampan ferry (30¢ per head).
Daily production of vegetables approximately 10 piculs

Livestock population:

Pigs	approximately 100
Chickens	approximately 3,000 (with 4 chicken farms)
Ducks	

Agricultural production:

Pineapples	100,000 trees [*sic*] (200,000 fruits a year)
Fruits, others	

Grass-cutting, for lime kilns and boats. Daily consumption 40 piculs.

TSING YI ISLAND: AN ELECTION (20 February 1951) (Paul Tsui)

Sometime about 2 weeks ago, two villagers from Tsing Yi Island came to my village and told me that they wished to have an election of Village Representatives and Rural Committee members for the Island. They said the existing Elder Mr Tang Hon Fung had expressed his idea of resigning from his position as the spokesman of the inhabitants of Tsing Yi. In fact, these two country gentlemen asked me if I had received his resignation already.

As far as I could remember Tang Hon Fung was never properly elected as a Chairman of any Rural Committee, nor could I remember there being any election on Tsing Yi at all. I remember in the early days of 1945–46, Tang Hon Fung was just accepted as the Elder. I have, from time to time, however, heard of gossip saying that Tang had been a veteran politician in various parts of China before, but was mixed up with certain rackets during the Japanese occupation. His being the unchallenged head of the Tsing Yi community has been more of a surprise to me rather than a personal favour of mine. His good points, however, are that he is experienced, he is careful and he is knowledgeable. I had therefore told the two country gentlemen that if it was the wish of the people living in Tsing Yi that they should have a proper election, I was all for it. I also said that I was looking forward to a properly organised Rural Committee, which I hope will be efficient. I was then invited to be present on the date chosen (which is today) so as to supervise the election. In connection with this, I also advised them to invite Mr Chan Wing On to be there.

I arrived there today in the company of Mr Chan Wing On. The meeting was held at the village school. The actual electoral meeting was preceded by a general discussion of an assembly of over 30-

odd villagers, including the list of persons who signed their names on a piece of paper I have provided. I call this assembly a General Assembly of all villagers who are interested in the village affairs. The discussion touched upon no less than 6 different subjects, including the ferry service, the provision of a pier at the end of Chung On Street, Police procedures, the powers of the Village Representative, the Rice Ration Depot and the Postal Service. I am dealing with these questions separately in the appropriate files.

I was presented with 8 sealed envelopes containing election ballots for the 8 different villages (or better described as hamlets in some cases). I was given the honour to unseal these envelopes, and read out the names of the elected persons. The results of the election were as follows:

No. of Votes for Village Representative

(8)	Tang Shiu Fung 鄧少封	to the Village Representative of Tai Wong Ha (upper), 大王下 上村
(10)	Chan Sai Cheong 陳世昌	to the Village Representative of Tai Wong Ha (lower), 大王下 下村
(14)	Tang Yuen Tai 鄧元泰	to the Village Representative of Ham Tin, 咸田
(5)	Chan Wai Tak 陳懷德	to the Village Representative of Chung Mei, 涌尾
(9)	Chan Sai Lung 陳世龍	to the Village Representative of Dui Mun, 對門
(5)	Tang Lup Fai 鄧立輝	to the Village Representative of Sai Shan, 細山
(8)	Leung Cho Sap 梁 cho 十	to the Village Representative of Kaifong, 街坊
(15)	Chan Sai Kam 陳世 kam	to the Village Representative of the Hamlets of Sun Urk, 新屋, Yim Tin Kok, 鹽田角, Fung Shu Wo 楓樹窩 (combined)

It was their wish that the above-named 8 gentlemen to be concurrently the committee members of the new Rural Committee for the Island; and the Chairman and the Treasurer for the Committee to be elected by them and from amongst them. The term of office will be for 1 year only; and a new election will take place again at the beginning of next year (presumably, shortly after the Chinese New Year).

The result of the separate election turned out to be:

Tang Lup Fai	Chairman
Chan Sai Lung	Treasurer

The newly elected Chairman is a young person (approximately 30 years of age), in contrast to the former elderly Tang Hon Fung who is already over 64 years old. The Treasurer appears to me to be a fairly sound and quiet person. The Chairman was elected with a marginal majority of 5 to 3, whilst the Treasurer choice was unanimous. The new Chairman happens to be one of the two who called on me a fortnight ago, whilst the Treasurer received a lot of compliments both from Mr Tang Hon Fung and Mr Chan Wing On. On the whole, the newly elected Committee appears to me to be well-balanced. Provided these gentlemen can exercise good control over the comparative young Chairman, I am quite happy.

Immediately after the election, I asked the newly elected Committee to work out a proper constitution of their organisation (which they have not got), I advised them to consult Mr Chan Wing On in connection with this question. I also asked them to provide me with a programme of their activities for my criticism at their earliest convenience, which they have promised to let me have within a month's time. Their immediate request was that the Government be persuaded to let them have the piece of land fronting the school to be the playground for the school. (It is foreshore, which needs reclamation, for which I have not made any promise yet.)

We shall wait and see what sort of a programme they are going to produce.

CHEUNG CHAU (Eric Hamilton)

Cheung Chau was always a very go-ahead village and the Kaifong were cooperative. It had been raided by pirates in 1912 and badly shot up. A new brick police station and quarters were then built on higher ground away from the waterfront. After 1920 Europeans started building weekend bungalows on the knob towards Hong Kong side. Donnelly and White (now merged in Dodwell Drinks Department) had one, as did Cartwright, editor of the *Hong Kong Daily Press* and Alabaster, a civilian barrister originally rather anti-government, but he became Attorney General in the thirties — after Kemp I think — and was knighted. The bungalows were not a very good idea, but then the south side of the island had not been opened up for residential use (malaria!).

I had a narrow escape from death once when, against my better judgement, I went over to Cheung Chau on a very stormy day. There was quite a bit of cash involved in the case. When I had finished the case, and set off home with a small safe full of cash, it blew like hell and perfectly torrential rain reduced visibility to about 50 yards. I got up in the bows and suddenly discovered we were heading straight for the cliffs south of Green Island. We clawed round and by the mercy of God managed to make Aberdeen where I turned over the safe to Martin Earner, Sergeant in charge there. A drowned rat had nothing on me. He gave me a half pint of neat whiskey laced with about one-eighth pint of ammoniated quinine. I had never drunk the mixture before or after, but although I had to wait an hour for a police car, I never even caught a sneeze out of it.

CHEUNG CHAU (S. H. Peplow)

Cheung Chau. Cheung — long, Chau — an island. Long Island, so named from its shape, long and narrow.

Pirates at Cheung Chau

One of the most audacious and dastardly crimes that has ever been chronicled in the history of Hong Kong took place on the night of 21 August 1912, when a gang of pirates attacked the Police Station at Cheung Chau. If the affair had stopped at plunder, it would have been bad enough, but when the murder in cold blood of three policemen has to be recorded, it is surely without parallel. The news reached Hong Kong in the early hours of Tuesday morning, being conveyed by Chinese who had been sailing nearly all night.

They fortunately had escaped the bullets of the pirates, and with all possible haste journeyed to Hong Kong bringing details. Cheung Chau is only some 10 miles distant from Hong Kong. It boasts of a big population, nearly all being connected with the fishing industry. The Police Station at that time was built practically on the water's edge, right in the middle of the village, and was in the charge of Sergeant Boulger.

Although of such importance, there were no means of communication between Hong Kong and Cheung Chau except by launch, and for this reason it was impossible for the authorities here to know anything about the affair until many hours had elapsed. The pirate gang consisted of some 40 or 50 men, and they were all armed. They arrived in boats which came close to the police pier, and the men were onshore before anyone in the station knew what was happening.

An Indian PC was on duty in the station, but he was immediately shot down, and the pirates commenced to ransack the place. The shot and the commotion aroused two other Indians, and they rushed to the entrance of the station, but they too were shot dead before they could even reach their rifles.

It was quite dark when the robbers landed, but rain fell about an hour afterwards, and it was perhaps due to this that the village itself was spared. There was over $1,000 in the safe at the Police Station, and it was taken away by the looters together with all the arms and, ammunition in the place.

The robbers undoubtedly knew that the only way by which news of the affair could reach Hong Kong was by means of launches, and they promptly put a launch which ran daily from the island to Hong Kong out of action. All the cylinder plates on the vessel were removed, and thus they were able to continue their looting until nearly one o'clock on the morning of the 22nd. Sergeant Boulger and his wife, who were living in a mat-shed some little distance away from the station were unhurt. Goods, money, jewellery and clothing to the value of $3,300 were also taken from the village. Two robbers were killed, the bodies being taken away by the pirates.

Such an occurrence as this would be almost impossible now as all the New Territory stations are protected by barbed wire entanglements. Far distant stations like Tai O and Cheung Chau are fitted, in addition to telephones, with wireless. Rifles and revolvers are issued to all ranks, and machine guns are also provided. The men are kept well up in drill, and the alarm bell is sounded several times during the month, so that there shall be no hesitation should an attack take place. Each man is allotted to a certain place, and on the alarm being given, he takes up his position with either his rifle or machine gun as the case may be.

CHEUNG CHAU (Walter Schofield)

One of the first questions I had to deal with was a request from the Sheung Tong villagers to make a grant from the small public works fund of $400 at my disposal to enable a footbridge to be built over the deep ravine dividing one side of their valley from the other. This was to be of granite beams, quarried in the Shap Pat Heung and carried up over 1,000 feet to Sheung Tong. I was anxious to get the village to contribute to the cost, as my vote for small public works was only $400 a year and the cost of the three granite beams, and their transport by coolie up the mountain, would have come to about $160; and a good slice of the vote was usually granted to the Cheung Chau Residents' Association for upkeep and extension of paths there. The villagers could raise no money; they could not furnish coolies for transport; and they would not consider laying anything so ill-omened as an even number of stone beams: so to my regret, I felt I could do nothing for people who could or would do nothing to help themselves.

In those days the Cheung Chau ferry was a large one-deck launch and passengers paid 3 cents each for a passage, but for 5 cents the Kaifong committee who ran it, largely in the interests of the fish industry, would give you a bamboo chair on the foredeck to sit on; and this ferry was what drew missionaries to settle on the island from about 1907 onwards and build themselves bungalows for summer holidays, so saving the high cost of a Hong Kong apartment. Its timetable rarely fitted my official arrangements, as by it I could never spend more than an hour ashore unless I got a night's lodging on the island; so I generally used my hired launch. In the thirties a guest-house was opened for visitors in a large bungalow not far up the hill from the police station, and after 1934 I went there two or three times with friends while working on archaeological sites on Cheung Chau and the nearby coast of Lantau. This police station was not built till 1913 or 1914: before then the police had used a large house near the Kaifong pier, about 150 yards south of the later concrete pier, as their station. In 1912 a junk came to the pier by night, the crew and passengers landed and carried the station by a sudden rush, as they were an armed pirate gang. The sergeant in charge and some police escaped and kept the raiders under fire from the slope behind, but they got away with their plunder, including some arms and ammunition. The Captain Superintendent of Police at the time, F. J. Badeley, a cadet officer, retired soon after, and the story went that the Governor, Sir Henry May, who came in July 1912 after about eight years as Colonial Secretary and two years in Fiji, took this opportunity to get rid of him because he was "persona non grata" to him. (There were said to be seven such in the Service.) The Government took the hint given by the pirates and built a new police station on a much more commanding site well inland, surrounded by barbed wire.

The increasing population and prosperity of the Colony caused developments at Cheung Chau. Building land was greatly in demand, part of the foreshore was reclaimed, and houses of reinforced concrete began to appear in the village, modelled on Hong Kong tenement houses. A great difficulty with this development was the problem of ensuring proper inspection of buildings of this type, as the Buildings

Ordinance of 1903 did not apply, and there were one or two rogue architects about who would run up such houses cheap, and make their profit by deviating from plans: swindles that can, as I saw in Hong Kong later, cost lives. The best way of controlling knavery of this sort is to refuse permits to erect any more houses to the architect responsible: that, I was told, is London practice.

The Cheung Chau Kaifongs, who in my time were led by a Mr Lo Yip, a prosperous shopkeeper, were certainly enterprising, and had not only started a ferry to Hong Kong on the funds obtained from the Pak Tai Temple at the north end of the town, but had renovated the temple and set up an electric light installation for the village on the raised ground in the middle of the isthmus. The Ferries Ordinance was passed about 1917 and replaced the ancient launches plying to Yaumati and Kowloon City by much more suitable craft, some of them second-hand Star Ferry boats, far less likely to turn turtle than the overloaded, overcrowded craft which daily imperilled their passengers in the old days, the disasters to which brought about the new legislation. About 1925 the Ordinance was applied to the New Territory, which meant that the existing ferries had to be thrown open to public tender and their boats brought up to a higher standard. The Cheung Chau Kaifongs were encouraged to bid, and as theirs was the only one, and not unreasonably, they got the concession. The old pier by the former police station had sometime before been supplemented by a new wooden pier some 150 yards further north, and this was the Cheung Chau Terminal of the ferry. The concession expired in 1928, and under my successor, Mr Wynne-Jones, new ferry concessions were made, which according to Mr Lo Yip had caused great trouble to the Kaifongs. The timetable was certainly improved from the Hong Kong point of view, and day trips to the island became possible.

An event which affected life in Cheung Chau in 1925–26 was the change of Governor in November 1925, during the great Communist-inspired but mainly Nationalist strike which started in June that year. Sir C. Clementi increased the District Officer South's public works vote from the absurdly small figure of $400 to $1,000 or more. This made it possible to repave some of the chief streets of the village with granite blocks set in concrete, which cost about $800. He also paid a state visit to the island in the summer of 1926. Flags and decorations were put up, and practically all the proceedings were in Chinese, including a lengthy prepared speech by His Excellency written out in characters by (I believe) Mr Sung Hot Pang. He walked through the main streets, keeping up conversation with the Kaifongs, and showing that despite a long absence in British Guiana and elsewhere his Chinese scholarship had undergone no observable decline. The chief impression of the day that remained with me was the heat!

No doubt there are many other places of interest, especially temples and their contents: one of the finest is the Pak Tai Temple in Cheung Chau, with its coloured relief showing the local ferry boat nearing the pier in Hong Kong harbour.

CHEUNG CHAU (Paul Tsui, written after he retired)

Off the coast to the southeast of Lantao lies the dumbbell-shaped island of Cheung Chau, a fishing base, always full of life and activities. However, all the business and social activities seemed to concentrate in the narrow isthmus where houses and shops were built very closely to one another with interwoven narrow and irregular lanes serving as access or communication.

The township has a land tenure unique of its own, in that all the socially and commercially usable land is the property of a private family by the name Wong Wai Tsuk Tong, whose claim to have owned from time immemorial the "bones" as against the "skin" of the land, had been positively affirmed by the Privy Council in an appeal from a long-drawn litigation between the family and the Hong Kong Government. As a result, when one buys a house or a shop from another owner in the small township, he would in effect be buying a sublease previously granted in perpetuity by the Wong Wai Tsuk Tong to the vendor. The purchaser therefore had to enter into two separate agreements, one with the Wong Wai Tsuk Tong for taking over the sublease with the express consent of the owner, the other with the vendor who sells his interests on that particular piece of land on which the shop or the house has been built. In consequence, both the instrument for the transfer of the sublease as well as the instrument for the sale of the interest in the land have to be registered with the Lands Office which then was the District Office.

For reasons unexplained the Wong Wai Tsuk Tong never claimed similar ownership of the land on the hills which had been declared a European Reserve within which only persons with the express permission of the Governor (in effect persons of European descent) would be permitted to own property or to live there. Shortly after the Second World War, this reserve along with similar reserves at Taipo and up on the Peak on Hong Kong Island were de-reserved and people of non-European descent were allowed to move in.

When the first delegation of Cheung Chau inhabitants came to the Peninsula Hotel, requesting an audience with the newly appointed District Officer New Territories, the delegation was courteously received. The delegation called themselves the committee members of the Cheung Chau Residents' Association. Little did we know then that in pre-war days, the District Officer South used to deal with the Kaifongs (i.e., elders), who spoke for the community on the island. Most of the personalities of the two organisations overlapped and as far as we were concerned, so long as they spoke in the common interests of the inhabitants, we would be quite happy to deal with them as they came. There was no such thing as a register of voters, nor were there any accepted criteria as to who would be entitled to vote and who would not. In so far as the inhabitants were concerned, anyone who would be given face by the administration would be considered acceptable.

The delegation who came to see the District Officer New Territories, John Barrow, in 1945, included Mr Wong Chung Hoi, the then managing trustee for the Wong Wai Tsuk Tong. In good faith, we granted them the recognition they sought. Little did we know then that there were inter-factional local politics beneath the surface. One of the problems which emerged later was the ownership of the Fong Pin Hospital (a communal trust property, not unlike the Tung Wah Hospital, but on a much smaller scale). The question, I believe, remained unresolved as late as the mid-1980s. Be that as it may, the Cheung Chau Residents' Association did a lot of good in the rehabilitation of post-war Cheung Chau. We cooperated with them and dealt with many local problems very smoothly and successfully.

In my time as District Officer South, there were already several thousand inhabitants on the island. The two temples there, particular the Pak Tai Temple, was very popular. However, their most popular religious celebrations had always been the annual "Da Chiu", a requiem for departed souls (popularly but wrongly referred to as the Bun Festival) when thousands of pilgrims from far and near would congregate to climb up tall stacks of buns so as to take home a few which would bring special blessings to the family throughout the following year.

Fishing boats congregate at Cheung Chau because it is here that they could de-barnacle their boats on the beach by fire from flames of burning grass (Tam Shuen). They also replenish their nets and other fishing gear.

For the disposal of the dead, Mr Barrow was very keen that a suitable site should be set aside for a properly managed cemetery.

Enlisting the help of an experienced Sanitary Inspector from the Urban Council, Mr Warburton, the site of Cheung Chau cemetery no. 1 was formed that year, and rules for proper burials were drafted which were promulgated the year after.

A proposal for a submarine pipe to conduct water from a reservoir on Lantao Island was also approved that year.

I took advantage of a meeting with Bishop Henry Valtorta in Sai Kung to bring to the bishop's attention the fact that the island of Cheung Chau was densely populated, and the community there appeared to me prosperous, but I could see no tangible sign of the Catholic Church's involvement. I suggested to the bishop that the Catholic Church might wish to establish its presence on the island. I was also glad to note that years later, Caritas operated an impressive holiday camp for youth and later a pre-vocational school in the peak area on Cheung Chau Island. Close by to it was also a retreat house operated by the Jesuit Fathers, and another institution not far away operated by the Salesian Fathers. The parish of Our Lady of Fatima on Cheung Chau Island has grown to become a very active and prosperous community.

CHEUNG CHAU (Austin Coates)

The second largest town in the Southern District, Cheung Chau is a small island situated 8 miles southwest of Hong Kong. The distance from the centre of Victoria (Blake Pier) to the town's main west-facing harbour is about 11 miles. It is well served by the Hongkong and Yaumati Ferry Company, which runs four direct passenger services daily, with extra services on weekends and public holidays. There is a daily fish ferry from Cheung Chau to Aberdeen, and a slower passenger service via Silvermine Bay and Ping Chau. The fastest direct ferry takes 65 minutes; the indirect service takes 2 hours, and the fares are low. The direct fares are $1.20, first class, $1 second class. From Cheung Chau to the District Office, Kowloon, takes 1 hour 35 minutes. The population is approximately 18,500, of which approximately 7,000 are fishermen permanently based at Cheung Chau and living aboard small boats.

Although this is not at first apparent, the island's importance is as a fishermen's anchorage, and the town has grown up on the peculiar relationship between land and sea people mentioned in the beginning of this section. There are thus several important temples, and religious festivals play a significant part in the year's main events. As with the other southern islands, human habitation is probably very old, although there have been no archaeological discoveries. The earliest historical traces are to be found at the Pak Tai Temple, of which the idol is said to have been stolen from Pinghoi. The sea-robbers were passing with it near Cheung Chau when they were shipwrecked and came ashore on the island. The first location of the Pak Tai Temple was near the present market, and the earliest tablet in the existing temple, dated Chien Lung, 48th year (1782), originally hung in the earlier building. The present temple was built in the 8th year of Hsien Feng (1856), and enlarged in the 30th year of Kuang Hsu (1899), the year when the British occupied the New Territories, when there was an outbreak of plague on Cheung Chau, in the control of which government officials considerably assisted. From family sources it is known that the earliest settlement consisted of five Hoklo huts situated near the market, in the zone known as Hoklo Hong. One of these families was Tsui; the earliest branch of the family to arrive has died out, but a branch that arrived later still survives and owns property on the island. The Wong family, which by 1905 owned the greater part of all the land at present constituting the town, apparently came about 1820 from Nantao. Mr Wong Chung Hoi, manager of the Wong Wai Tsak Tong, is aged 61, of the 5th generation at Cheung Chau. The Hoklo settlement is fairly close to the market, where the original site of the Pak Tai Temple is said to have been; and it is just possible that the pirates referred to in the story of the temple's foundation were the same people as those who settled at Hoklo Hong. At any rate, the temple tablet, and the bell dated the following year (1783), indicate a fair degree of prosperity by that date, and it seems safe to say that Cheung Chau probably began to be permanently settled about the middle of the eighteenth century.

The earlier name of Hoklo Hong was Ng Kau Uk, the Five Old Houses. In the southwest part of the island is a cave associated with the notorious pirate Cheung Po Tsai; whether in fact this pirate was connected with Cheung Chau is not known, though he may well have been. He was subdued by the Portuguese in 1810, and subsequently became an anti-piracy official of the Chinese government.

Cheung Chau's principal function is as a shopping and money-lending centre for fishermen, who visit the place in their thousands at certain seasons, chiefly at the Chinese New Year, the Pak Tai festival (3rd day of the 3rd moon), the Tin Hau festival (celebrated in Cheung Chau a week before the proper festival, which is the 23rd day of the 3rd moon), the Feast of the Hungry Souls (a moveable feast held usually in the 4th moon), and the Dragon Boat festival (5th day of the 5th moon). After this there is a lull until the autumn, when the "wong fa" (yellow croaker) season brings many fishing boats into the waters between Cheung Chau and Tai O. Good and bad commercial years in Cheung Chau are gauged almost entirely by the numbers of fishermen who visit the island during the year, the size of their catches, the price of fish in Hong Kong, Kowloon, and to a lesser extent in Kwangtung, and the amount of money fishermen have to spend.

The island's second function is as a market for a large number of villages on Lantao Island. For all the villages from Shek Pik on the west to Shap Long in the east, Cheung Chau is the natural market; there is considerable trade with Silvermine Bay, and even with the Pak Mong group. The Soko Islands are also linked commercially with Cheung Chau.

There are several important industries. Cheung Chau is probably the most important place in the New Territories for shipbuilding and repairs; there are five shipyards. There are soya factories, fruit preserving factories, and a leather factory. The island's beancurd is well-known, and considered as second in quality only to that of Shatin. Salt fish of high quality is prepared by many individual families, and many people earn a few extra dollars by making match-boxes for sale to the match factory in Ping Chau.

A good deal of money is also made out of weekend holidaymakers from Hong Kong and Kowloon. Although this element results in a high standard of food in the principal restaurants (Yuen Long and Cheung Chau serve the best restaurant food in the New Territories), the income seldom amounts to more than 15% of any restaurant's annual takings. Tung Wan is a popular swimming beach, and the ferries sometimes bring as many as 2,300 people over from Hong Kong on a Sunday. Other smaller swimming beaches are Kun Yam Wan (411891), Lam Tam (409886), and the more remote Pak Cho Wan (396879), which is the best swimming beach and rewards the person willing to walk the 40 minutes needed to reach it.

The island consists of two small hilly masses, north and south, with a low thin strip, hardly more than a sandbank, between. The formation has earned it among Europeans the name of Dumb-bell Island. The town is situated on the low strip. On the northern hill there has been very little development, apart from a small Protestant cemetery (where incidentally there is one foreign grave, that of an American woman missionary from Canton), and a goat farm started in 1953. On the southern hill, which until 1941 was a European Reservation, about twenty different missionary organisations had small holiday houses, all of which were looted and seriously damaged during the internment of their owners during the Japanese occupation of the colony. In 1945, when the various European Reservations were removed by law, some of these mission properties were put up for sale, and since then the Peak, as it is called, has been the most significant area of development on the island. The morning and evening ferries are so timed as to make it feasible to travel into Hong Kong to work each day, and between 20 and 30 people are now doing this. Many former mission properties have passed into Chinese hands, the houses being rebuilt in more modern and cheerful styles, while overlooking Lam Tam cove there has been extensive new development. The path that formerly led from the town to the Peak has now become Peak Road, with small single-family houses on either side; and those parts of the Peak nearest to the town (and the ferry) are nearing capacity development. The Peak now includes a Jesuit seminary, where Irish fathers learn Cantonese; a Protestant American Bible School training Chinese students for service as pastors in (presumably) China; a seminary and rest-centre for the (Italian) Catholic Mission, which provides for Hong Kong's Catholic parish priests (here the Italians learn English from the Irish Jesuits); a private hotel run on European lines; and numerous family residences. This urban population is partly responsible for the high standard of the island's shops, which are probably the best in the New Territories.

On the highest point of the Peak there is a Royal Navy observation station; other evidence of Government on the Peak is a weather reporting station, reporting to the Royal Observatory, Kowloon, and to Kai Tak Airport, a Government-controlled cemetery, and a new urn cemetery, overlooking Pak Cho Wan. In the town there are two Government piers, a Police Station, two doctors and a large nursing staff running St John Hospital, the largest hospital in the New Territories, a Marine Licensing Office, a Post Office, a resident Health Inspector and scavenging staff, a large Government School with resident teaching staff, a Fire Station and an office of the Cooperative and Marketing Department. The St John Hospital, given to the St John Ambulance Brigade by the late Aw Boon Haw and Aw Boon Par, is rented from the Brigade, staffed and maintained by the Government.

In addition to the private hotel on the Peak (which is patronised equally by Europeans and Chinese), there are 4 hotels or boarding houses in the town. There are 6 restaurants, 3 tea-houses, 3 goldsmiths and 1 pawnshop.

The indigenous population lives in widely varying standards of housing. Some of the wealthier residents live on the Peak in European-style houses; others live over their shops, or else in houses of traditional

Chinese design. On Tung Man there are about 200 people living in grass huts, while near the Pak Tai Temple at the north end of the town, and around the bottom of Ching Hing Street at the south end, there are numerous families living in boats lying on land, supported by stones or on stilts. This squatter population is unfortunately increasing, and if it is not to become uncontrollable, definite action will be required fairly soon to deal with the problem of housing for the lowest and poorest levels of the community.

In this, as in almost all matters concerned with the further development of the island, the Government is seriously impeded by the fact that the town is almost entirely privately owned, by the Wong Wai Tsak Tong. The drainage of the town is unsatisfactory; the road levels will have to be radically altered if the drainage is to be improved; there are grave fire risks which can only be removed by the demolition of various buildings and the widening of roads; public lavatories and bath-houses are urgently needed, etc., and against all of these stands the private ownership of land by the Wong Clan. It seems to me that if Cheung Chau is to be changed from a tightly packed village into the modern town which it ought to be, some resumption of Wong clan land will have to be effected, although this is certain to arouse serious opposition to the Government from some of the island's chief vested interests with the widest ramifications. Such resumption might start with all uncultivated agricultural land.

The island's greatest need, a good water supply, is now being met by the Government. A reservoir has been built near Shap Long, on Lantao Island, from which water will be piped under the narrow channel between Lantao and Cheung Chau to a service reservoir situated on the highest point of the northern hill on the island. It is hoped that water will be laid on for the town by the autumn of this year. Later it may be possible to extend the supply to the Peak. For the moment, the supply will reach as far as the Weather Reporting Station, near the top of Peak Road. Most of the Peak houses in any case have large water storage tanks, and there are several wells. Previously, during dry periods, water was carried over from Lantao by sampan and sold by the bucketful.

Public lavatories and a bath-house, incinerators, and a large increase in the scavenging staff are really the island's chief needs. One large lavatory and two incinerators are provided for this year, and at this rate it should be possible to effect improvements sufficient to meet the needs of the population by 1960.

The District Office is guided by the Owen town plan, made in 1935, laying out what planning is possible in the already heavily built-up town zone. The plan is not particularly satisfactory, but it is safe to say that nothing short of pulling down a section of the greater number of all the buildings standing could hasten the implementation of any plan, however well thought out.

In spite of having a respectable number of schools, giving a reasonably high standard of primary education, it is estimated by the elders of the people that there are still between 500 and 600 children not receiving any education at all. These figures are obtained from the numbers of candidates for annual entry to the Public Free School, and are a fair assessment of the island's educational needs. The principal educational institution is the Cheung Chau Government School, which has six Chinese primary classes, and three lower-middle classes. From this school, boys and girls can mount to King's and Queen's Colleges in Hong Kong, and there is at least one free scholarship per year to King's College. Most of the Cheung Chau men who today occupy prominent positions in the island's affairs, or work in business or in government service in Hong Kong, studied at this school. Since the war there has been much activity in increasing the size of the premises of several of the island's other schools. The most prominent improvement is the construction of the Kwok Man School, which is actually a group of three schools situated together overlooking Tung Wan. The three schools are the Po On, Tungkun and Waichiu, and they reflect the composition of the land population. In its small way, Cheung Chau resembles Singapore more than Hong Kong, in that it still has clearly definable quarters in which the people of one Chinese province or district predominate over others. In the centre of the town these distinctions are being obliterated; but the north end of the town, the Pak She Street quarter is still mainly Hoklo (Waichiu), and it will take a long time yet before these divisions finally disappear. The other important new building is that of the Cheung Chau Public Free School, which is maintained by a Government subsidy and by funds raised by the Cheung Chau Residents' Association from rents on market stalls and other properties.

Although Cheung Chau seems to be a comparatively settled community, there are few families who do not look upon some other place as their home, and if conditions were even to become desperate, about nine-tenths of the population might be expected to leave. The Chinese districts that have mainly supplied the population are, as the former paragraph indicates, Po On, Tungkun, and the Swatow region. People from the last-named area were most of them at one time fishermen living ashore, several generations back in the case of

Cheung Chau's Waichiu people. They maintain their special Waichiu Association and School. It would be impossible at a quick glance to distinguish them as an entity of population. Of the other districts that have fed Cheung Chau, the most important is Shuntak, which may in the next few years contribute the fourth building in the square forming the Kwok Man School. There are also elements from Chungshan and Toishan. The people are overwhelmingly Cantonese, with the important Waichiu element. Very few New Territories people have migrated to Cheung Chau; men from Shap Long and Tai Long work as labourers in Cheung Chau, chiefly as house-builders and factory workers. The only prominent person from another part of the New Territories is Mr Chow Li Ping, a former Chairman of the Residents' Association and at present Vice-Chairman of the Cheung Chau Chinese Chamber of Commerce, who is a native of Lamma Island. Once again, the fundamentally unsettled and migrant nature of the population (when the Chinese frontier was open, there was a constant movement of families in both directions) is explained partly by the fact that 90% of the people do not own land, but are tenants of the Wong Wai Tsak Tong.

The Cheung Chau Residents' Association, the principal public body with which the District Administration deals, was founded during the war, and is actually a sad remnant of the Greater Asia Co-Prosperity Sphere, although none of its members would consider themselves complimented if one reminded them of it. It holds annual elections for its committee of seven, on a basis of universal franchise, with no electoral role and no literacy test. Voters are asked to produce either their rice ration card or their identity card to prove Cheung Chau residence, and the elections are freely manipulated by the more ambitious citizens, principally by the simple method of paying large numbers of the illiterate fishing community to vote for them. It was a mistake made in the balmy democratic days of the British Military Administration to have allowed the Association to pass virtually unsullied from war to peace, and it has caused District Officers a number of headaches since. Mercifully, the politics of Cheung Chau are unconnected with Peking and Taipei. Schools were being used in 1949 for the formation of Communist cells, but due to prompt action by the Government this growth was nipped in the bud. The strife is thus one of personal ambitions. It is an honour to be the Chairman of the Association, and at certain times there are perquisites to be made. On this matter, however, it should be added that there is considerable expenditure on entertainment, and if the Chairman makes a little on the side, provided it is not too much, it is well to turn a blind eye, because he probably needs it. For example, it may happen that the District Officer visits Cheung Chau on Monday, the Divisional Superintendent, Marine Police, on Tuesday, the Assistant Director of Education on Wednesday, the Chief Engineer, Port Works, on Thursday, and the Medical Officer of Health on Friday. If prior notification is given, as it sometimes has to be, to enable local discussions to be held, this means, by local etiquette, a substantial restaurant lunch every day of the week, which for the Chairman is a considerable outlay.

Then on top of it all, the entire Police Band may arrive on Saturday and expect to receive light refreshments after playing for an hour on the waterfront. This aspect of Rural Committee work is not fully appreciated by the Government, but in 1953 one of the best Chairmen in the District had to resign due very largely to the expense involved.

Each annual election has to be carefully watched by the District Officer. The first serious crisis occurred in December 1951, when Mr C. G. M. Morrison was District Officer, and Mr John Barrow OBE was District Commissioner, New Territories. As a result of measures taken then, this Administration has posted independent supervisors at every polling booth. The following year, with Mr Morrison's approval, indirect elections were substituted, in an endeavour to counteract the influence of bought votes; the public elected a college of 55, who then in private session elected the committee of 7. The leading participants were, however, quick to learn how to manipulate this too; in fact, the troubles have increased with their increased experience of election techniques. A climax came in December 1954, when a group of residents, encouraged and guided by an outside advisor, Wong Kam Fan, a former minor Nationalist official from Nantao and a distant relative of the Wongs of Cheung Chau, irregularly altered the constitution of the Association 17 days before an election was due. The new Constitution was not submitted to me for approval, as it should have been but I allowed the election to take place provided certain public statements were made (and these were made), while reserving judgement on the new Constitution. The main reason for this was that arrangements were already in train for the election, and its cancellation would have meant a loss of over $500 to the Association. In the event, two citizens who under the old constitution had no right to become members of the Committee (having served the constitutional maximum of three consecutive years, after which a member must stand down for a year) were re-elected and publicly announced their membership of the 1955 committee, ignoring a letter from me suggesting a compromise solution. In January 1955, with the approval of the District Commissioner, Mr K.

M. A. Barnett, ED, and in response to a petition signed by 200 people objecting to the constitutional amendments, I declared the elections temporarily invalid, pending enquiry, and ordered that all further connection between Government Departments and the Association be suspended. At the time of writing, this order is still in force.

What is now needed is the dissolution of the Residents' Association as the principal body administering funds raised from property leased by the Wong Wai Tsak Tong to the Kaifongs of Cheung Chau, and its replacement by a Cheung Chau Rural Committee similar to, although probably organised on slightly more democratic lines than those of Tsuen Wan and Sai Kung. The Residents' Associations of Tai O and Ping Chau should be similarly transformed. The zone of the Cheung Chau Rural Committee should include the South Lantao villages from Shek Pik to Shap Long, and the Soko Islands, and the Full Session Committee should include one representative from each village and a number of town members worked out on a basis of approximately two members per thousand of population. This would give about 36 town members and 11 country members (Fan Pui would probably prefer to be under the Tai O Rural Committee, and Pui O would probably send two members to represent their whole group). This Full Session Committee would elect the Executive Committee, which should consist of 9 members, 3 of whom should retire each year. General Elections to the Full Session Committee should be held every three years. Any person having served three consecutive years on the Executive Committee should be ineligible for election for one year, although eligible for election to the Full Session Committee. An electoral roll must be compiled, and a literacy test imposed, under the supervision of a Justice of the Peace of such Government officials resident on Cheung Chau as the District Officer may approve. The meeting of the Full Session Committee at which the Executive Committee is elected should be attended by the District Officer, and conducted as at Tsuen Wan. This will not entirely cleanse the electoral arrangements, nothing will, because Chinese people, although liking the outward show of a fair election, find uncomfortable the social relationships arising from any situation of uncertainty in leadership. Hitherto there has been no candidature for election; there should be. Candidates should be nominated by responsible citizens and registered ten days before the election. A Presiding Officer should be appointed, a Justice of the Peace or some other responsible party. There are at present three Justices of the Peace resident on the island, Dr Khoo Keng Wah, Medical Officer in charge of St John Hospital, Mr So Pak Sui, Headmaster of the Government School, and myself.

In various small valleys among the southern hills about 70 families of farmers grow vegetables, most of them for sale in Hong Kong. They have a Vegetable-Growers Union, which is about 50 years old and should be brought more into touch with the Cooperative and Marketing Department. With the introduction of water latrines, obtaining night soil is going to become more difficult, and careful thought will have to be given to this. At present the farmers obtain night soil from the earth latrines at the bottom of Peak Road and on Tung Wan; Cheung Chau also exports night soil to villages in South Lantao.

The Cheung Chau Electric Company provides electric light and power. The company, founded in 1913, and at present owned by Shanghai residents of Hong Kong, enlarged its 12-hour service to a full day and night service in 1950. The rate is high: $1 per unit for light. As a result of representations made by this Office, a slight reduction has been made for Government buildings only, and negotiations are being pursued, with a view to extending this to the general public. The principal users are the cinema, the restaurants and soya factories.

The Hongkong Telephone Company provides an internal service, and a service to Hong Kong which is by way of radio telephone, the cost of a call being $1. At hours of peak load the radio service is not at all clear, but it is better than nothing. The telephones were installed in 1952.

At Tai Shek Hau (402887) there is an important breaming beach, owned by a branch of the Wong clan, which claims that when the New Territories were leased to the British Government it was accorded breaming rights. I can find no reference to any such rights on what is, by the laws of Hong Kong, Crown foreshore, and the operator of the beach, Mr Wong Wai Fong, will be asked to produce evidence for this statement. The prices charged for grass are probably the highest in the Colony. Further round the coast, at Saiwan (396883) there is a private reservoir where Mr Wong Shing Yip, a former Chairman of the Residents' Association and a member of then Wong clan of Cheung Chau, sells water to fishermen. Wong was imprisoned for a short time in 1950, under the Emergency Regulations, for extorting fees from junks passing between British and Chinese waters. Had he not been able to prove that he was born in the New Territories, he would have been deported. He has since been behaving himself, and his relations with the Administration while he was Chairman (1953 and 1954) were excellent, although he is not a good organiser and therefore in some ways an inefficient chairman, often failing to consult his fellow members. His chairmanship marked the rise to influence of his distant cousin, Wong Kam Fan, referred to earlier, who acted as his private advisor, and who unwisely piloted Wong Shing Yip into the constitutional amendments which led to the interdiction of the Association.

As occasion offers, Cheung Chau makes large profits out of smuggling. In general, the Colony's smuggling activities are centrally controlled from Victoria. Those who appear to be big shots in the New Territories are actually only small fry acting, with a certain amount of independence, within the framework of a ring embracing Hong Kong, Macao, Nantao, Shekki and Kongmun. Commodities vary, but there is always something worth smuggling from somewhere to elsewhere. In 1948 it was gold; in 1950 fuel oil; and so on. Even rice sometimes figures high on the smuggling scale. Cheung Chau, Ping Chau and Tai O all participate in the smuggling, which alters in nature according to the laws, restrictions and controls enforced by the Chinese, British and Portuguese governments at the mouth of the Pearl River. Lesser smuggling centres are Silvermine Bay and Yung Shu Wan, and when the Hong Kong preventive services are being particularly watchful, smuggled goods enter the Colony from Macao and Nantao even through Sai Kung. Other places which I suspect of being involved in a small way are Po Toi and Tai A Chau, the more southerly of the Soko Islands.

Cheung Chau smugglers made a great deal of money out of fuel oil smuggling in 1950 and 1951, and many of the large European houses on the Peak rest on kerosene foundations. Since that date, although there have been occasional flutters, such as that in rice from Macao in 1953–54, these activities have been so restricted that they can no longer be said to count as a salient factor in the island's economy. From the heyday of 1950 the island had notable benefits, including donations to the building of new schools, the construction of many new roads and the repair of old ones. The timely discovery by the then Chairman of the Residents' Association that a subordinate member of the Police Force was heavily involved in the smuggling led to the latter being obliged to fork out large sums, all of which went on road-building.

Great social changes have taken place in Cheung Chau since the 1930s. Some of them are illustrated in the photographs round the walls of the Residents' Association Office. When Sir Cecil Clementi (Governor 1925–30) visited Cheung Chau, none of the elders accompanying him wore European clothes; on such an occasion today there might be one or two Chinese long coats in a group of fifty, and black silk jackets (ma kua) have gone completely out of fashion. Cheung Chau is in fact even more slapdash in its manners than Hong Kong.

Before the war, few of the local people indulged in swimming as a recreation, and Cheung Chau was noted for its smugglers, opium smokers and missionaries living in exclusive isolation on the forest-clad Peak. During the Japanese occupation nearly every tree on the island was cut down, the main exception being large good luck trees in the town area. Although about 10,000 trees are planted per year, grass-cutters, children and holidaymaking vandals account for the loss of many, and typhoons for more, and it is only in one or two places that the skyline is gradually beginning to be more verdant. Swimming is now a popular village recreation, and in the summer evenings, between six o'clock and sundown, Tung Wah will have as many as 500 local people, including small children, enjoying themselves in the water. Most of the schools wisely encourage this, and so do the better employers of labourers in the town. There are opium and heroin seizures from time to time, but I would say that the post-war generation has far less interest in drugs than its forerunners, and far more interest in sport, the cinema, and pictorial magazines, many of which are pornographic, judged by European views on morals.

There are a number of unregistered doctors (herbalists), dentists and midwives, and all the hotel-keepers apparently act as procurers. At weekends and holidays, a small number of prostitutes come over from Hong Kong and Kowloon.

Urn burial, along lines similar to those followed in the urban area, is being introduced this year, and this autumn there will be a large removal of graves which are technically illegal and all more than ten years old. The entire southwest sector of the island will be affected, west of a line from Tai Shek Hau to the Jesuit seminary. As a special concession, any grave for which reasonable proof can be produced to the effect that it existed prior to the lease of the New Territories will be exempted by order of the District Commissioner.

As will be seen from what is written above concerning the Residents' Association, Cheung Chau is a place of incipient petty quarrelling. This has hindered all cooperative Kaifong work ever since anyone can remember. In the early days after the First World War the Kaifong attempted to run its own ferry junk, but the attempt failed as a financial loss due to internal disagreements. The Kaifong started the Electric Company, but again because of bad management and constant friction in the management they had to sell out. There is no sign whatever that this trend is coming to an end. The principal contestants in Cheung Chau politics, Messrs Chow Li Ping, Chu Pak Sang and Wong Shing Yip, are in fact all fairly young; so disputes may be expected to go on for years. The first two gentlemen named received the Coronation Medal in 1953.

CHEUNG CHAU[7] (*Gazetteer* pp. 89–90) (James Hayes)

Whereas Austin Coates' account of this important island is voluminous, I have very little information in my notebooks. *Gazetteer* gives a land population of 15,100 for 1960, mostly Cantonese, whose livelihood was mainly centred around the fishing industry and its related trades. Its home-based and visiting fishing fleet was the lifeblood of this generally flourishing and interesting place.

Cheung Chau had limited cultivable land, and in 1957–58 this was now used only for vegetable cultivation. There was a vegetable growers' guild, the Choi Yuen Hong, which was established before the war. I spoke to its vice-chairman, Yu Kai-hang, born in the island's Sai Wan, a cultivator himself, with some fields of his own there. There are, he says, over 100 members, either independent farmers or their employees. Some have been in Cheung Chau for a long time. The majority rent their land from owners or on annual permit from the District Office.

There is no government-organized Vegetable Marketing Organization depot on the island, and the guild is the sole body concerning itself with the vegetable-growing community and its welfare. The guild arranges collection, transportation and sales in the Hong Kong wholesale market at West Point. Around 30 piculs a day are sent over, usually by the Kaifong's cargo junk.

There are yearly elections of office-bearers. The guild operates a burial fund among its members, but there are no services like a school, clinic, or the like.

Another member was present. This is Fu Yun-fuk, a man in his thirties. Born in Outer Ling Ting, in Chinese territory, he came to Cheung Chau after the war. He is a cultivator, but also engages in the fish trade.

There are also four pages of jottings made on board a government launch before or after a meeting with the leaders of the Cheung Chau Chamber of Commerce, led by its charismatic, capable and energetic chairman, Mr Chau Li-ping. This was in regard to securing agreement on elections for the new Cheung Chau Rural Committee. The Chamber was a stop-gap substitute for the Cheung Chau Residents' Association, from which the District Commissioner had, in 1954, on the D.O.'s advice, withdrawn his recognition as the island's representative body following electoral irregularities. Securing agreement on a replacement was an involved business, and the rural committee was not established until the end of 1960.[8]

Chinese Theatre and Performance (S. H. Peplow)

In Hong Kong there are several well-built Chinese theatres; to mention a few, the Po Hing, Tung Hing, Tai Ping and the new Lee Theatre attached to the Lee Gardens. But, should a performance be required in the Territory, or on any of the larger islands, such as Cheung Chau or Lantao, a mat-shed is erected. These structures are wonderful affairs. Large enough to hold from one to two thousand people, they are erected in about two days, and only about a quarter of this time is required to demolish them. The materials used for their erection are bamboo poles and matting. No nails are used, each pole being tied to the other with strips of rattan, about two or three feet long and half an inch wide. The framework is erected first, the higher the building, the more poles tied together. After the framework is finished, it is covered with strips of rattan about 6 feet by 3 feet, roughly sewn together with thin strings of rattan a little narrower than that used on the poles.

In addition to the theatre, a shed is erected for the players who travel from village to village to give their performances.

The play, although lasting three or four days, with short intervals for sleep and meals, is a performance made up of several short plays or sketches, lasting about four or five hours each.

The scenery is conspicuous by its absence. For instance, suppose an act calls for an episode wherein the heroine has to escape over some high mountain. The mountain will be represented by placing several chairs on top of each other. Over this perilous path she will assist herself by clutching at imaginary trees and boulders.

In fact, one has to use a great deal of imagination when viewing a Chinese performance.

Men dressed as women take the female parts, and pitch their voices in a high falsetto to make the illusion greater. The players wear a certain "make-up" so as to be known at once. The villain of the piece always appearing with his nose painted white; the hero always young and handsome, and the father as an actor wearing a long, flowing beard, etc. There is no need to issue programmes; everyone will recognise who is who at once.

Should a fight be staged between two of the actors using either swords or daggers, there will be none of the "all or nothing" style of the Western theatre. The duel is performed by the combatants taking up various postures. They will face each other and at each slight action or movement, will yell what they intend to do later. There is no real tussle or throwing about, for, in the first instance it may hurt, and secondly the costumes have to be taken care of. Very gorgeous robes are sometimes worn, especially if the play is an historic one: these robes contrasting greatly to the poverty and bare appearance of the stage.

Mention must be made of the stage hands and general assistants. These people wander about at will. No one takes the least notice of them. If a table or chair is used in a scene during the performance and there is no further use for it, one of these gentry will walk on to the stage in full view, of the audience, and remove it; and as the dress of these people during the hot weather is limited to a pair of thin trousers, one can imagine the contrast between them and the actors dressed up in their beautiful costumes.

To one unacquainted, it is very hard to give a description of the band. There are generally four or five players scraping on native fiddles, and a few others with drums and clashing cymbals, all the lot from a European point of view, playing as they like. Sometimes, one of the principal actors will be speaking his part. At the end of every sentence, crash will go the cymbals, and a little tune on the fiddles and drums will follow. Then another sentence and the same deafening noise. A few minutes of this sort of thing is quite enough to satisfy the average European, but the Chinese will sit through it for hours. They always go to the theatre well provided with melon seeds, roasted peanuts and cakes; and will sit, eat and chat; or watch the play, whichever appeals to them most. Hawkers, with baskets on arm, will walk about shouting out their wares, and should one of the audience catch sight of a friend in another part of the theatre, he will give a yell to attract his attention, no notice being taken whatever.

The visit of a theatrical company to a certain village is looked upon with mixed feelings. Relatives from other villages consider the occasion an ideal time to visit their uncles or aunts, brothers or sisters, as the case may be, resident there. Extra food has to be provided, and in many cases the show attracts other visitors, in the shape of pickpockets and thieves.

Cheung Chau Island: A Theatrical Performance (17 August 1946) (Paul Tsui)

It was the usual practice in the past for the Kaifong of Cheung Chau to sponsor theatrical performances, but they appointed a sub-committee to manage the production. The sub-committee, as a rule, was chosen by casting the "holy dice" before the altar for the "approval of the Gods". This year, the people of Cheung Chau predicted that the performances were going to be a losing business proposition and so they found that voluntary candidates for the sub-committee were scarce. The Kaifong (or Residents' Association) therefore decided to appoint the "Youth Group" (少年團) to be the sub-committee for managing the performance (as they were the only body of men who were willing to undertake the risk).

It was also the usual practice for the sub-committee to run its finances independently. The Kaifong Committee usually does not provide any subsidy out of its current public funds, but it applies for a subsidy from the Chinese Temple Committee, which was how it was getting $1,000 in the past.

The sub-committee usually starts by raising a sum as initial capital; it then goes round to all shops and households for donations and subscriptions and then puts up the performance. When the show is over, the sub-committee will then collect the subsidy from the Chinese Temple Committee, which will be treated as repayment to the sub-committee members for the initial expenses. Very often, the performances turn out to be a profitable racket and the sub-committee makes a handsome profit out of it from the by-products, namely gambling stalls, food stalls, etc.

This year, the performances turned out to be a losing proposition, so instead of three performances, they only gave two. Messrs Wong Chung Hoi and Au Hop Chuen told me that the "Youth Group" was the recognised sub-committee for this year's performances, and they were the right people to get the Chinese Temple Committee's subsidy if there was going to be any.

I recommend the Chinese Temple Committee's subsidy to be paid accordingly.

Cheung Chau Festivals (Austin Coates)

Being a fishing centre, and being dependent on the regular visits of fishermen, Cheung Chau is notable for religious festivals. The first four months of each Chinese year, from the New Year to the Dragon Boat Festival, are in fact so taken up with one festival after another that it is seldom possible for the Administration to have any serious dealings with the Residents' Association before the middle of the fifth moon. Usually the largest annual congregation of junks is seen at the Chinese New Year. The Pak Tai Festival is celebrated with theatrical performances for four or five nights in a very large mat-shed theatre specially erected near the temple. The capital needed to finance the show is raised from the townspeople; part of what each person provides is a gift, the rest a loan. Admission fees are charged, and if anything is left over after the troupe has been paid, it is spent partly on charity and partly in repayment to the promoters. This is the only annual theatre in the Southern District at which the most famous stars of the Cantonese opera are to be seen. The performances are sometimes very good. The Tin Hau Festival, which follows shortly afterwards and is given before the proper date so as not to clash with similar festivities at Po Toi and several other places in the District, is held at Saiwan. Here the capital is raised from among the fishermen themselves, admission is free, and the troupe is third or fourth-rate.

By far the most interesting festival of the year is the Festival of the Hungry Souls. Although this feast is observed at different times and by different methods in most Chinese communities, it is celebrated nowhere in the Colony so fully as at Cheung Chau. The purpose of the various ceremonies is twofold: to drive away evil spirits, and to offer food to the spirits of all those who have died leaving no descendents to sustain them in the nether world by the performance of the traditional ancestral rites. Such spirits, the hungry souls, can sometimes be malevolent, and in any case their condition is pitiable, and to care for them is a public duty.

The Feast at Cheung Chau differs in several respects from those held elsewhere in the New Territories. To begin with, it is organised by the Hoklo community; secondly, it is annual, whereas at other places it occurs every five years or so, at Tai O every ten years; thirdly, whereas elsewhere the feast is held for the known dead, at Cheung Chau it is held for the unknown dead. Here we get indications of the island's history prior to the Hoklo settlement described earlier. It appears that in the early days of settlement it was found that the Tung Wan side of the island was littered with partially buried bones, those of the victims of the pirates who had earlier made Cheung Chau one of their lairs. Within living memory, parents warned their children never to go near Tung Wan after dark, because of malevolent spirits, and it is for this reason that the village grew up on the other side of the island, facing west. It is thought that the name Tong Yan Shek (Kill-by-a-Knife Rock), which stands on Tung Wan in front of St John Hospital, is a surviving reminder of pirate atrocities. At all events, the Feast of the Hungry Souls has been celebrated by the Hoklo community for as long as anyone can remember, and is considered important for the island's spiritual welfare.

The elders of the Hoklo community traditionally meet within the first fifteen days of the Chinese New Year to fix a day for their ceremonial visit to the Pak Tai Temple to obtain the God's instructions regarding the organisers of the feast and its date. The visit usually takes place during the 2nd moon, on a date fixed to suit convenience. In the temple the God is consulted by the throwing down of two wooden cups, a favourable fall being one up, one down ("shing pui"). The name of each elder is called out one by one, and the cups thrown ten times for each name. The person getting the largest number of favourable falls becomes the organising chairman, the second the vice-chairman and the third the treasurer. The God is then consulted about the date. Days are called out, beginning with the 1st day of the 4th moon, and the cups thrown three times for each day. The God indicates his wish by causing two favourable falls ("shing pui") and one fall with both cups down ("po pui"). The first day called on which this occurs is the date of the Feast, which, being a four-day affair, consequently begins three days before the favourable day. From the first day of the 4th moon until the end of the Feast, the chairman must eat no meat or fish.

The Feast is held on Tung Wah, in the open space between the Catholic Church and the Kwok Man School, close to the original place of Hoklo settlement and to the place where tradition asserts that the bones of the murdered seafarers were discovered. A mat-shed theatre is erected, the show lasting four days. On the first day, all the idols from all the temples on the island are brought in procession and installed before the theatre. The same evening meat and fish eating stops throughout the island; none is served in restaurants, and none is available for sale in the market. On the second day, the theatre and the fast continue. On the third day a procession is held; it is a fairly recent innovation, originating from Fatshan district, and consisting of a series of tableaux of small boys and girls, in couples, representing anything of social interest or amusement. Most of the tableaux are of similar basic design, contrived by means of a light iron frame. For example, Mr Hong Kong, in long coat and with large spectacles, is lighting a cigarette. He holds the box of matches in his left hand and in his right holds a huge match, on the top of which sits Mrs Hong Kong, complete with handbag and powder compact. It is one of the most ingenious and pleasant sights in the annual New Territories round. On the evening of the 3rd day, the feast begins. 36 places are laid for the ghosts, and there are separate washrooms provided for both sexes of ghost, where a new suit of clothes is burnt, so that, despite the horrible way in which many of them died, none need come to the feast unclad. Lantern-bearers come from all parts of the island, including the Peak, to show the spirits the way to the feast, and a hell-god presides at the feeding. Among the other food provided are high cones of buns, rattan affairs shaped like a maize corn, and at about 3 a.m. the end of the feeding of spirits is signified by the officiating priest, whereupon the youths of the town seize the buns off the cones, scaling their sides in any attempt to grasp the luckiest prize, which is the bun on the top. On the fourth day, the ceremonies end with the return of all the idols to their respective temples.

Thus, the two main objects of the feast are carried out. The buns feed the hungry souls, and the passage of the idols through the streets clears the town of evil spirits. Until the Central People's Government came to power, it was the custom for a Taoist priest to come from Swabue to perform the principal ceremonies. These days, a Buddhist priest from within the colony is invited.

On the 5th day of the 5th moon, the Dragon Boat Festival is celebrated with races in the main harbour. There are three dragon boats, each manned by between 60 and 70 paddlers. The boats are kept in a specially long boat-house just north of the Pak Tai Temple.

PO TOI ISLAND (Eric Hamilton)

I recall one very interesting show as regards Po Toi Island. A Chinese had a great idea of an enormous fishpond up in a hollow in the hills which drains out into the only landing on that island, south of Cape D'Aguilar and west-southwest of Waglan Island. It is undoubtedly a very old crater of an extinct volcano. His idea was to dam the stream and stock the resulting reservoir with freshwater fish from Hong Kong. I had a look at it, then took a Public Works Department bloke out. The figures for the amount of concrete required for the dam ended it, to my relief, because there would have been endless trouble over water rights. I had considered it as regards water storage, but I felt pretty sure the storage capacity would be nowhere big enough to justify a pipeline to Tai Tam.

PO TOI ISLAND (Paul Tsui, written after he retired)

Off the coast to the south of Stanley lies the Po Toi Island, where a few families occupied the few houses in a small village in a bay on the island. The island was not very accessible, as one would normally have to charter a sampan or a small junk to take you there from Aberdeen. The livelihood of the villagers had always been closely tied to the sea, fishing and seaweed gathering.

POI TOI (Walter Schofield)

The careful investigation of applications to use land was more than once impressed on me by experience. Desire to develop apparently unused land may mislead a District Officer into sanctioning the spoliation of an object of natural beauty, the monopolizing of an area in common use by a village community, or such damage to hill slopes as to cause villages or fields to be flooded with mud and soil wash, or the erection of a gimcrack structure of bad concrete, instead of a brick or stone village house in harmony with its surroundings.

Proposals for forest development may turn out to be schemes for evicting villagers from areas where they hold forest rights: though proper forest lot maps should make such schemes impossible. An instance of an application designed to monopolize an area already used by villagers occurred in 1931, when a man applied for a mat-shed permit for a small area in the middle of the beach at Tai Wan village on Po Toi. I took a launch there to see the place and found he had picked the centre of an area on which were a large number of poles used by the villagers to support bamboos for drying nets and similar purposes: so after a few enquiries I told the applicant he could not have that place. (That was the day I found a fine shouldered stone adze-head on the path above the village at the 150 ft. contour.)

PO TOI ISLAND (Austin Coates)

Situated 1 hour from Aberdeen by motor launch, 1¾ hours from Queen's Pier, Victoria. Calm seas can only be expected in the 2nd moon, and between the 5th and 8th moons, inclusive, only large-sized vessels can make the journey in safety.

Population 110, Cantonese, surnames: Chan, Yeung, Cheung, and 2 families of Man, who are Hakka. Commercial and other connections are almost entirely with Aberdeen. Po Toi presents in its worst form the practice whereby coastal Cantonese batten on the requirements and superstitions of the fisherfolk to make a living. The entire village lives off the fisherfolk, and if they ever stopped coming, the land inhabitants would be obliged to evacuate. The harbour provides excellent shelter and there are good breaming and net-varnishing facilities. The villagers' principal means of livelihood is as middlemen, purchasing the catches of the fisherfolk and selling them at Aberdeen. They hire one purse seiner to transport the fish, but this boat is too small to make the journey during the southwest monsoon, and they accordingly asked me if the Government could provide them with a junk. The reason given for this request was so that pregnant mothers could be brought into Aberdeen to the clinic there. It is true that, for fishermen and villagers alike, a regular junk service would be of great assistance; but it would be useless to recommend a loan for the purchase of a junk unless a full proper scheme can be worked out to guarantee a return of the money. Ng Yung Kan, of Yung Shu Wan, Lamma Island, tried to run a regular junk ferry service between Yung Shu Wan, Aberdeen and Po Toi during 1954, but had to close it down as a financial loss. Other sources of income to the villagers are the sale of grass and water, the use of varnishing kilns, and the sale of sundries. There are two tea-houses and a number of ramshackle establishments just recognisable as shops.

On my first visit to the island, in the spring of 1954, I was able to interest the Departments concerned in providing pigs and two cows under the Kadoorie Agricultural Aid Association scheme, and in getting a one-room school in the Estimates for 1955–56. The pigs have been an almost total failure, partly due to ignorance in rearing them (and in the difficulty of giving instruction by regular visits), and partly, I suspect, due to the ease of making money by other means, including smuggling and gambling. Great care will be needed in starting the school, and the assistance of the Registrar of Cooperative Societies is essential, the school being for land and sea children. This view has been made known to the Education Department. The teacher will have to be carefully chosen, and will have a difficult job. The temple is of some importance to the fishing community, and we have here in raw form the annual theatre performance based on the traditional mixture of commerce and superstition.

The fisherfolk consider it important for luck that the performance be held; the villagers know that unless the temple is maintained and the performance given the fishermen will go away and they will be ruined. Each community thus puts up about $3,000, and the show is given free. If the villagers would dare to hold no performance for one year their junk problem would be quickly solved, but they are so afraid of losing the prey on which they feed that they dare not even introduce a small admission charge; the show is always free. There are a number of cultivated fields, growing vegetables for local consumption and sale in Aberdeen whenever there is a surplus. In the last years of the Ching dynasty a number of wine distilleries existed here, but they all went out of business about forty years ago. The Village Representative is a woman; the 6th generation of the Chan clan to reside there. Since March 1946 the population has risen from 74 to 110, but the hill villages which were occupied by one or two families at the end of the war have now been deserted.

PO TOI ISLAND (*Gazetteer* p. 107) (James Hayes)

Po Toi is an island south of Hong Kong. Its connection with Hong Kong is by boat, from Aberdeen on the south side of Hong Kong Island.

I have a record of the inscription of the bell in the Tin Hau Temple, which is dated in 1892.

In regard to Local Public Works, I promised materials for some work they want to do now, and some they wish to carry out in September 1958. The work to be done includes two sections of footpath, two sets of steps, and a new well. They also want to construct a hut that can be used by the doctor when he comes to visit the island.

There was also a request to be passed to the area health inspector, for DDT, and for spraying to be carried out. There are some requests for post-registrations of birth to be passed on. A 50-year-old man requested assistance in finding employment.

Gazetteer mentions that there are around 180 villagers, living in three villages, and maintaining a "precarious living by farming and a little fishing", and that there were three abandoned villages on the island. The population is likely to have been mixed, and mostly part Cantonese-part Hakka.

NORTH SOKO ISLAND (Walter Schofield)

In the outlying islands are three interesting structures: one is on the North Soko island, where in a small valley on its south coast are two converging lines of megaliths. The other two are on Sha Chau (north of Tai O), one a stone burial chamber on the south isthmus in the form of a "kistvaen", the other a ruined guard station on the flat area northwards of the chamber, with an earthwork protecting the landing place to eastward.

THE SOKO ISLANDS (Austin Coates)

This group of small islands, of which two, Tai A Chau and Siu A Chau, are inhabited, is one of the remotest parts of the New Territories, neglected and difficult to care for. They are situated at the southwest corner of the Colony, south of Shek Pik, the south coast of Tai A Chau being about 2 miles from the shore of Chi Chau, one of the islands of the Lapsapmei group, which is Chinese territory. Their trade connections are almost entirely with Cheung Chau, from which they are situated just over an hour by launch, 3 hours by sailing boat with a good wind.

Both islands insist that they have no connections with Lapsapmei, and that boats from there never enter the waters round the Soko Islands. In 1952, however, there was an incident in which a police launch surprised three Chinese junks sheltering between the islands. There was an exchange of fire and one of the junks, a dumb vessel, was captured. As a result of this, the District Administration asked for more Royal Navy patrols in the waters south of Lantao, and there have been no more incidents. The villagers of Tai A Chau complain, however, that sea-robbers, who they say are from Macao, sometimes come to the islands at night and have stolen pigs and chickens. They have asked me to obtain rifles for them from the police. As Fanlao and other villages in the vicinity possess rifles, and on occasion have had to use them, I think this request should be supported.

One or two of the inhabitants have a little schooling, but the majority are illiterate, even those who have been to school being only able to read a few characters. The most important contribution the Government can make here is to start a school. Once there is a resident teacher on the islands, some progress in various directions can begin. The Education Department will be consulted about this. A school can only be started, due to the low population, if the fishermen's children also join, but with the help of then Registrar of Cooperatives this should not be impossible.

Tai A Chau (Austin Coates)

Population 31. Surnames: Ng and Yeung. The people describe themselves as Cantonese of Hakka origin. They came from Im Tin, in Po On district, and a man of 25 is of the 3rd generation, giving a date of settlement of c. 1905. It is possible that the islands have been occupied from time to time, as is the case with other islands (Shek

Wu Chau and Chau Kung), and that these are by no means the first inhabitants. Certainly the place has been a fishing anchorage for quite a long time. The bell in the Tin Hau temple by the shore dates from the 8th year of the Tao Kuang (1828).

The principal sources of livelihood are pig-breeding and fishing, the products being sold at Cheung Chau. The village owns one small sailing boat (which goes to Cheung Chau about three times a week) and 3 sampans. There is a small amount of rice land, and more could be brought under cultivation. The island is fertile and there is sufficient water. The villagers need instruction in the proper keeping of pigs, and the Kadoorie Agricultural Aid Association could be of great help here. They have hitherto been using cross-bred pigs, but with the lack of success which is widely seen in the District. What is now required is a new start, the pigs being segregated from the pigs which the villagers have now, many of which are diseased. There are breaming facilities on the beach and evidence that these are well used and a source of income to the Ng family.

Siu A Chau (Austin Coates)

Population 37 (10 families). Surname: Chow. These people also describe themselves as Cantonese of Hakka origin. A man of 57 is of the 4th generation on the island, and they come from Shataukok. Chief occupation is fishing; there are 8 boats and the waters around the Soko Islands are well known as a good fishing ground. There is one shop on the island, run by a merchant from Cheung Chau, who acts as agent for the villagers, and who probably makes a tidy profit out of everything they produce. They sell shrimp paste and occasionally papayas. There are 8 taochungs of rice land, 10 taochungs of hill land and some pigs of low standard. The people need a loan to build proper pigsties and to receive instructions in pig-breeding.

It is difficult to know what to recommend for this group of islands. It seems undesirable that within shouting distance of communist territory we should have two almost entirely illiterate villages, but a school is only possible provided the two islands will combine, and provided fishermen's children participate. Both of these provisos present difficulties. There is obviously smuggling going on at Tai A Chau, and my general impression is that the character of the Siu A Chau people is better than that of the more southerly islanders. But the people are as a whole unusually tight-lipped and one would probably have to take up residence there for quite a long time in order to find out what really goes on. A substantial building was erected illegally near the breaming beach on Tai A Chau in 1954 and is described innocently as a dormitory, though for whom and under what circumstances is not made plain.

The Soko Islands (Tai A Chau and Sai A Chau)[9] (*Gazetteer* p. 79) (James Hayes)

The two islands of Tai and Sai A Chau (shown as Siu A Chau on the map) are, like many outlying ones, big and small, between Hong Kong and Macau, inhabited by landsmen who are as much fishermen as farmers.

The Hong Kong official gazetteer of 1960 gives populations of 30 and 35 respectively for the two islands. I suspect, probably of mixed origin.

I was told that besides a few latecomers, Tai A Chau had four families of Ng and three of Yeung. They had come to the island around the same time, about 120 years before, say, in the 1830s, and in 1957 farmed about 20 taochung of paddy and twice that area of dry cultivation. Water for irrigation is clearly a problem.

Sai A Chau is inhabited by the Chau family. They must have been there for generations also.[10]

There are 25 boats at Tai A Chau, and 10 at the other, of which only three have engines, but my notes do not say whether they were owned and operated by the landsmen or by boat people. Some may well have been, since fishing was one of the mainstays of livelihood in the islands.

There seem to be many children, and upon enquiry was told that there are 15 between ages 7 to 12 (primary school age), with 5 over 12, 10 or more between 4 to 7, and another 7 under four years of age. I have a vague memory of seeing a classroom with a portrait of Chairman Mao on one wall, and I think we set about getting a proper school building thereafter, with the help of the Education Department. At any rate, my note states that I received a request for a school, to be shared with the Sai A Chau people.

The small temples on Tai and Sai A Chau each have an inscribed bell, presented by devotees in 1828 and 1773 respectively.[11] The former was donated by the operator of a stake net (*tsang pang*), a type of fishing from a land-operated heavy drop net on poles, still commonly found across the Hong Kong region in 1957–58.

They said that a small footbridge and a well had been completed recently on Sai A Chau. Our Assistant Inspector of Works had visited the two communities a few months ago.

Finally, why "Sokos"? This was because the older Chinese name for this group of mostly uninhabited small islands was Sok Kwu Kwan To (*Gazetteer* p. 79), meaning a group of Fishing Net Islands. It has long been brought into Hong Kong English usage as "The Sokos".

Ma Wan (S. H. Peplow)

Ma Wan. A small island near Kap Shui Mun. Ma — a horse, Wan — a bay. The bay or bays of this island are supposed to be shaped like the body of a horse.

Notes

1. See the chapter on Peng Chau in *Rural Communities of Hong Kong*, pp. 12, 187–191.
2. I had overlooked this scribbled note when writing about another of these rocks in *JHKBRAS* 17 (1977), p. 173.
3. From this brief account, it would seem that the Peng Chau community took good care to ensure that it was well-protected.
4. Their removal is not mentioned in the District Commissioner's printed annual departmental reports for this period, nor is there any mention of what was done for them.
5. This is the story taken down in 1958, overlooked when I was preparing the narrative given at p. 89 of *The Great Difference* in 2006. Based on what was said by other informants, it differs in some particulars from the later version: principally, in that the decline in population seems to have begun much earlier than stated here.
6. I seem to recall an illustrated article in *Reader's Digest*, entitled "Welcome to the Isle of Hope", and featuring a photograph of Mr Borgeest taken on the island. Soon after taking up my post, I had visited Yi O on northwestern Lantao with Mr Borgeest to the same end (not pursued) and also the Cheung Shan plateau in East Sai Kung, where another resettlement project had already been achieved.
7. See the chapter on Cheung Chau in my *Hong Kong Region*. Also my article, "Notes and impressions of the Cheung Chau community", in David Faure and Helen F. Siu (eds.), *Down to Earth: The Territorial Bond in South China* (Stanford: Stanford University Press, 1995).

8. See my article, "Notes and impressions of the Cheung Chau community", in Faure and Siu, cited above, especially at pp. 89–90.
9. The notice for an unexpected RAS outing there in 2008 brought back vivid memories of my first visit in the winter of 1957–58. Who, then, could have dreamed of today's intended liquefied petroleum gas storage facility on Tai A Chau, or the Vietnamese boat people's detention camp there of the early 1980s?
10. I found out later that one member of the family had bought land at Shek Pik, and later sold it. The deed of sale in the 1860s is among surviving land documents from there.
11. By the time the Museum of History produced its three-volume record of *Hong Kong's Historical Inscriptions* (published by the Urban Council, Hong Kong, in 1986) the bells had gone, but their inscriptions had been preserved. The only other remembrances from those times are two photos: one of some rather unusual huts on Sai A Chau, and one of our visiting group with some local people.

Sai Kung Market, c. 1960. Courtesy of Tim Ko.

The anchorage at Po Toi O, 1935. Photograph from the collection of the late Mrs Joan Fuller, now in the possession of Jason Wordie.

A village brick house in Sai Kung, 1935. Photograph from the collection of the late Mrs Joan Fuller, now in the possession of Jason Wordie.

The Tin Hau Temple at Rocky Inlet, 1935. Photograph from the collection of the late Mrs Joan Fuller, now in the possession of Jason Wordie.

The deity of a Tin Hau Temple in Cheung Chau, 1930s. Courtesy of Tim Ko.

Worshippers at Tin Hau Temple in Joss House Bay, 1935. Photograph from the collection of the late Mrs Joan Fuller, now in the possession of Jason Wordie.

An earth god shrine at Hebe Haven, 1935. Photograph from the collection of the late Mrs Joan Fuller, now in the possession of Jason Wordie.

The abandoned Kung Clan school/ancestral hall in Yi O, 1990. Photograph by Tim Ko.

6
Sai Kung Peninsula

SAI KUNG PENINSULA (Paul Tsui, written after he retired)

Prior to the Second World War, Sai Kung was virtually inaccessible. There was one small ferry boat running only once a day, between Shaukiwan and Sai Kung market. The ferry had to take the route via Lyemun, Fat Tong Mun, crossed the open sea by Clear Water Bay, through Port Shelter to Sai Kung Market. Off the coast round Fat Tong Mun, the sea could be very rough, and it took over an hour each way. Seldom was the journey attractive to picnickers, particularly if they were not good sailors. Alternatively, travellers would have to take the route via Ngau Chi Wan, climbed the steep hill across the Tseng Lan Shu valley, then down the slope bypassing Wong Keng Tsai to Ho Chung, and then along the coast to Sai Kung Market.

That route would take over two hours, and could be exhausting. In my schoolday picnics the furthest I ventured was Rennie's Mill down the slope from Tseng Lan Shu. I could only gaze at Sai Kung from a distance across the valley from the ridge behind Rennie's Mill. Crew members on board picnic launches going as far as Clear Water Bay pointed vaguely at the direction of Sai Kung.

To my pleasant surprise there already existed after the war a well-paved motor road running from Ngau Chi Wan via Tseng Lan Shu, Tai Po Tsai and the whole length of the Clear Water Bay Peninsula, to Tai Wan Tau, beyond Tai Au Mun village. It was along this quiet but smooth Clear Water Bay Road, where I learnt and practised how to drive a jeep. It was here too, where I practised horse-riding with the help of the Commando troopers. I combined such pleasant pastimes with my official duties, by visiting the villages along Clear Water Bay Road on my way, and got to know some of them well enough to be on first-name terms. Apart from Tai Po Tsai, most of them were Hakkas.

From a junction near Tai Po Tsai, the Japanese army, using prisoner-of-war labour, had started constructing a rough but "jeepable" path, down the treacherous steep slope via Wo Mei, Ho Chung, and Pak Sha

Wan to Sai Kung Market. The Commandos, on taking over, did a good job in trying to improve the road, and named it the "Hiram's Highway". Thus when I came to the scene, it was possible to gain access to Sai Kung Market by jeep, driven by skilful driver.

The Commando Unit garrisoned in Sai Kung Market had established its command post including an officers' mess in a two-storey schoolhouse near the Tin Hau Temple. It was here I had luncheon with General Festing. It was here also, where Jack Cater and I spent a night, when a typhoon frustrated our attempt to visit Tai Long, Tap Mun, Lai Chi Wo, Sha Tau Kok, Kat O, and Peng Chau Island to assess the potential of a fishing industry. We laid on a Harbour Defence motor launch to take us from Sai Kung. The naval lieutenant in command of the launch had never experienced a typhoon before. When the typhoon signal was lowered, he insisted that his launch could withstand any rough sea. So we set sail eastward, towards High Island. As we were heading towards the open sea off Long Ke, the waves were so high that our launch nearly capsized. The coxswain, who apparently was a seasoned sailor, was not happy with the situation, but the young naval lieutenant was rather insistent. After a few tosses, I told the skipper that I considered it to be too dangerous to continue, and Jack Cater agreed with me. So we turned back.

What impressed me most in my round trip walking all over the Sai Kung Peninsula was the number of Catholic Churches and Chapels all over the place. At the eastern end of Sai Kung Market, almost right next door to the Tin Hau Temple, was a Catholic school for girls. About 50 steps up the hill slope behind the same Tin Hau Temple was a substantial Catholic Church, with a fairly elaborate school complex attached to it. There was a Catholic Church at Yim Tin Tsai and Chapels at Wo Mei, Chek Keng, Tai Long, Sai Wan, Leung Shuen Wan and halfway up the hill behind Sai Kung Market.

I had a shock, when accompanied by two commando troopers, to have been confronted by a gang of some thirty young men and women, some of whom were armed, camping in a disused Catholic Chapel at Chek Keng Village. I was equally shocked to find that the entire village of Sai Wan was deserted and almost all of its houses were in ruins. I later learnt that the Japanese army had raided the village and burnt down all the houses, because it was being used by the guerrillas as their operational base. All the villages except Tai Long and Leung Shuen Wan were at that time thinly populated. The villagers, however, gave no convincing reasons as to why. To me, it seemed obvious that the reason was that it was a guerrilla area.

The next day, while I was still lingering in Sai Kung, I learnt that Bishop Henry Valtorta was resting at the Catholic Church halfway up the hill behind Sai Kung Market. I called on him and over a cup of tea we had a long chat. I told the bishop what I had seen in my recent visits to the peninsula and made the point that if the Catholic Church would not take steps to re-establish its presence, the entire Sai Kung peninsula would soon be totally communist. I urged the bishop to think about the problem and see if he could do something about it. I left the matter at that, and I was glad that some years later, the Catholic Church established a secondary school at Sai Kung, the first secondary school in the district. A few of the Churches, including the one at Yim Tin Tsai, have also been rehabilitated.

I was glad that I made that recce, for the first-hand knowledge that I gained was most useful when one day in August/September 1946, I was asked to "attend before", His Excellency the Governor, Sir Mark Young, along with all the members of the Executive Council, supported by the Commanding Officer of the Royal Air Force and the Director of Civil Aviation, on board a 4-engine Sunderland aircraft, on a flying trip over the New Territories in search of a suitable site for a replacement airfield. It must have been the first time that most (if not all) of them had flown over the New Territories. Many questions were asked, and I was the only one in the party, who could provide the answers, both up-to-date as well as historical. They were all, I believe, impressed by my masterly knowledge of the place. As I could tell without hesitation the name of every village, the links of every footpath, and the story (if any) relating to a particular village.

SAI KUNG PENINSULA (Austin Coates)

The Sai Kung region is predominantly Hakka. In the section that follows, each village mentioned is Hakka, unless otherwise stated. The three principal exceptions are the three large villages of Sha Kok Mei, Pak Kong and Ho Chung which are Cantonese. The region is entirely agricultural, with a small amount of fishing. Rice is the principal crop. Pig-keeping is widespread, and in villages nearer the town it is a principal source of income namely, in the Ho Chung valley and along the main road. Vegetables are beginning to be grown, under the influence of Government departments, but the alteration is coming slowly, and only in places near the main road.

In the entire region, consisting of 72 villages, the town of Sai Kung, and a total land population of approximately 6,100, there is only one road. It was built by the Japanese during the war. After the war it was improved and surfaced by the British army. It is known as Hiram's Highway. It is for one-way traffic only, with a passing place near Ho Chung. It opens for traffic every 40 minutes, and stays open for 8 minutes. That is to say, if one arrives at the head of the highway 8 minutes after the time of opening one can just get through without colliding with the traffic coming from the opposite direction. If one enters the highway at the exact time of opening, one has a 10-minute wait at the halfway stop near Ho Chung. When His Excellency the Governor visited Sai Kung in 1953, he informed the Sai Kung Rural Committee that while it might not be feasible to double the size of the highway, making it suitable for two-way traffic, the question of enlarging it section by section annually might be examined. His Excellency subsequently gave no directive on this; but in view of the work which the Public Works Department is now doing to knock away hillside from the more dangerous blind corners on the highway, it would seem the appropriate moment to raise the question of widening, the work of which has been much simplified by these makeshift arrangements.

During the war the assistance which British people received from the inhabitants, who helped them escape through the Japanese lines to unoccupied China, was outstanding, and the Government's recognition of the fact was later signalled by the award of a British Empire Medal and many Certificates of Merit, at least one of which can be found today in almost every village of any importance. The certificates are used as passports to preferment, quite rightly, I think. It is nice to have an award being of material use.

The region suffered extremely in the later years of the war. Fields were sold, houses demolished, to pay for food; there was a tendency for profiteers in the three Cantonese villages mentioned above to enrich themselves by purchasing fields owned by smaller, chiefly Hakka, villages. The horror of the famine of 1944 and the deplorable condition to which the country had been reduced by the middle of 1945, when the Japanese were finally defeated, has left its mark on the people. Mentally, they are thankful for the return of British government. Materially, they have not yet recovered from their losses during the grave years of war. There was a momentary swing of opinion towards the present Government of China in 1949 and 1950, induced possibly by the idea that it was necessary to take precautions lest the new regime occupy the New Territories. This phase has, in my opinion, passed. The Catholic Church authorities inform me that there is still a good deal of communist sentiment, and even activity, in the region but my staff and I are agreed that we see little significant evidence of it. Any undercurrent of hostility towards us as servants of the Government is entirely absent from our widespread and intimate contacts with the people.

It is to me remarkable that the British should have received the wartime help that they did from the people of Sai Kung, considering that until 1945 the Government of Hong Kong had, to all intents, done nothing for them whatever. The Japanese road, and the Government's wise decision to improve it, started things moving; and it can be said that along the road there is evidence of governmental activity, in forestry, pig-breeding, vegetable-growing, the supply of fertilisers and advice on animal health, and, very gradually, in improved irrigation.

Away from the main road the process has hardly started, and I am constantly surprised by the ignorance of the Government Departments concerned with these more remote areas.

However, nothing is more obvious than that the District Administration must in all things give the lead; where this is done, there are results; where it is not done, there is the likelihood that the improvements will be either piecemeal or concerned with inessentials. It should be the keynote of our policy to stimulate the interest of other Departments in the outlying areas about which they have only scanty information, much of it unreliable.

The first and most important need in the region, now that a permanent Government doctor has been stationed at Sai Kung and a health centre is shortly to be constructed, is the extension of Hiram's Highway to open up the most remote and inaccessible section of the entire mainland area of the New Territories. This road (the same width as the present Hiram's Highway would be adequate for as long as can be foreseen) should start at Sai Kung and pass west of the Police Station to the foot of the hills, thence turning to the coast and passing through Sha Ha, Taiwan, Wong Chuk Wan, just south of Tai Mong Tsai, coastally round to Tsam Chuk Wan, crossing the river at Pak Tam Chung, and mounting thence through Pak Tam, over the hills on the main track to Chik Hang (Chek Keng), thereafter following the main track sloping gently up to the pass leading over to Tai Long. There it should leave the path, which is too steep, and descend gradually to the north side of the valley, past Lam Uk Wai. This important development is the prerequisite to all improvement in the area, just as the Japanese road was to the area it serves. The matter is being investigated in detail, and will shortly be submitted to the Rural Development Committee for urgent consideration as a Colonial Development and Welfare scheme.

Lesser needs, affecting individual villages, will be found in the substance of this summary.

The region has been unaffected by the influx of refugees into Hong Kong, and there are very few strangers settled in any part of it.

There is a tendency for the larger villages to gain control over the smaller ones by enlarging their schools, to the elimination of the smaller village schools. While educationally this is not a bad thing, the profiteering experience during the war makes it necessary for the matter to be watched carefully by the Education Department and this Administration.

The Catholic Church is powerful throughout the region, chiefly in the remote parts. Through their middle school at Sai Kung they are the principal educational institution, and in view of the care given to pupils subsequent to their studies (in finding employment) they rival the Government in the services they offer. In the two years that I have been District Officer I have noticed that increased activity by several Departments is gradually giving the Government the ascendancy.

Fishing, grass-cutting and cattle-grazing are considerably interfered with by military firing. The Port Shelter Range, used by all three services, prevents all traffic throughout the greatest part of Port Shelter every day of the week except Saturdays and Sundays, with a minor relaxation on Wednesdays. The Port Shelter Artillery Range prevents access to large areas of hillside along the border of Taipo and the Southern Districts. Calf's Head Range takes up another area, admittedly seldom used, in the hinterland of King Kong Shek, and southeast of Wong Chuk Shan (Taipo District). Various complaints have been received from many villages, and complaints from the fisherfolk are so regular as to be almost seasonal. The office has taken the line that while there can be no relaxation of the present arrangements while the Garrison remains at its present strength, the Government will not allow the firing hours to be extended any more than they are now.

SAI KUNG (James Hayes)

With the exception of the large Cantonese villages of Ho Chung, Pat Kong, and Sha Kok Mei, the area is totally settled by Hakkas. There were also some Cantonese shopkeepers and businessmen in the market town of Sai Kung.

HO CHUNG GROUP OF VILLAGES

Ho Chung (645077) (Austin Coates)

Population 459. This is one of the few villages whose population has declined since 1898. 25 of the population are abroad, 30 in Hong Kong, 15 in Kowloon, and 15 live and work at the main road bus stop. A Cantonese village surrounded by predominantly Hakka hamlets. Surnames: Wan, Wong, Lam Cheung, Lok Lau, Tse, Lee, Tsang, Chan, Lai, Fung, So. The village has a large number of fields, and during the war, when some of the surrounding hamlets were selling their fields to buy food, Ho Chung people purchased these also. Other occupation: pig-breeding and grass-cutting. It is perhaps significant of the prosperity of Ho Chung that only women cut grass, whereas in the hamlets men do this work. There are two shops, at the bus stop, about 5 minutes' walk from the village. There is a good school, and the youths are of comparatively high standard in education and appearance. Requests were received for help in finding employment, but the jobs required are in the Police, public utilities, etc., or as mechanics. The more menial jobs that are acceptable to lads from remote villages would be unsuitable here.

500 bags of cement were requested for improvement to irrigation; a Land Bailiff will inspect later and advise on the work, which should be done in the autumn. A petition was presented for a grant of $20,000 for the construction of an additional classroom at the school, but the figure given for school-age children was exaggerated and has had to be queried; it included children from the Hakka hamlets, some of which had already informed me that they preferred to send their children to Sai Kung, where the teachers are Hakka, rather than send them to the Cantonese surroundings of Ho Chung. The village also makes money in a middleman capacity, as agent for rice barter and the sale of grass from the hamlets. It is proposed in the autumn to make a concrete road from the village to the bus stop. There was a request for a pillar box at the bus stop; this is being referred to the Postmaster General. 10 bags of cement were authorised at once for the repair of wells at Nam Pin Wai. I am also approaching Messrs China Light and Power regarding a supply of electric light.

Ho Chung (*Gazetteer* p. 139) (James Hayes)

One of the largest villages in the Southern District, with 574 persons, all Cantonese.

There are over 10 families of Cheungs, settled here for over 10 generations, and many Wans, here for about the same length of time. There are also Li and Chan lineages.

Around 20 men work in Hong Kong and Kowloon, and the same number are overseas.

They reported 244 children aged between 2–15 years. There is a village school with three classrooms and four teachers, accommodating 96 children in one session. They said that over 80 other children from the village had no education. They had applied to the Education Department in 1955, to help them build extra classrooms. However, they had been advised to operate their school in morning and afternoon sessions, as many schools in the urban area and elsewhere in the New Territories are doing at present.

There seems to have been some discussion about the school charges, which in their case amounted to $2.50 per child. They claimed that other schools were only charging $2.00 ($1.50 plus 50 cents). Why were they being required to charge more? This, like the large number of children unable to go to school, has to be taken up with the Education Department.

They have over 500 taochung of paddy fields, and a further 80–100 taochung of vegetable land, but these fields have very little water. They have 90 brown cattle, buffaloes are not required for ploughing here, and there are currently 20 sows.

In regard to Local Public Works needs, they would like materials for paving the streets in the village. They also wish to do work on irrigation channels. I said that I would ask our Inspector of Works to visit, and discuss further.

Shek Pok Wai (638069) (Austin Coates)

A Cantonese hamlet, subsidiary of Ho Chung. Population included with Ho Chung.

Nam Pin Wai (647072) (Austin Coates)

A Cantonese hamlet, subsidiary of Ho Chung. Population included with Ho Chung.

Kai Ham (634080) (Austin Coates)

Population 117, of whom 7 are abroad and do not send regular remittances. This figure included Wang Tse, a subsidiary village of

Kai Ham. Surnames: Tse (Kai Ham), and Ip (Wang Tse). The Village Representative is not entirely satisfactory. He joined the Police about 1951 but was dismissed after one month, before completing the course. His cultivated appearance suggests that he is a racketeer of some kind; he was selected for work in Seria oilfields in 1954, but the Police advised against it and he was struck off the list. The families have lived here for 8 or 9 generations. They cultivate over 100 taochungs. Principal occupations: Farming and grass/wood-cutting. Like most of the New Territories villages, they produce high quality paddy which they barter at Ho Chung in exchange for an equal weight of lower grade rice.

50 bags of cement are needed for a small bridge, a dam and to irrigation channels. They also wish to surface the area in front of the school. They asked for 2 rifles to guard against pig stealers, but in view of the Village Representative's past history I do not propose to pursue this. All public works approved after inspection by a Land Bailiff will have to wait until the autumn.

Their grass and wood is sold partly at Ho Chung, the rest is carried on foot over the hills for sale at Ngau Chi Wan, in the urban area. More than one mouth per taochung is not nearly enough to keep the village alive. The economy of the village is thus balanced by the women working the fields while the men cut grass. When there is not enough money, they cut more grass. There are 30 able-bodied men, and more women than are really needed in the fields; thus women also do a certain amount of grass-cutting. While there is clearly a need for these villagers to find employment in town, the need is by no means as great as the villagers say. Every one of the 30 men, I was told, would like to leave and work in Kowloon. It is in fact at Kai Ham that we have an extreme example of the lure of the city, as described in the first section of this memorandum. If employment could be found for 10 of them, it would help. It would also be an advantage for the Kadoorie Agricultural Aid Association pig scheme to be extended here, and I have addressed the Director of Agriculture about this. It is a pity, in view of the loan basis of the scheme, that the Village Representative is unreliable. I consider that he is himself a hindrance to the improvement of conditions here.

Kai Ham (*Gazetteer* p. 139) (James Hayes)

They are not sure what the place name means.

102 persons, including Wang Che [*Gazetteer* also mentions another associated hamlet, Chuk Yuen]. The Village Representative is Tse Leung-

man, a new appointee. The villages in the area all have new Village Representatives, I was told.

There are 25 children of school age. The village has a one-classroom school, providing Primary 1–3 classes. Only 6–7 attend from this village, the rest coming from the small hill villages in the area. Some of the bigger children go to school in Sai Kung.

They have 30 taochung of paddy, but very little vegetable land.

Two or three men are working abroad, and a few are earth coolies in Hong Kong, on casual jobs from week to week.

Drinking water is reported to be a real problem. They have only one well. They would like water pipes or materials to build an open channel. I said that I would pass their request to Assistant Inspector of Works Mr Chan, who will arrange with them for an inspection and discussion.

Their forestry licence is up for renewal, they said, and they will be putting up some names for post-registrations of birth.

Wang Tse *(Wang Che)* (637080) (Austin Coates)

A subsidiary village of Kai Ham. See above.

Tai No (632087) (Austin Coates)

Population 75, of whom 15 are working in Hong Kong. This village, Tai No Sheung Yeung, and Tin Liu are under a single Village Representative. The surnames are Tse and Tsang (Tai No), and Yau (Tai No Sheung Yeung). Tai No and Sheung Yeung are comparatively remote, 45 minutes' climb up a steep but well-paved track from Kai Ham, and thus about 1¾ hours from the bus route. Sheung Yeung is about 10 minutes' walk further up.

They own a number of fields scattered in cultivable spots up the valley through which the path runs to Heather Pass. Before the war they owned 45 taochungs, but during the war they sold 25 to Ho Chung people, from whom they now rent them. There is a longstanding dispute arising from this which the Sai Kung Rural Committee has tried unsuccessfully to solve. I have called for a report from the Chairman. The principal occupation is grass-cutting, with farming as a subsidiary. Grass is carried either to Ho Chung or Ngau Chi Wan for sale. If employment could be found for 4 men, the economic position would be eased. There are 8 families at Tai No, 3 at Sheung Yeung. Repairs are needed for a stream embankment.

Tai No (*Gazetteer* p. 139) (James Hayes)

The village is said to be several hundred years old. The Village Representative is Tsang Sang, who is also Village Representative for Sheung Yeung and Tin Liu. This village has these three parts, of which Tai No itself is at the highest location, and is also the oldest settlement.

Tai No was settled from Ng Wah in Canton province. The clan record (Tsang's, I think they said) was lost during the Japanese occupation. Tsang and Tse came about the same time, and the Yaus arrived not long after.

There are currently 50 persons in ten families at Tai No proper: five of Tsang and five of Tse.

At Tai No Tin Liu, a 100-year-old offshoot from Tai No, there are another 30 persons in eight or nine families, all or mostly Tsangs.

Sheung Yeung was also settled from Tai No. This was 74 years ago, known with certainty since one of the original settlers is still alive, and had moved there with his family when he was 4–5 years old. Thus the date of settlement is in the early 1880s. There are 25 persons in four families, named Yau here, owning 25 taochung of rice fields.

There has, in fact, been a continuing and gradual move to the lower, more convenient locations. They said that more village families moved down from Tai No about thirty years ago (late 1920s), and others a little later still, during the Japanese occupation. Two families had moved to the urban area, one before and one after the war.

They mentioned that many houses in Sheung Yeung had been knocked down in a recent typhoon, including the Yau family's ancestral hall. This usually indicates that they were in a poor state of repair.

At Tai No the village families own about 20 taochung, with 22 taochung of vegetable fields. However, only 2 taochung of the latter are cultivated. This was said to be partly because wild boar and deer come to eat the crops.

At Tin Liu about 10 taochung are owned by a lineage trust (Tong) named for the Tsang first ancestor. There are no Tongs in the other villages.

There are supposed to be around 40 children of school age here, but only 15 go to the school at Kai Ham. In this hill area, with unsurfaced footpaths, children are usually between 7 and 8 years old before they attend school.

Three men from Tai No are in the United States, all of them over 70, and only one of them now sends a remittance home. No one is overseas from Sheung Yeung or Tin Liu.

Four young and not so young men, aged 21, 26, 28, and 30, had asked for employment in government service.

Potable water is sufficient in Tai No and Tin Liu, but not in Sheung Yeung. When the well there is dry, they have to fetch water from Tai No. They say there is no use deepening the well or digging another, and it is better to install about 1,000 feet of piping to bring water from a reliable stream.

Tai No Sheung Yeung (630091) (Austin Coates)

Subsidiary of Tao No. See above.

Tin Liu (636079) (Austin Coates)

A subsidiary of Tai No. Population 30. Surname: Tsang. The villages here are in a chessboard pattern, Tin Liu being situated 45 minutes' walk from Tai No, in the lower part of the Ho Chung valley, between Kai Ham and its subsidiary Wang Tse. Repairs to a bridge are needed.

Man Wo (633073) (Austin Coates)

Population 42, which includes the subsidiary hamlets of Chuk Yuen and Ngau Fan Tun. Surnames: Yuen, Chan, Wong, Ho, Tai. The present generation is said to be the 12th, which would make the date of settlement earlier than other hamlets in this area (c. 1650). Principal occupations: grass-cutting and farming. Paddy produced is bartered at Ho Chung, grass is carried to Ngau Chi Wan for sale. The population is Cantonese. An important Local Public Works job to be done this autumn is the construction of a stepping-stone (stone and concrete) crossing over a stream on the main track from the village to Ngau Chi Wan. Repairs are needed to a wall and a dam, also an autumn job.

Man Wo (James Hayes)

31 people, in six families. The Village Representative is Yuen Tin-fuk.

There are five families of Yuen and one of Yip. The Yuens came here from Lo Wu in the border area. They said they had been settled in Man Wo for 12–15 generations [but this really refers to their lineage's length of settlement in the parent village, I was told later].

They have 15 taochung of rice fields, and 3 taochung of hill slope land (dry cultivation). There are 7 brown cattle, and 5 sows.

With exchange, they only have enough rice for 3–5 months.

Two men are working in the United Kingdom, and one in Kowloon.

They say they do not have enough potable water, and want to build a well, and want also to build two small bridges over a footpath. They can do the work in the autumn, if given the materials by then.

I omitted to ask about children of school age, and the number at school.

Chuk Yuen (633075) (Austin Coates)

Subsidiary Cantonese hamlet. See Man Wo.

Ngau Fan Tun (634068) (Austin Coates)

Cantonese subsidiary of Man Wo. See above. Surname: Chan.

Tai Nam Wu *(Tai Lam Wu)* (632068) (Austin Coates)

Population 40, including 2 families at Ngau Liu. Surname: Wan. A Cantonese village, the clan being an offshoot of the Wans of Ho Chung. The present inhabitants claim to be of the 8th generation to inhabit this particular spot. As might be expected, they rent some of their fields from the earlier settled village of Man Wo. Principal occupations: grass-cutting and farming. Grass is carried to Ngau Chi Wan for sale. About 50 bags of cement will be needed for repairs to a dam and an irrigation channel.

Tai Nam Wu (*Gazetteer* p. 139) (James Hayes)

There are 49 people, of whom 20 are children. The Village Representative's name is Chan Hap.

This is said to be a Punti village, but speaking Hakka. The men are Punti and all the wives are Hakka.

There are 11 families of Wan, and only one of Chan. However, the new Village Representative is a Chan. The Wans came before the Chans, but do not know how long they have been settled there, nor where they are from. Both clans say they lost their records during the Japanese occupation.

I noted, "They don't seem to know much about themselves, but are very hospitable!"

Two men are working in a Chinese restaurant in England.

They farm 70-plus taochung, but say they have very little paddy, most of the land they farm being dry or hill cultivation. They buy necessaries in Ho Chung, Sai Kung and Kowloon.

Nine children attend the primary schools at Ho Chung and Kai Ham. Two go to school in Kowloon.

They have enough drinking water, but say the supply is too far from their houses: being 20 minutes' walk away. They need cement to improve the footpaths and to repair the irrigation channels. They do not have enough water for irrigation purposes anyway, even when the channels are in good repair. I said that Assistant Inspector of Works Chan Chi-nin would come in three weeks, and would let them know when in advance.

Ngau Liu[1] (*Gazetteer* p. 139) (James Hayes)

A very small hamlet, with only two families, one of Wan and the other Chan. Both Punti, but curiously enough they each claim a settlement of 200 years. This is perhaps because they are offshoots of the lineages at Tai Lam Wu, under which this hamlet comes.

I was asked to arrange for a change of name on a birth certificate. The name on it is Wan Kwai-kam, but the requested entry is Wan Tai-hing. Date of birth was 20 January 1935, the name of the mother was Fu Yue-mui, and the address/place of birth was recorded as an unnumbered hut at Ngau Liu.

Kau Tsin Uk (*Gazetteer* p. 139) (James Hayes)

Apparently a new hamlet, as the people moved there during the Japanese occupation. Population included with Tai Nam Wu, from which presumably they had come here.

Wong Keng Tsai (631061) (Austin Coates)

Population 15, of whom 8 live and work in Hong Kong. Surname: Lau. Situated high up in a beautiful, well-wooded valley only 10 minutes' descent on foot from Customs Pass Road. Occupation: grass-cutting. Grass sold at Ngau Chi Wan and Sai Kung. 15 taochungs of paddy land and some tangerines. 10 bags of cement needed to repair a leaking irrigation channel. This would be an excellent area for fruit trees.

Wong Keng Tsai (James Hayes)

Only two families totalling 5 persons are at home at present. The Village Representative is Mr Lau Tin-sung, and his son, Lau Fong, 24, is at home helping his father.

15 people are away in Hong Kong or Kowloon, and two men are abroad. They then recalled another two persons who are away, a husband and wife. He is a bus conductor with Kowloon Motor Bus.

Their branch of the Laus was at Ha Yeung and Hang Hau, before coming here, perhaps a 200-year local settlement in all, but they don't know for how long in this place.

They suffer from insufficient potable water, and asked if I could get them a piped connection from the new Butterfield and Swire's senior residential housing complex recently built at nearby Customs Pass.

They have 16 taochung of paddy fields, of which only 80 per cent is cultivated at present for lack of water, and one taochung of vegetable land. They have 5 water buffaloes [surely these must be brown cattle] but the younger ones are not yet trained to the plough. They have one sow and a few pigs.

They buy everything they need at Ho Chung, because by the time the buses to Kowloon from Clear Water Bay reach their stop, they are crowded. It takes 45 minutes to walk to Ho Chung, and an hour to return, uphill.

They would like to build an "Irish" bridge along the footpath to Ho Chung. Assistant Inspector of Works Mr Chan will visit.

There was a confusing tale about the owner of fixed pitch market stall No. 5 in the Tsuen Wan Market, with money owing to one of the Laus, because (the owner was alleging) of non-payment by customers for goods supplied. The Lau seems to have been on a monthly wage of $170 per month, and the arrears to date amount to $800. The stall operator was trading under the name of Wong Wing Kei.

Wo Mei (652070) (Austin Coates)

Population 142. A Christian village, some Catholic, some Protestant, with a Catholic church and a Protestant school. The Catholic church has just received a grant of Crown land for a school of its own, so doubtless that will be checkmate. Surnames: Tse, Cheung, Ho, Yue, Chan, Lau, Wong, Mak. A hard-working village with good prospects.

Occupations: agriculture and pig-breeding. The village has recently turned over the greater part of its fields from rice to vegetables, and has received assistance and advice from the Department of Cooperatives and Marketing. Some of the vegetables were damaged in the frost, particularly the cabbages, but I think that nothing but good can come from this change. It seems to me that all the villages on or near the highway (Nam Wai, Pak Wai, Pak Sha Wan, Ho Chung, Ta Ho Tun,

Chiu Hang and Pak Kong) would be well advised to make the same change as Wo Mei has made. But we must first judge carefully for one year before pressing this suggestion. A request for loans for the purchase of pig food was forwarded as a Kadoorie Agricultural Aid Association item. Rice and vegetable fields were damaged here in last year's typhoons, surprisingly far inland, and as assistance is being given elsewhere it may be necessary, in fairness, to give a small amount here.

They also asked for cement to enlarge two wells, for a waiting shelter at the bus stop, and for erection of conventional signs for a school.

Wo Mei (*Gazetteer* p. 140) (James Hayes)

Population 140. There are six families each of Tse, Cheung, Yue and Ho, with one or more families of Chan who are the latest arrivals, though not stated when. All are Hakka.

Twenty children are attending school in the village, but there are said to be six or seven of school age whose families are said to be "too poor" to send them to school.

There are two men working in Chinese restaurants in the United Kingdom, and two others are on ships. Three men work in the urban area, one or perhaps two of them as tailors.

More young men want to work in the United Kingdom, and half of the 8–9 requests received on this visit for post-registrations of birth are for this purpose.

They have, in all, some 80 taochung of rice fields, and 30 of vegetable land. There are 24 brown cattle, but no buffaloes. They gave the cost of a working cow as $500, and for a buffalo $700.

Currently there is one boar, 86 sows, and 400 porkers. They valued the boar at $400 and a sow at $200.

There is a Vegetable Marketing Organization (VMO) depot here, and others at Pak Kong and Sai Kung, but they said that the vegetable-growing business has not been very good for the past few years.

They requested Local Public Works assistance for two projects they have in mind. These are both intakes for large and small dams, needed to improve irrigation flow.

They asked for traffic signs to be put up outside the school, and at the entrance road into the village, to slow down vehicles.

Mok Tse Tse *(Mok Tse Che)* (650067) (Austin Coates)

Population 64. A Cantonese subsidiary village of Wo Mei.

Pek Uk *(Pik Uk)* (646061) (Austin Coates)

Population 50 (8 families). Surname: Shing, a branch of the same clan at Mang Kung Uk. A man of 54 is of the 5th generation at Pak Uk, giving a date of settlement c. 1825. A hill village situated at the extremity of the Ho Chung basin. It belongs to the Wo Mei group. It is fairly near Clearwater Bay Road, from which the villagers in 1954 constructed a dirt road running two-thirds of the distance, down steep slope. They have asked for the loan of wheelbarrows this autumn when finishing the road, and for cement to construct a small reservoir and filter for drinking water.

Their cows have sometimes been forcibly driven off the fields near the road where the Kowloon Dairy are growing elephant grass on permit; in order to avoid further trouble, they have asked me to request the Dairy to fence in their fields. They have one man working in town. Occupations: rice-growing (the village owns 175 taochungs near the village, 19 taochungs near Ho Chung, and rents 19 more near Ho Chung); pig-keeping, and grass-cutting, the grass being carried by women for sale at Ngau Chi Wan.

Pik Uk (*Gazetteer* p. 140) (James Hayes)

Population 35, in eight families of Shing. They moved here from Mang Kung Uk, "before the port of Hong Kong was founded by the British". The Village Representative is Shing Yuen-wan.

There are, in addition, seven more families of newcomers who live higher up in a separate area. They came here after the war, at least one from Tai Po, the Village Representative said.

From both groups, 17 children out of 20 of school age are attending school at Nam Wai.

The Shings have 80 taochung of rice fields, which give (with the usual exchange) a six-month supply. Also four taochung of vegetable land.

I described them as being "a go-ahead lot". They told me how they had opened a rough road for lorries to come down from Clear Water Bay Road into the village to bring pig food, with a turning circle; how they had themselves provided a piped water supply for each house from a stream intake; and how they had plans to pave the fronts and lanes of the houses if we would supply materials. I said I would ask our Assistant Inspector of Works Mr Chan to visit and discuss, and also check on the slopes and whether their road needed any protective work.

No men are overseas, and only one man works in Kowloon.

Nam Wai (657073) (Austin Coates)

Population 350. Surnames: Yau and Shing. Probable date of foundation c. 1760. The Shings here are related to those of Mang Kung Uk; the Yau family have relatives at Sha Tau Kok. Although this village has a fair number of people working in Hong Kong or abroad, it still has a large local population and 30 under-employed youths, many of them of good physique. The village complained that they had formerly engaged in fishing, their ground being between Kau Sai Island and Sharp Island, but that the opening of the Port Shelter Range had put an end to this occupation. In fact, it appears that Nam Wai's fishing activities had been impaired prior to the start of the range. Before the war they owned 30 boats, and by the end of the war this number had dwindled to 10. Now they have 3. I have, however, raised with the Director of Fisheries and Registrar of Cooperatives the possibility of revitalising this industry.

The village at present subsists on remittances, rice and pigs. There is a Kadoorie Agricultural Aid Association battery of styes, and the Agricultural Department is prepared to establish more provided private land is available. A shelter is required on Hiram's Highway near the village, chiefly for goods awaiting lorry transport. It would be a good idea to make it also a tea stall, which the villagers could let out on tender and make some money for village uses. This is being followed up.

A good deal of work will be needed here in the autumn on improving irrigation channels, and repairing roads and a bridge.

300 taochungs of paddy land.

A native of this village is Mr Yau Pak Tung, now a press translator in the Government Public Relations Office, and formerly (during the war) in Chinese Ministry of Information at Chungking.

Nam Wai (*Gazetteer* p. 140) (James Hayes)

Over 350 people. Village Representative is Shing Chung-on.

They said there were more people in the village before the war, between 400 to 500. They also mentioned two small subsidiary hamlets at O Tsai Tsuen and Chuk Kok Tsuen, both coming under Nam Wai.

Not many fields, their rough estimate was 80 to 100 taochung of paddy at the most, and 50 to 80 of vegetable land. Only enough, with exchange for cheaper rice, to give three months' supply during the year. Mostly with shops in Sai Kung and Kowloon.

No fishing takes place now, as there are few fish to catch, but it was a major activity before the war.

Nowadays, pig-breeding has taken over. There are 100 sows, and each family rears 2–3 pigs. The current sale price is $140 per picul.

There is a one-room school in the village, operating in morning and afternoon sessions, with five primary forms and three teachers. About 80 children attend school.

We had hill tea to drink here.

Among the men employed outside the village, two are drivers for Kowloon Motor Bus, one is a lorry driver, and 2–3 are at sea. Before the war, many more, an estimated 40 to 50, went to sea.

One of the villagers, Mr Yau Fo, showed me a Certificate of Honour, which was presented to him for assistance given to escaped Allied prisoners of war during the Japanese occupation.

Apparently our Assistant Inspector of Works, Mr Chan, is advising them about repairs to a village road, and a well.

Wong Chuk Shan (*Gazetteer* p. 180) (James Hayes)

There are reported 133 persons in the village. All are Hakka. The Village Representative's name is Chung Kwun-ma.

There are 25 families of Chung, and one of Sham. Curiously enough, despite the great disparity in numbers, the forbears of both families came together and at the same time, 9 generations ago, from Hing Ling of Cheong Lok District, Eastern Kwangtung.

The villagers [mainly the women] farm 150 taochung of paddy and about 10 of vegetable land. They have 40 cows and 3 sows.

There is no forestry lot or orchard, just a few tangerine trees, and the Village Representative said they would like to apply for a "Kadoorie Orchard" from the Kadoorie Agricultural Aid Association.

Two men are in Borneo, and 4 are working in Hong Kong or Kowloon.

However, apart from helping with farming, the main occupation of the village men seems to be the construction business. Several of them employ their own relatives and other male villagers from the Sai Kung area, and go round other parts of the New Territories building houses. This is a traditional rural industry, often concentrated in several villages in a district, as here.

The village school holds 27 pupils. There are 19 children from Wong Chuk Shan in this school, plus 4 more from Mau Ping, another hill village but located within Taipo District.

There are, they said, five hill villages in this area, three of them within the Taipo District boundary. A new two-classroom school is being built for them, about a 15-minute walking distance. At present, 15 of the bigger boys from these villages attend school at Ma On Shan, going and returning daily.

They have proposed to install a better water supply, building two storage tanks and requiring 900 feet of piping to bring water from a suitable intake. They applied for 500 bags last year. They got 200 then and would like the rest any time. I said that our Assistant Inspector of Works would look into it, adding that since they were contractors they should have no difficulty doing a satisfactory job.

They would also like to apply for a Kadoorie Orchard from the Kadoorie Agricultural Aid Association, and will do this through the Agriculture and Forestry Department.

I must have been asked about successions to property in this village, as there is a note on the procedures, which are as follows.

After filling in an application form in the District Office, notices of the proposed successor(s) are posted by the land staff in the village in question for two weeks, calling for objection. If, as is usual, there is none, the application is then transferred to Mr Lo Cho-chi who checks the landed property with our land registry and obtains a valuation from the land staff. If the valuation of land and houses is more than $2,000 in value, the papers go to the Estate Duty Commissioner who collects the duty assessed by the District Office. After payment is made, the papers come back to us. A memorial of succession has to be prepared in the Land Registry, for signature by the applicant with a witness, and passed to the District Officer for him to approve and sign.

If the deceased died before 1937, no estate duty is required, because that Ordinance was not applied to the New Territories until that year.

Pre-war, it was often, in fact, usually, the case in outlying places, that people didn't bother to update succession until it became necessary. Succession went by custom, and everyone in the village knew who was/were the successor(s). It was only when the government wished to pay compensation for land taken for a new road or other public work that successions were required, and then the villagers thought that having to make application for this was a great nuisance!

Ngong Ping (*Gazetteer* p. 180) (James Hayes)

Two families of 24 persons named Kuet. They came 5–6 generations ago from Fukien, and are Hakkas. One man is working in England, and another at Ma On Shan; otherwise, the men are all at home. They said there had "always" been only the two families here.

They have 30 taochung of paddy land and 10 taochung of vegetable land. I noticed several taochung of abandoned fields. The Village Representative who is in his 60s said they had already been abandoned when he was a small boy. His name is Kuet Fuk-lei.

Six children are over 6 years old, of whom 4 go to school at Ma On Shan each day. A new school is planned for this and other local villages, and its location will be closer to this village. I saw the site on this visit.

In the dry season, there is not enough water in their well. They could deepen it, they said, and would need about 25 bags of cement, which I said we would supply. There are transport costs besides, because of the remote hill location.[2]

Mau Tso Ngam (*Gazetteer* p. 146) (James Hayes)

Population 58: Village Representative is Cheng Fuk. *Gazetteer* gives 65 persons, Hakkas, mainly surnamed Cheng.[3]

Lo Shue Tin (*Gazetteer* p. 146) (James Hayes)

Population 10: Village Representative is again Cheng Fuk. *Gazetteer* states that the population is included with Mau Tso Ngam.

Fu Yung Pit (*Gazetteer* p. 146) (James Hayes)

Population 20: Village Representative is Yau Fuk-tin. *Gazetteer* gives 25 persons, Hakkas, with the family name Yau.

My notebook entries for these three villages in the Kowloon Hills provide no other information, but the *Gazetteer* entries given above help out a little.

Perhaps I should add that, after fifty years, I am not entirely sure whether these three hill villages were in the old Southern District, or in Tai Po District.

Chuk Kok (660066) (Austin Coates)

Population 10. Surname: Ho. A subsidiary of Nam Wai. A hill village.

Tseng Lan Shu (S. H. Peplow)

Tseng Lan Shu. Tseng — a well, Lan — a fence or low wall, Shu — trees. From this we find that the village has a well, with a low wall or fence around, situated in the midst of a thicket or small wood.

Tseng Lan Shu (637051) (Austin Coates)

Population 425 (76 families), of whom 12 live in the United States of America, sending remittances, and 29 work in Hong Kong and Kowloon as drivers, mechanics, labourers, etc. There are 60 refugees settled in the village zone of grass-cutting; all of them are Cantonese, chiefly from Punyu and Sunwui. The indigenous population is Hakka, surname Yau. Occupations are vegetable-growing, pig-breeding and grass-cutting, and Tseng Lan Shu's nearest market is of course Kowloon. A man of 54 is of the 7th generation, giving a date of foundation c. 1775. This is a prosperous village, and because of the vegetable growing, it employs more of its men than do other villages. It has only 35 under-employed, which for its population is below average. There are 284 taochungs of rice land, 45 taochungs of vegetables and the same for sweet potatoes. Repairs are needed to wells and paths.

Tseng Lan Shu (*Gazetteer* P. 128) (James Hayes)

Population 450, all surnamed Yau. They say they are in their 12th generation at this place, and have two ancestral halls. The older of the two is over a hundred years old.

They have 350 taochung of rice fields, and 110 taochung of vegetable land. The majority of families exchange their rice. The situation is different for each family, dependant upon its own land holdings, whether it rents, family size, etc.

About one-third of the younger men work outside the village, but very few men are overseas.

Many of the vegetable fields are now rented by outsiders. They say there were 150 of these when Mr Coates was District Officer, but the number increases from year to year, and there are now about 200.

The villagers have 40–50 brown cattle, and three water buffaloes. There is a pig-raising cooperative in the village.

There are a reported 132 children aged between 6–16. The existing village school was built before the war by the villagers, and is now inadequate for the demand. It operated primary classes 1–6 in morning and afternoon sessions, with 45 children in each, up to capacity. A new and larger school is required. Fortunately, this has been approved by the Education Department, and a site has been found by the District Office.

The village has enough potable water, and for irrigation. Their Local Public Works needs, they say, include more all-weather footpaths to replace the existing unsurfaced ones, including two new footpaths to the new school site.

They have a request currently under consideration by the District Office and the Forestry Officer to fell several old camphor trees, to help with the village contribution to the building cost of the new school.

Pak Shek Wo (642053) (Austin Coates)

A subsidiary village of Tseng Lan Shu, a short distance further from Kowloon and situated south of Clear Water Bay Road, below road level. It is occupied by the same clan as Tseng Lan Shu and its population is included in the above. A new well will be needed here, and a bridge to link two parts of the village and enable children to reach the school at Tseng Lan Shu.

Pak Shek Wo (*Gazetteer* P. 129) (James Hayes)

Six local families of Yau with 36 persons.

Au Tau (644045) (Austin Coates)

Another subsidiary of Tseng Lan Shu, occupied by the same Yau clan and its population included in that of the chief village. Au Tau is situated some distance from Clear Water Bay Road on the south side, south of Tseng Lan Shu.

Au Tau (*Gazetteer* P. 128) (James Hayes)

Au Tau with Au Tau Tsai has about the same number of Yau families, as well as over 50 newcomers.

Tai Po Tsai (661055) (Austin Coates)

Population 346, of whom 37 are working either abroad or in Hong Kong, and 14 are working locally for the army (10 of them only casual workers). 30 under-employed youths. Surname: Wan. This clan has relatives at Ho Chung and has a considerable family history. The tradition is that there was a Wan, of Tungkun District, who was an officer in the Sung army and fought in the last battles of Sung Ti Ping in Kwangtung. Taipo Tsai was founded in 1629. They own 227 taochungs of rice land, and rent another 60 taochungs from the Shing clan of Mang Kung Uk. The village is on the main road, favourably situated outside the Military Closed Area, and is, in its rough way, one of the most prosperous places in the District, presenting the curious mixture of sophistication and simplicity which reaches its apogee on Lamma Island. There are two shops. The village is a central point of rice barter for the small villages of the Hang Hau peninsula. The population is Cantonese.

Pig-breeding is an important and expanding occupation. There are Kadoorie Agricultural Aid Association batteries of pigsties here, but the villagers do not entirely favour the design of them, besides which the Army's habit of firing off heavy guns in the immediate neighbourhood has scared the pigs (understandably). They now ask for more Kadoorie Agricultural Aid Association loans, repayable after 8 months, but will build their own sties.

Other requests were for cement to improve drainage channels, and for a "Go Slow" sign on the main road. Army drivers are the particular offenders, apparently. The village wishes to knock down the present school building, erected in 1925, and erect a new school. A formal application will be made in June. This is another instance where a Cantonese major village is seeking to expand its school to serve children from smaller surrounding Hakka villages, in this case Yau Yu Wan and Hang Hau. The explanation given by the village for the peculiar fact that the Wan clan should rent fields from the Shings of Mang Kung Uk, who did not arrive in the region till 70 years after the foundation of Taipo Tsai, is that about 1865 there were serious disputes within various branches of the Wan clan, involving litigation for which money was urgently required by the Taipo Tsai branch; they accordingly mortgaged the fields in question to the Shings. When the region became part of the New Territories of Hong Kong the Shings declared that they owned the fields: the Wans did not dispute this, and thus ever since the latter have paid rent to the former.

Tai Po Tsai (*Gazetteer* P. 129) (James Hayes)

Gazetteer has: "A prosperous farming village on the Clear Water Bay Road. Population 340, with all villagers surnamed Wan."

A Cantonese village, with over 400 inhabitants.

90 children attend school in the village. A small number come

from Hang Hau, Yau Yue Wan, and Pik Uk. Six classes are held in two classrooms. There are 4 classes in the morning, and 2 in the afternoon.

They have 170 taochung of paddy fields, producing 3 months' supply for exchange. This is done here, or with the Wah Hing Lung shop in Ho Chung.

For the last two years there have been no geese or ducks, but only some chickens.

40 men work outside the village, producing (they said) a very good income from remittances. Most of them work on ships.

Local Public Works needs are improvements to 2, and possibly 3, wells, and construction of a small footbridge inside the village and a path to one of the wells.

No widows are as yet registered for assistance from the Kadoorie Agricultural Aid Association. The Village Representative was asked to send in any names to the District Office.

The village received a Kadoorie Orchard some time ago. We looked at it, and it seemed in a poor state. Will ask the Agricultural and Forestry Department to send an officer to advise.

PAK SHA WAN GROUP OF VILLAGES

Pak Sha Wan Old Village (659087) (Austin Coates)

Population 70. Surnames: Yau, Lee, Lam, Tseng, Tse, Wan, Wong, Shek, Yuen. Situated on Hiram's Highway, with a shop at the bus stop. Occupation: fishing, pig-breeding, and grass-cutting. The village owns 8 pairs of small fishing boats, whose activities are somewhat hampered by the Port Shelter Range. In the autumn two wells situated on the sea-shore will need repairing. The village has an unemployment problem, but it does not appear to be serious. There are very few fields, and the whole crop of sweet potatoes was destroyed in the frost. As a substitute pig food, they are using seaweed. They have asked to join the Kadoorie Agricultural Aid Association pig scheme, and I have asked Director of Agriculture, Fisheries and Forestry what the position is.

Pak Sha Wan Old Village (James Hayes)

Population 140 persons. There are now families of 10 surnames, all Hakkas. The oldest among them are the Lams, who came from Sha Tau Kok about 40 years ago, say around 1910. The next oldest are families of Tse.

There is a Village Representative, and also one for the New Village (see below).

The village has only 7–8 taochung of cultivated land, and no paddy fields.

Many men work outside, and people at home rear pigs, with currently 20 sows. There is also some fishing, done with 7–8 sampans.

There are 16–17 children aged between 6–16.

A new, one-classroom, school was built in 1952. It is full to capacity (45 places). The children come from this village and the New Village, but half the local children go to school in Sai Kung.

Pak Sha Wan New Village (659088) (Austin Coates)

Population 100. Surnames: Shek, Wong, Cheung, So, Fan. This is the former village of Kau Sai, which was moved in 1950 to Pak Sha Wan because its grass-cutting activity on Kau Sai Island and Jin Island were impeding the full use of the Port Shelter Rage. A new model village, designed by Mr C. A. V. Hall, was erected for them, situated on Hiram's Highway. Occupations: pig-breeding and grass-cutting. No fields. The village is slowly settling down, and there has been a definite improvement in the health of the pigs since 1953. The Kau Sai pigs were superb specimens, but at Pak Sha Wan the villagers suffered many losses, due to various animal diseases. They are experts at pig-breeding, however, and have been greatly helped by the Department of Agriculture and the Kadoorie Agricultural Aid Association.

There has been a request made by the Government that part of the grass-cutting and forestry land north of the village be reserved for the possible resettlement of urban cultivators. I have not hitherto opposed this, but in view of the importance of grass-cutting to Pak Sha Wan, I propose to ask that the request for reservation be cancelled and that the whole of this available area to be included, as was originally intended, in the New Village's forestry lot. The two villages have also asked for assistance in making a small pier. The best site for this would be the old pier formerly used by a lime kiln, in front of the village, which closed down about 1949.

Pak Sha Wan New Village *(Kau Sai San Tsuen)* (James Hayes)

Also Hakkas, totalling 164 persons. There are six surnames in the New Village, namely So, Shek (2), Cheung, Fan, and Wong. The village has its own Village Representative.

This village was removed here from Kau Sai, an island and boat anchorage in Port Shelter, in 1952, to make way for the Port Shelter Artillery Range.

The villagers have 15 taochung of vegetable land, but no paddy fields.

There are 22 children aged between 6 and 16.

There are no boats and no fishing here. Pig-rearing is a main occupation, with 23 sows and 100 pigs. Six men work outside.

Pak Wai (655081) (Austin Coates)

Population 79, of which several live and work in Hong Kong. Principal surname: Lau; two families of Chan. The Laus are an offshoot of the clan at Mang Kung Uk. Occupations: pig-breeding and grass-cutting. A few fields scattered up the narrow valley leading to King Kang Shek. The village is situated just below a bend in Hiram's Highway, and the Village Representative asked for white posts to be placed along the road and a small stretch of embankment built to protect the houses from the danger of cars crashing over the edge of the road. This is being referred to the Roads Office.

In discussing pig-breeding one villager explained that he made about $1,800 every 6 months from the sale of 10 pigs; it cost him $80 to feed one pig for 6 months. The profit was, however, less than it looked, because he was obliged to buy piglets, which were now expensive. The price of piglets had in fact gone up so much that he had now started breeding, and others were doing the same. Cultivation is done entirely by women, and most of the grass-cutting too.

The villagers are comparatively prosperous and very self-reliant. 10 young men are seeking employment in town, but if 5 could be found jobs, it would help. This village, because of its situation on the highway, is less connected with Ho Chung than some of the others. Ho Chung would like Pak Wai's children to come to their school, but Pak Wai people prefer to send their children to school at Sai Kung.

Pak Wai (James Hayes)

Village Representative is Lau Man-fong. Population 80, with 21 families.

The Lau lineage is the most numerous, with 15 families, here for 9 generations, and coming here from Mang Kung Uk. There are also three families of Chan, settled here for 4 generations. Three families have just moved here from Kowloon, Wong, Tang, and Chui.

There are 18 children aged between 6 and 16. Twelve attend school, all going to Sai Kung Public School.

They have 30 taochung of paddy fields, and 10 of vegetable land. There are 3 buffaloes and 3 brown cattle.

12 men work outside the village, mostly in Kowloon, with six working in the Wah Lun textile factory near Kowloon City Pier. There is a man in Borneo, and another at sea.

There are no sampans, and they don't fish. The village used to have fishponds, but these were sold to a Hong Kong man, Lam Pak-chuen, after the war.

They had nothing immediately in mind for Local Public Works, but said they would discuss and let us know.

King Keng Shek *(Hing Keng Shek)* (651088) (Austin Coates)

Population resident in the village, 35 (6 families), and there are another 3 families in Hong Kong. This village is incorrectly described on maps as Sam Fai Tin, which is the local name for some fields situated above the village. It is a poor and remote spot, about 30 minutes' walk uphill from the bus route. Various small public works are required in the autumn and the Village Representative thinks he can raise the money to do them.

Hing King Shek (*Gazetteer* p. 140) (James Hayes)

I have no note. *Gazetteer* gives population 40 for this hamlet with two other locations.

Ta Ho Tun (664094) (Austin Coates)

Population 67, of whom 12 reside in the urban area. Surname: Lo, with one family of Yuen and one of Lau. A short village about 150 yards from Hiram's Highway. The village has 30 taochungs of paddy and sweet potatoes, one small fishing boat, and about 30 pigs. They appear to manage quite well, probably thanks to remittances from town. A pier needs repairing, and a fairly long path leading to the principal well. These jobs can be done in the autumn (300 bags of cement were requested). The Village Representative stated that 6 men were out of work.

Ta Ho Tun (*Gazetteer* p. 135) (James Hayes)

Again, I seem to have no note, but *Gazetteer* gives population 55, and the people as Los. The omission is curious, as my notes refer to "the Ta Ho Tun group of villages", and list Tsiu Hang, Che Tuk, Che Keng Tuk, and Ta Ho Tun as its member settlements.

Tse Kang Tuk *(Che Keng Tuk)* (670094) (Austin Coates)

Population 25. Surnames: Lau, Cheung. Situated just beyond Chiu Hang. Access to these villages is through the Sai Kung Agricultural Station; they are about 15 minutes' walk from the main road.

Che Kang Tuk (*Gazetteer* p. 135) (James Hayes)

This comprises four locations at different points along the coast, all known under the same name, though two of them are called "Chiu Hang Hau" by the Chiu Hang people. My note says, "This is a poor little place."

The Village Representative, Cheung Kwun-yau, is a new man, and has still to be registered as a Village Representative with the District Office and the Police, and get his identity card.

The total population is 70–80. This may be too high, since the number of children of school age is only reported as eight (of whom four go to school in Sai Kung). On the other hand, maybe they are only reckoning in terms of boys. The notes add that more provision for education is very necessary in this area.

One man is abroad, and two work in Kowloon.

There is a Cheung Ancestral hall here, but it seems just like an ordinary house, and is being used as a store room. [I note that *Gazetteer* has the villagers as being Laus, with no mention of Cheungs.]

They have 80 taochung of rice fields, and 25 taochung of vegetable land. Some families have land, and some not. It was stated that 30 taochung of paddy are owned by a number of families who have enough land to provide (with the usual exchange) themselves with a six-month supply.

A little coastal fishing is done, but they have no boats.

Ma Nam Wat *(Ma Lam Wat)* (674077) (Austin Coates)

Population 70, Surnames: Chung, Wan. 10 taochungs, all owned and cultivated by the Chung family. 4 men working in Hong Kong, 1 in

North Borneo, pig-breeding and fishing. The Chungs own 3 sampans, the Wans 4. Three children from the village go to school in Sai Kung. Village founded c. 1870.

Ma Lam Wat (James Hayes)

70 persons. The Chung lineage has been here for 3 generations, and the Wans also, but the Chungs came first, ten years before them. We were met at Pak Sha Wan by Mr Wan Tin who came with us by boat.

There are 15 children of school age, but only four attend, going daily to Sai Kung.

They have 10 taochung of paddy fields, all owned by the Chungs, and 30 taochung of vegetable fields, half owned by Chungs and half by the Wans.

The Chungs have 7 brown cattle, the Wans have none. There are currently 10 sows and 30 pigs.

Some families are still fishing; we saw them drying nets. In the 3rd lunar month, they begin to catch crabs and abalone until summer, but not before or after. They use nets from small sampans, but not stake-nets.

One man is in Borneo, 2 are at sea, one man works on the Kowloon Canton Railway, and one is a government electrician.

They wish to improve their potable water supply; the Assistant Inspector of Works will visit.

When we visited, we took one of the village sampans, crossing over from Pak Sha Wan. To save time, it was taken in tow by our government launch. Neither party was skilled in towing or being towed, and on the way it turned over, and we were all thrown into the water. It was a beautiful day, and very calm, I remember, so it was good for a laugh all round!

PAK KONG GROUP OF VILLAGES

Pak Kong (658104) (Austin Coates)

Population 265, of which 220 are actually resident, the rest being either abroad or working in Hong Kong and Kowloon. A Cantonese village, although the hamlets in its group are Hakka. Surnames: Lok, Tseng, Lau, Leung, Lee. The eldest family is the Lau, a 23-year-old man being of the 12th generation, giving a foundation date c. 1680. The population figure includes 10 villagers, recently settled near Uk Cheung, an amateur farming family with a shop in Kowloon. The village suffered severely during the war, when it was reduced to starvation. Several houses were demolished during this time, the inhabitants selling the materials to buy food. It is the home of Lok Kau Kee, a prominent member of the Sai Kung Rural Committee and a keen hunter, who has done a good deal to keep down the wild pigs which have been a nuisance to villagers throughout the whole region from here to Tai Long.

Although this village is more prosperous than the more remote places, it can still not be said to have recovered from its losses during the war. A dirt track connects it with Hiram's Highway, and lorries can come within 100 yards of the village. The track should be improved, and I intend to consult the Chief Engineer, Road Works, to obtain technical advice on making, as a local public work, a cement wheel-way job with aggregate between. The army has constructed an excellent full-sized football field near the highway, and they share the use of it with the village. Now that the Battle School has closed down, I will raise with the Army the question of issuing a permit to the village for the use of the field, subject to future Army requirements being met.

Local Public Works cement was granted to control the overflow drainage from the football field and to make a temporary improvement in the dirt track approach to the village. 90 bags of cement will be required in the autumn for repairs to irrigation channels, and there are two large dams which could be constructed to the great improvement of the general water supply. These jobs are, however, larger than the villagers can handle, and the Irrigation Engineer will have to be consulted. Flies are very troublesome here. I authorised an immediate issue of Local Public Works cement for repairing two small reservoirs.

There was formerly a sugar industry at Pak Kong which came to an end around 1925 largely, I gather, through neglect. The place is conveniently near the urban area, 15 minutes' walk from the bus route, and doubtless the case of finding jobs in Hong Kong and in ships drew people (and money) away. Large cows are needed to operate the grinding machines, made of stone, and some subsistence from the Government would be required in the first year while fields formerly used for sugar industry. 12 taochungs of land formerly cultivated are now without water, and could be opened for sugar, should villagers be in favour of this. There is an unemployment problem, which could be solved by restarting the sugar industry. The young men, however, are of the type that prefers urban jobs. No priority for such jobs need, in my view, be given to this village. Nearly 100 taochungs of sweet potatoes were lost in the frost. The strangers near Uk Cheung grow sugar, and sell molasses through their shop in Kowloon.

The school operates in a single room of a private house and is unsatisfactory. A school building is in the Estimates for 1955–56.

Pak Kong (*Gazetteer* p. 135) (James Hayes)

Population 250, all Cantonese. The Village Representative is Mr Cheng Chung-shing. He is of the 10th generation at Pak Kong, and his grandson the 12th.

There are 37 children in the school (primary 1–3 classes only), plus some others from the temporary structures along the Sai Kung road. Another 15 attend school in Sai Kung. "It is quite easy to get them into Sai Kung schools," they said, "and they all have cheap fees, at $2 per month."

They have 500 taochung of paddy fields, inclusive of vegetable land. There is a large village forestry lot, but no hill tea bushes.

There are 80 brown cattle, all brought in at night, and consequently no deaths reported. About 20 sows and 200 pigs currently.

With exchange, they have about 8 months' supply of rice in a normal year. The majority of families exchange their own rice with shops in Sai Kung.

There are 5–6 men working in United Kingdom, and over 10 are at sea. Another 10 work either in Sai Kung, Kowloon, or Hong Kong.

There is a Tai Wong shrine in front of the village, and a Tin Hau Temple. There are family altars in private houses (branch ancestral halls) but no separate structures.

Their main Local Public Works project is for a motorable road into the village, in the form of two 2 x 2′6″ concrete strips. "This," they stressed, "is very important, and high priority."

Another need is for piped water and two stand-pipes, after the last flood destroyed a well, leaving only one behind. Also, for an "Irish" bridge on the footpath leading to the Tai Wong shrine, to be done before the start of the 1958 wet season.

They applied to the China Light and Power Company for an electricity supply two years ago, but have not got it yet, although (they say) Ho Chung and Sha Kok Mei already have been connected.

Speaking about the Tin Hau temple, they mentioned that they used to have brass temple ornaments, but after these were stolen during the Japanese occupation, they were replaced with pottery ones. As to its founding date, they say it is "as old as the village", but the only dated object is an internal screen of 1872 (Tung Chi 11th year). White ants are always destructive, causing roof collapses, and the need for periodic repairs and renovations.

There had previously been a rubble wall surrounding the village.

Pak Kong Au (661111) (Austin Coates)

Population 75 (17 families), with a high natural increase, 20 since 1945. Surname: Lau. The land cultivated is partly their own, partly leased from the Lok family of Pak Kong. No school; the children go to school at Sai Kung. The village is fairly high up, 35 minutes' walk to the bus route, then 5 minutes by bus to Sai Kung. Pak Kong hopes that when their new school is built the children will go to it, but there is a definite tendency to prefer schooling at Sai Kung where, although the medium of instruction is Cantonese, the teachers are Hakka and can explain difficult points to the children in their own dialect.

A sugar industry formerly existed here too, but died out around 1925. Here the causes are more specifically known. The Lau family had a serious quarrel at that time and split into two groups; it has since splintered still more. The consequent diminution in capital backing made it economically difficult to keep the sugar going, particularly in the matter of buying the large cows specially needed. I think too that the villagers never troubled to breed these cows, and probably followed their usual practice of letting the bull run with the herd. When necessary they would buy new cows, the calves bred at the village being useless for sugar grinding. In the level fields above the village (661112) peanuts are grown as a dry crop on land ideal for sugar, and a number of fields formerly cultivated are deserted. Subsistence would be required if men are to bring these fields under cultivation again, and I have raised the question with the Director of Agriculture, Fisheries and Forestry. It would be useful to this village if sugar-growing could be started again, and failing this, if the neglected fields could be utilised in some other way.

It has to be remembered, however, that this village is sufficiently near the urban area for the lure of the city to operate; an economic boom in Hong Kong in the future, and it might be found that work spent of making full use of the land was quickly wasted. The village wishes to improve its path from Pak Kong, but I said this could best be done by themselves without assistance from the Government; a stone path is all that is needed. Forestry licenses are so long outstanding here (10 years) that the total sum owing in fees ($180) is too much for the village to find. The Forestry Officer has been asked to advise on this. 5 bags of cement were granted immediately for repairs to a well.

Pak Kong Au (*Gazetteer* p. 135) (James Hayes)

A Hakka village. Over 80 people, all of the Lau clan. The Village Representative, Lau Shang, is of the 7th generation here. They have just repaired their ancestral hall.

Their children attend the Sai Kung Public School, in the township.

They have 20 taochung of paddy fields, some rented from Pak Kong people, and a little vegetable land. They could farm more land if water was available. I said I would ask our Assistant Inspector of Works to investigate.

The fields in question were abandoned from before the Village Representative was a child. There were 12 taochung up the hill, and about three near the village. Some had been planted with sweet potatoes during the Japanese occupation.

They had previously been used to grow sugar cane. This had been a flourishing industry, and we were shown several abandoned grinding stones beside one of the village wells. It had stopped perhaps thirty years ago.

They have rice only for 3–4 months in the year, and must exchange and buy for the remainder. There are 8–10 trees of hill tea.

It then turned out that, of the 16 families in the village, only seven have fields to cultivate. Among the village men, many belong to the traditional building trade, and are bricklayers, masons, plasterers, and carpenters, known collectively under the names "nai shui", or otherwise "sam hong". They work in the Sai Kung area. Six or seven of them are skilled artisans, the rest as builders and labourers. Two men are in England.

The Local Public Works requests relate to two wells, for which materials were requested. None have yet been supplied, so they built one themselves. They also want to make small repairs to the footpath connecting them to the Sai Kung Road, along which their children walk to school.

Uk Cheung (650108) (Austin Coates)

Population 19. Surname: Lau. A remote hill village, 30 minutes' walk, carrying a load, from Pak Kong, and 40 minutes from the main road. The village is an offshoot of Pak Long Wan, in Hang Hau peninsula. The Village representative, aged 57, is of the third generation at Uk Cheung, giving a migration date c. 1875. He appears to have leprosy, and the Medical Officer of Health is being notified. The remoteness of the place, and the peculiarity of a small branch of the Lau clan wishing to leave Pak Long Wan, a prosperous place, suggests that leprosy may also have been the cause of the migration. There are 17 taochungs of rice land; Kadoorie Agricultural Aid Association loans for pig food would be of great assistance. The Village Representative is a village contractor, and his son works as a foki in the nearby farm run by strangers.

Uk Cheung (*Gazetteer* p. 135) (James Hayes)

I have no note on this place. *Gazetteer* gives the population as 20, and the people as being of the Lau surname.

Chiu Hang *(Tsui Hang)* (669097) (Austin Coates)

Population 39. Surname: Chan. 4 under-employed youths. The clan comes from Tamshui District. Chiu Hang was founded c. 1830. They have no one working abroad or in Hong Kong, but have a good standard of prosperity (and idleness) breeding pigs, at which they are expert. They asked for cement to build a sea-wall, but the number of fields to be protected is small, and it may be difficult to supply the 150 bags asked for.

Tsui Hang (*Gazetteer* p. 135) (James Hayes)

Population 46, all Chans, and with an ancestral hall. They say they came from the mainland, but forget where. They have been five generations here, a man of 55 being in the fourth generation. The Village Representative is Chan Yuet-shing.

Two men are working on ships, and another two are working in Kowloon and Hong Kong. The Village Representative is employed with a fish dealer in Sai Kung.

They have 25 taochung of paddy fields, giving enough with exchange for a four-month supply of rice, and there is another 25 taochung of vegetable land. There are eleven brown cattle, six sows, and 50 porkers.

There are over ten children aged between six and sixteen. Five of them walk to school in Sai Kung, and it appears that the smaller children would go if communications were easier. The village women all came in and joined in the discussion over schooling, which is unusual. The four small villages in this group (see above) are asking for a school of their own.

They say they have enough potable water, and also that water for irrigation is usually sufficient.

There is no village temple here.

The people at Che Tuk would like materials to build an all-weather footpath. They said they were offered cement in 1956, but couldn't take it then. They are keen to do the work next September, and I will ask Assistant Inspector of Works Mr Chan to visit and discuss with them.

SAI KUNG TOWN (Austin Coates)

Population 1,597, the principal market town of the Sai Kung region, the fifth largest town in the Southern District. Its population is actually a good deal larger than the figure given here, but this is because many villagers from the region own subsidiary houses here; as such people are included in the numbers given for their respective villages, they are not included in the market town's population. The daytime population of Sai Kung is something like 3,000. Whereas Tsun Wan grew up as a group of nearby villages which gradually expanded until they formed a single town, Sai Kung developed solely as a market, its founder inhabitants being villagers from neighbouring places such as Sha Kok Mei, and its earliest appearance being probably no more than a group of mat-sheds and market stalls, occupied by day, but left almost unattended at night. The same phenomenon will be observed further on in this memorandum, in respect of Hang Hau and Yung Shu Wan (Lamma Island). The only definite statement regarding the age of Sai Kung which can be given without further detailed enquiry is that it is not as old as Sha Kok Mei, which was founded about 1580.

Apparently the villagers of Sha Kok Mei were among the first settlers at Sai Kung. The town is ideally placed as a market for farmers and fishermen, and has grown considerably since the lease of the New Territories. In 1898 there was little to choose between Sai Kung, with a population of 800, and Ho Chung with 600. Since then Ho Chung has shrunk in population, while Sai Kung has more than trebled in size. Neighbouring Sha Kok Mei, in the same period, has doubled its population. The inhabitants are chiefly Hakka, but Cantonese influence is strong; there are 4 restaurants, 2 goldsmiths, 2 rice mills, and 3 small shops making handles and other wooden parts of agricultural implements. There is no pawn shop.

In front of the Tin Hau Temple an important annual theatre performance is held. Because of the large number of similar performances on the Queen of Heaven's birthday, Sai Kung holds its festival, incorrectly, on the 15th day of the 4th moon. The oldest tablet in the temple dates from the 22nd year of Tao Kuang (1842), which suggests that Sai Kung benefited almost at once from the British connection with Hong Kong Island and Harbour. From 1830, when British ships began sheltering regularly in Hong Kong harbour, a good deal of the farm produce purchased for the crews came from the Sai Kung region, most of it sent by sampan to Nam Tan Sha, thence over the hill to Hang Hau, and from there by boat direct to the harbour. (See also under Hang Hau.)

Until 1945 Sai Kung was somewhat neglected by the Administration, due to its remoteness. Since that date there have been numerous improvements. Under the British Military Administration, the Japanese road was surfaced and improved, a new Police Station was built, with the help of Government subsidies various schools have been enlarged, a resident doctor, midwife and nurses have been provided, and a Health Centre is shortly to be built. The Cooperative and Marketing Department has a rented office on the harbour; there is an Agricultural Station just off Hiram's Highway near Pak Kong; and a small resident health staff works in the town under the Health Inspector for Tsun Wan. There is a Government market and public lavatory, and this year the harbour has been dredged by the Port Works Office. The China Light and Power Company provides electric light, and it is intended that the District Office should re-open negotiations as soon as possible with the Hong Kong Telephone Company for public lines to be laid; at present only the Police have telephonic communication with Kowloon. Main water has been connected this year, the source being a large stream behind Lung Hei, in the Sha Kok Mei group of villages.

It cannot be claimed that the appearance of Sai Kung reflects any remarkable sign of Government activity, but at least a respectable start has been made to obliterate 40 years of neglect. The Sai Kung region has a Rural Committee, with an office in the Tin Hau Temple. The Committee's Full Session (Toi Piu Wui) consists of 8 town members and a representative from each village. All members are theoretically elected, although it does not do to enquire too deeply into methods of election. The Chairman must be a town member. The present Chairman is Mr Li Shiu Yam, a Sai Kung shopkeeper, who is concurrently the Village Representative of Lan Nai Wan, north of High Island. The Committee is satisfactorily active, and there is a good deal of younger blood which is decidedly promising for the future. The Headmaster of the Sai Kung Public School is an advisor to the Committee and has Communist leanings, but it does not appear that his influence in this direction is particularly serious. The Education Department is aware of the position. In the Rural Committee Office is the large silk banner, with Chinese inscription, presented to the people of the Sai Kung region by Major-General Festing after the war, as a gesture of thanks for their notable services and loyalty.

SAI KUNG TOWN (*Gazetteer* p. 136) (James Hayes)

I did not take notes on the township. It was a market town and boat people's anchorage, with many shops and businesses, including those servicing the fishing fleet of the Port Shelter area, and those connected with the wholesale fisheries industry. The population was mixed.

SHA KOK MEI GROUP OF VILLAGES

Sha Kok Mei (*Sha Kok Mi*) (670115) (Austin Coates)

Population 526, of whom as many as one-fifth are either abroad or shortly going abroad. Their main destination overseas is London, where they run restaurants, laundries, etc., in the East End. They have an interest in the well-known Hong Kong Restaurant in Shaftesbury Avenue. A Cantonese village surrounded by Hakka hamlets. A multi-clan village; principal surnames Tse and Lau. The first family to settle here was Ip, but they have died out. They were followed in two or three years by the Tse and Lau, the former from Tungkun District, the latter from Ng Wah. A man of Lau aged 69 is of the 14th generation, giving a probable date of settlement c. 1580.

The place is among the most prosperous places in the District, although formerly there were more men employed as seamen than there are now. Remittances are the principal source of income, far exceeding anything else. In fact, the village has no products for sale outside as a regular source of income. Rice, pigs and grass are all used for home consumption. A Colonial Development and Welfare irrigation scheme is due to be carried out in the next phase of Hong Kong's irrigation scheme, but as it appeared that the Irrigation Engineer had not made a thorough examination of all dam sites, his attention was drawn to two in particular which should be included in the project. Other requests included cement for the construction of wells (Kadoorie Agricultural Aid Association), a bridge between the village and Nam Shan Long Mei, and two paths connecting the village with the main road. 100 bags were included in the Kadoorie Agricultural Aid Association grant for strengthening a stream wall which threatens to break and flood rice fields. Many unemployed men, mostly former sailors. Not much room for extended cultivation.

Sha Kok Mei (*Gazetteer* p. 136) (James Hayes)

This is a large Cantonese village, with 608 inhabitants. Included in the total is the hamlet of Kap Pin Long, with 30 persons in 8 families, all belonging to the Tse lineage of Sha Kok Mei.

The Tse lineage is the most numerous in the village, with 39 families. Wai, Lau, and Cheung lineages account for 13, 12, and 10 families respectively. There are also Tsang, Li, Wong, Leung, Poon, and Chu with 3–5 families each, Tang, Lok and Laam with two families each, and a single Chiu family. All were settled here from before the war.

The Tse lineage came here from Tung Kwun, and are in their 20th generation, say about 370 years. The Wais are a branch of the Wai lineage of Tai Wai, Shatin, and have been here for 16 generations. The Village Representative, and those other men we met during the visit, were not sure about the other clan's length of settlement.

The total farmed area is around 400 taochung, including vegetable land. Kap Bin Long has about 50 taochung. The majority of the villagers own their fields, with a minority renting land from other villages. There are 200 brown cattle, again, no buffaloes are needed. Currently, there are 20 sows in the village.

The village school is their own, not a modern building. It was built and owned by a Yick Wo Tong, which seems to be a "whole village" organization. It is probably the building in which a man of 77 told me he had studied for 5–6 years in the 1890s. Presently, 87 children study in its five classrooms, mostly from Sha Kok Mei. Over 10 of the village children do not attend school, but this is not because there is insufficient room for them. The school seems well run, though with only two and a half teachers.

There are a large number of men away, of whom 48 are seamen, with others in the United Kingdom, and 5–6 in the USA. Very few men work in Kowloon or Hong Kong, and only one person is employed in Sai Kung, in a grocery shop.

The village water supply is from 3 wells, all with clean water and providing enough to drink.

Some of the village fields do not have enough water for irrigation in the dry season, and they would like to see what might be done to improve the supply. Mr Chan, our Inspector of Works, will be asked to visit.

In regard to Local Public Works needs, they want materials to surface the village paths, "and the sooner, the better", they said! They also want to start on the first part of a wide footpath connecting the village to Sai Kung Road, not far from where we left our Land Rover.

Long Mei (*Long Mi*) (666119) (Austin Coates)

Population 40. Surname: Ho. Closely linked with Sha Kok Mei. A bridge required (see above) connecting the two.

Long Mei (*Gazetteer* p. 136) (James Hayes)

A small village close to Sha Kok Mei. The Village Representative is Mr Ho Man.

There are 6 families of 40 persons: five of Ho and one of Wong. The Hos have been here for 4 generations, and are Hakkas coming from Hing Ning. The Wongs arrived only three years ago from the Mainland.

There are 7 children of school age, all of whom attend school.

They have 30 taochung of paddy fields and four and a half taochung of vegetable land.

One man is in Nauru, another in United Kingdom, but there is nobody working in Hong Kong or Kowloon. Three men have no jobs.

They would like to have 50 bags of cement to construct a path in September.

Lung Mei (668124) (Austin Coates)

Population 16. Surname: Kong. The village near which the Waterworks have dammed a stream to provide water for Sai Kung. On my advice a free supply is being installed for the village, and the Village Representative's son will be caretaker of the water installation. Anxiety is expressed lest the damming deprive the rice fields of water; this will need watching. In the worst of the drought a special supply had to be diverted to Sha Kok Mei fields.

Lung Mei (*Gazetteer* p. 136) (James Hayes)

A hillside village, a half-mile north of Sha Kok Mei, with 38 persons in four families, all surnamed Kong. They are Hakkas, coming from Pak Mong Fa in Wai Chau, and have been here for four generations.

They have 25 taochung of paddy fields and 3 taochung of vegetable land, with 7 brown cattle and 3 sows.

There are 13 children of school age, of whom 11 attend school.

They want to strengthen an embankment along the stream, but with only 8 men at home, all of them employed by the Water Supplies Department, say they have not enough manpower to attempt the work. There are also 2 men away at sea.

Nam Shan (664118) (Austin Coates)

Population 160. Near this village are the two dam sites referred to above, which it is hoped to include in the Colonial Development and Welfare irrigation project.

Nam Shan (*Gazetteer* p. 137) (James Hayes)

Ngau Liu (675122) (Austin Coates)

Population 16. Surname: Lau. 24 taochungs of rice land, 9 of vegetables. One member working in England. This village derives water from the same stream as Lung Mei, the stream that has been dammed for the Sai Kung supply, and there may be disputes between the two villages, Lung Mei being able to cut off Ngau Liu's supply.

Ngau Liu (*Gazetteer* p. 136) (James Hayes)

This must be the Ngau Liu entered at *Gazetteer* p. 136 where its population is given as 20 persons, but without stating of which lineage. However, they are probably the Lau families credited with living at the other Ngau Liu, which clearly they do (or did) not.[4]

I seem to have missed them during my visits.

TAI WAN GROUP OF VILLAGES

Tai Wan (680124) (Austin Coates)

Population 160. Surnames: Wong, Ho, Lau, Chan. The age of this group is not known, but it is younger than Sha Kok Mei. It was reported that this was the first time a District Officer had visited for several years. A good spirit and evident pleasure at Government interest. 2 wells and paths leading to them are required (Kadoorie Agricultural Aid Association). The village seawall was damaged in one of the 1954 typhoons, but has been partially repaired. Employment is a problem, and there seems to be little land available for opening.

Tai Wan (*Gazetteer* p. 137) (James Hayes)

Population, 160, with four lineages. The Wongs and Hos have been here for 9–10 generations, they say, the Laus for four, and the Chans for three. The first three each has an ancestral hall.

There are 35 children aged between 6 and 16. Twenty-five of these study in the village school. The other 31 come from surrounding villages. There are primary classes 1–4, with two and a half teachers, meaning that one teaches part-time in the Tai Wan school.

The villagers farm 90 taochung of rice fields and 30 of vegetable land. They say they mostly have to rent, as they do not own many. About 10 taochung of fields were abandoned after the war, owing in part to insufficient irrigation water, and because wild pig and porcupine ate crops on dry cultivation.

There are over 20 brown cattle, and currently 20 sows and 80 pigs.

There are five men in the United Kingdom, but no one at sea. Over ten men work in Hong Kong and Kowloon.

The elders said that the village had been prosperous before the war, doing much business in the transportation of lime, lumber, crockery and miscellaneous goods needed for small village stores in the area. They were especially connected with the flourishing lime kiln business conducted by a non-villager, which did not resume after the war. The Japanese had not allowed them to transport goods, and had destroyed their boats [this would have been part of an overall effort to restrict movement and prevent smuggling of people and goods, including escapers from the prisoner-of-war and internment camps in Hong Kong and Kowloon].

Tai Wan was part of a local federation known as the Shap Hung (the "Ten Villages"). There was a small Tin Hau temple near the lime kiln and its pier, but it was damaged during the occupation, and there is no money to repair it.

The Village Representative said they have asked permission to cut down some camphor trees, to raise funds to repair the school playground. The District Office requires villagers to apply where such valuable trees are concerned. After a check by the Agriculture Department, and subject to its OK, the District Officer will approve.

The Village Representative requests materials to improve the path from the village to the school, and also for constructing a teacher's quarter on Crown land, as there is no available accommodation in the village. He explains that it is necessary to have a quarter "in a remote place like Tai Wan" [this was before work began on the extension of the Sai Kung Road to Tai Mong Tsai, not long after] to attract and retain teachers, who invariably come from the urban area. Two of the teachers stay in the village but return to Kowloon at the weekends, and the other rents accommodation in Sai Kung at present. The desired quarter will measure 25 ft. by 15ft.

Sha Ha (*Sa Po*) (677118) (Austin Coates)

Population 30. Surnames: Wong, Lau (2 families). This village will lie on the future road and some small exchanges of land will be required. A new well (Kadoorie Agricultural Aid Association).

Sha Ha (*Gazetteer* p. 136) (James Hayes)

Population 32. There are two lineages in the village, Lau and Wong. The Lau families came here from Ngong Wo, and the Wongs from Shan Liu, both upland villages at no great distance.

They have 12 taochung of rice fields and some vegetable land, clearly not enough for their needs. They have only enough rice for 3–6 months of the year. They have a village orchard, but say they have not enough hands to tend it properly.

There are currently nine brown cattle, four sows and 26 pigs.

There are 12 children of school age, between 6–16, but only three boys are at school in Sai Kung, the others helping in the fields. I noted that the Village Representative didn't seem to know the details and had to get answers from the women.

One man is in the United Kingdom, but there is no one either at sea, or working in Kowloon, or even Sai Kung. One is working as a labourer with the contractor building the new road.

They have three wells, but still not enough water. They cannot be deepened because there is rock underneath.

All in all, they seem rather poor and ignorant people.

Long Kang (684130) (Austin Coates)

Population 9. Surname: Wong. Bunds need improving.

Long Kang (*Gazetteer* p. 184) (James Hayes)

This hamlet has only two families, one with over 5 members, the other being a one-person household. Only one man works outside the village. This is the Village Representative, who is a sanitary coolie at Sai Kung Market, i.e., a government employee.

The two families have 2 taochung and 2 cows. There are no pigs at present.

There is one widow, and I was going to check if she needed assistance. No children were at school, perhaps because there was none of school age in this tiny place?

As at Wong Chuk Wan, I was told that "the fields here are owned by people from many many villages, including some in Tai Po District" [probably from the adjoining Shap Sze Heung, otherwise known as North Sai Kung].

Long Keng has the distinction of having probably the oldest and largest banyan tree in the Southern District. There were also camphor trees and papayas.

The villagers pointed out two other trees: one whose leaves were fed to cows, whilst the other supplied wood shavings which were made into a paste used to dress women's hair. This was said to be an old practice.

Au Tau *(O Tau)* (682132) (Austin Coates)

Population 17. Surname: Ho. A fairly large dam near the military track passing between this village and Ngong Wo is required. Irrigation Engineer's report requested. Cement and steel will be needed. I approved, subject to Forestry Officer's agreement, the cutting of trees for sale of wood to meet expenses of minor public works.

O Tau (*Gazetteer* p. 184) (James Hayes)

Population 21, with four families of Ho and one of Wong. Six children go to school, 3 to Tai Wan and 3 to Sai Kung.

The Wong family owns 4 taochung of rice fields, but the Ho family has none, and when I asked why, I was told that they used to have fields but an uncle had sold them [perhaps during the wartime occupation, when many were being sold in this area]. They currently rent two taochung. There are a few taochung of vegetable land.

I seem to have noted that the Hos had been here for six generations, but the pencilled note is faint.

One man is overseas (place pronounced as "La Lo", said to be in the South Seas, the Nam Yeung). There are three working as apprentices in a rattan ware shop in Kowloon or Hong Kong, probably the former.

There is no well, and they get their potable supply from a stream.

Looking around the place, I seem to have considered them to be "very poor", which is no wonder reading this account, and had scribbled a note about trying to help them out with a few CARE parcels.[5]

At the same time, I was very taken with the Village Representative, Mr Ho Tso, who seemed to be in his 50s–60s. A most courteous and pleasant man, he sat us down, produced tea, entered into conversation, and left me with a lasting impression of the good upbringing and easy manner of even the poorest Chinese village families.

Tso Wo Hang (692122) (Austin Coates)

Population 50. Surnames: Lee, Tseng, Yeung. New well (Kadoorie Agricultural Aid Association). Post-registration of births requested.

Tso Wo Hang (*Gazetteer* p. 184) (James Hayes)

The official *Gazetteer* has "Tso Wo Hang: population 50, Hakkas, with many villagers surnamed Lei. San Tin Hang: population included with Tso Wo Hang". The latter is an upland offshoot or attachment to Tso Wo Hang, but I have no information on this place.

Regrettably, I have no separate note on Tso Wo Hang either. This may be because I had a lot to do with them during construction of the road from Sai Kung to Tai Mong Tsai, then proceeding, and did not bother to take a special note for this small village.

However, a number of scattered entries in my note book show the difficulties we experienced in getting the new road past this village. I had similar problems on Lantao (see under Tong Fuk) but somehow it sticks in my memory that none of the other villages affected by road works in Sai Kung were as temperamental or as difficult as Tso Wo Hang. This is certainly borne out by my notes.[6]

The Village Representative says that 5 villagers would like the District Office to get them employment as earth coolies with the road contractor, "as they can't get the jobs on their own". But even when they were offered work as earth coolies by the contractor, the Tso Wo Hang folks were choosy about it. The work is paid at the rate of $3.80

a day for shifting 10 cubic feet [1 cheung] daily. They do not consider this to be enough and so have not taken up the offer. On the other hand, Kowloon people reportedly work for $3 per day, and normally a contractor does not expect to pay more in the country than in town.

A request was also made on behalf of a Cheng Pin-cheung who wanted to join the Roads Office. He was 23, and had been a casual worker for the Health Office in Sai Kung; a street sweeper probably, but the Village Representative was not sure.

I have a note on Local Public Works. The villagers say their drinking water comes from the stream. They have nothing particularly in mind, and will let us know if they think of anything.

The Tso Wo Hang villagers are coastal fishermen as well as farmers. I recall a particular man, aged about 60, who had only stumps instead of hands and forearms. He had lost them as a result of an accident whilst using dynamite for "fishing". This was formerly a common (but illegal) practice.

Nam A (678130) (Austin Coates)

Population 36. Hill village, surname Kuet, related to the villagers of Ngong Ping, on the border of Taipo and Southern Districts.

Nam A (*Gazetteer* p. 184) (James Hayes)

I described this upland settlement as "a really nice, almost English, village", and thought the Village Representative, Mr Kuet Po-shing, as "a good, intelligent chap who came all round with us", a reminder that, time permitting, I used to like poking my nose into all parts of a village!

Population 31, in five families, all named Kuet. They claim ten generations here, coming from Wai Chow, and have an ancestral hall. Another family has gone to live in Sai Kung recently. Two men are in the United Kingdom, and five work in Hong Kong and Kowloon.

Eight children of school age all walk to school in Tai Wan.

They farm 30 taochung of rice fields, all their own, and three taochung of vegetable land. With exchange, they have 4–5 months' supply of rice.

There are nine brown cattle, two sows and 12 porkers at present.

They said their well usually supplied enough potable water, and they also had enough water for irrigation.

We had excellent hill tea to drink here, from a few tea bushes planted twenty years ago in the hills behind the village. They can pluck the leaves four times a year, but the best crop is usually at Easter. They get enough for a six-month supply. The leaves are kept in a drawer for two months, and then rolled. Best brewed in porcelain, with boiling water added. Apparently, cows like to eat the leaves.

Wo Liu (S. H. Peplow)

Wo Liu. Wo — rice, Liu — hovel or hut. A collection of huts forming a village near rice fields.

Wo Liu (688128) (Austin Coates)

Population 16. Hill village, surname Lee.

Wo Liu (*Gazetteer* p. 184) (James Hayes)

There are only 18 persons in the village, all of the Li lineage. The Village Representative's name is Li Hei.

They have been here, they say, for ten generations, coming originally from Fukien [I think the ten generations are more likely to be "in Kwangtung", rather than in this place]. There is a Li ancestral hall.

There are ten children, of whom five go to school in Sai Kung, and one to Tai Wan.

The village has five taochung of rice fields, and two of vegetable land. Their rice supply lasts for up to 4–5 months. They have four cows, one sow, and 12 pigs.

No one is working at sea, or in United Kingdom, but one man is working in the Marine Department, and another is a lorry driver.

Ngong Wo (686133) (Austin Coates)

Population 85. Surname Lau. The highest and most remote of the hill villages of this group, clearly neglected by the Government. I have requested the Irrigation Engineer's advice on a general scheme of improvement of waterworks; gave permission for the construction of a new stone road connecting the village with Taiwan, and, subject to the Forestry Officer's agreement (which is being sought), the felling of trees to meet expenses. There was an old school here before the war, but none since; I have asked Education Department to examine early the prospect of rebuilding and reopening the school. The first ancestors arrived here c. 1800 and there is a fine ancestral hall of the low-roofed type reminiscent of buildings further north. Forthright people of good physique, a number of whom have been to Ocean Island and Nauru. Portuguese and other huntsmen sometimes stay here.

Ngong Wo (*Gazetteer* p. 184) (James Hayes)

I described this place as "an upland village, very pleasant and with splendid views". The Village Representative, Lau Yik-pan, "a young chap" according to my note, says there are 74 persons, all Laus, in some 19 households, living here. They have been here for over a hundred years, coming from Hing Ling in eastern Kwangtung.

Another 7 Lau families from this place are resident at Tai Wan village, and 3 more are now living in Sai Kung Market. There are another 20 persons living in Wong Chuk Wan, down below, an offshoot from the main settlement which was established four generations ago, and therefore is nearly as old as Ngong Wo itself.

Among the families in residence at Ngong Wo, none of the men work outside the village, save for one in England. Ten children are in the 6–16 age group, but as there is no school here, have to go elsewhere to study.

The families at Ngong Wo are farming 40–50 taochung, and another 4–5 taochung of vegetable land. They have enough rice, with exchange, to give them four to five months' supply. The Wong Chuk Wan families are worse off, having only enough for 3 months, even with exchange, and having to buy for the rest of the year. These low figures were not unusual, as the 90 persons at nearby Shan Liu also had only a three-month supply.

There are 25 brown cows, 4 sows and 25 porkers at Ngong Wo. They have sufficient potable water, but not enough for irrigation needs.

There is a fishpond in front of the houses, and they want 160 bags of cement to make some repairs to it.

At the time of the visit, the villagers had just finished the lower part of an intended "road" leading from behind Wong Chuk Wan and up to the village, probably along the line of the existing old path. It had taken them 2 months, and they said they had spent $700 on food for the workers and their families in that time. In answer to my question how they had financed this, they said they had sold trees. This would explain the tremendous amount of cut wood stacked round the houses. My note says: "Very commendable. Get them to do more, and get some CARE parcels for them, to help eke out their livelihood during the work."

They explained that wood-cutting was their major activity in winter, and complained about the Port Shelter Firing Range. This takes in the hill land above the village, and prior to its establishment, they had been used to cutting firewood there 7 days a week in the winter months. Nowadays, it is closed to them for 4 days a week. Moreover, trees are being destroyed by firing from the larger guns, so that they have less wood to sell. They mentioned the magic word "compensation", but did not pursue it.

Shan Liu (S. H. Peplow)

Shan Liu. Shan — hill, Liu — hovel or mat-shed. A collection of hovels on the hillside forming a small village.

Shan Liu (675125) (Austin Coates)

Population 79. Surname: Wong. Although a hill village, this place is fairly prosperous. A jeep track gives access to within a few yards of it. It has 2 members in USA, 2 in Borneo, several working in Army jobs in the Colony, and one working as a Kowloon sub-contractor, presumably employing a number of others from the village. There are several well-built houses of the semi-Western-style now becoming popular. The village is an offshoot of Taiwan. A man of 55 is of the 5th generation, giving a foundation date c. 1825. Complaints were received about the Army ranges, but examination showed that these were exaggerated. Grass and wood is sold at Sai Kung; pig-breeding. Only a few fields.

Shan Liu (*Gazetteer* p. 139) (James Hayes)

The Village Representative, Wong Hang-fai, who met me earlier at Nam A Village, said there were 90 persons in 19 families, all named Wong (King). They seem to be Punti, from Tung Kwun District, and say they have been here for 10 generations. Nine men are working outside, in the Nanyang, in Hong Kong and Kowloon, and in Sai Kung Market. 12 children attend school, 8 in Tai Wan and 4 in Sai Kung.

They farm 22 taochung of their own land and rent 5 from others. They also have 10 taochung of vegetable land. Their rice, with exchange, is only enough for 3 months of the year. There are 14 brown cattle, 6 sows and 30 porkers.

They have enough water to drink, from a new Kadoorie Agricultural Aid Association well, and apparently for irrigation too. Their main need is to improve some bad footpaths, especially the one leading to the new school. They seemed delighted when I told them they could have materials for this improvement work. Assistant Inspector of Works Chan hasn't visited this place yet, and when he does, he can say what is required.

Interestingly, this hill village had complained to Austin Coates a few years before about the Port Shelter Firing Range, but he records in his Summary Note on the District that their complaints were exaggerated. They said nothing on this subject to me on this visit.

Wong Chuk Wan (689127) (Austin Coates)

Population 19. Surname: Lau. A coastal village.

Wong Chuk Wan (*Gazetteer* p. 184) (James Hayes)

The Village Representative, Lau Yuen-cheung, said there were 20 persons living in this offshoot from Ngong Wo. He repeated that they had been here for 4 generations. One man is in the "South Seas", the Nanyang is probably meant, but no specific country was named. Two other men work in Kowloon, and a third in Sai Kung Market.

The villagers farm 3 taochung of paddy, and even with exchange have only enough for 3 months and have to buy for the rest of the year. My note says: "There is a tremendous patchwork of shared fields in front of the village. People from five villages own land here, including villagers of Sha Ha near Sai Kung."

On the Local Public Works side, they said that an irrigation channel was leaking and needed repairs, and they wished to improve their drinking supply.

TAI MONG TSAI GROUP OF VILLAGES

Tai Mong Tsai (705123) (Austin Coates)

Population 105. Several fairly large bridges are needed in this valley. Steel is needed, and works therefore cannot be contemplated before autumn. A small pier was requested last year, but my Local Public Works vote ran out before the request could be met. It will be included in the autumn list. The valley is peopled by families who are resourceful and industrious; a good spirit. Always very hospitable, with excellent meals provided, even when sternly ordered to provide nothing. Surnames: Tseng, Law. New wells and dams needed (Kadoorie Agricultural Aid Association).

Tai Mong Tsai (*Gazetteer* p. 184) (James Hayes)

This is the largest of the villages in the valley, and is the lowest-lying, located beside the main stream.

The reported population is 105 persons: namely, 22 boys, 14 girls, 39 men and 31 women, in that order. The Tsang, Law, and Wong clans have been settled here for about three hundred years, they thought.

All possess clan records. There was much argument over one of them, the Tsangs I think, because, unexpectedly, they couldn't find it when they wanted to show it to me.

The villagers farm 150 taochung of paddy land, and grow some vegetables for their own use. They exchange two-thirds of their rice in Sai Kung, and have to buy rice for six months of the year. They have 20 brown cattle. Currently, there are 50 pigs, including 10 sows. They sell all their porkers, which currently sell at $160–170 a picul. They also sell eggs, firewood and a little grass. The village has a forestry permit.

A number of men are away. Two work as coolies in Sai Kung, and six are in Kowloon, all working in a rattan furniture making shop in Sham Shui Po, introduced by a villager who had been the first to have found work there. Two men are policemen, one of them a sergeant in Wanchai. Another man is a wireless operator in the Post Office. Five or six went to England before the war, but haven't come back, and another left for the United Kingdom last year.

Some widows were mentioned. Wong Chan Shiu, aged 62, has been widowed for 12 years. She has no children or close relatives, and relies on 1 taochung of paddy and half a taochung of hill cultivation, plus a few chickens for her livelihood. Two others were said to be in need of help, but no details were given: Tsang Tang Shi and Tsang Cheung Ng.

Another request concerns a young man, Tsang Kin or Kin-hei, aged 20. He has an open sore on his leg but can still walk, though he doesn't look too good. They asked if I could get him into Kowloon Hospital for treatment.

There is a request for 1,300 feet of water piping to bring clean drinking water into the village.

Among other points of interest are the seamen's papers shown by Tsang Hing, aged 60, who had served on British and Dutch ships in the 1930s and 40s, and now asks for help in getting work. I was also shown some land papers signed by Sir Cecil Clementi as Assistant Land Officer.

I visited Tai Mong Tsai again on 23 February 1958, probably in connection with the new road works which were extending the motor road from Sai Kung Market through the coastal villages to Tai Mong Tsai.

I was told there were 110 people in the village (up from 105) with more specific information on the various clans. There are 14 families of Tsang, 8 of Lo, and 2 of Wong, "all coming about the same time", they thought. However, this time, they said their family records had been burned during the Japanese occupation.

Again, I was told they owned 150 taochung of their own land, and did not rent from others. They reported having "over 10 cows", instead of 20. Either way, these figures seem low for the amount of cultivated land.

They mentioned the road works. The earth fill for the small bridge at the end of the road will interfere with irrigation, and they request that work on an alternative new channel start now. The Village Representative also asked if some village men could be employed on the future road maintenance gang.

Among the other requests made on this visit was one for something to be done about the porcupines that were eating their sweet potatoes. I said I would speak with Mr Common, the Inspector of Police at Sai Kung, who might send some men to help deal with them.

They said they had earlier asked for a dam and channels for improving the local irrigation. It appears that the Irrigation Engineer in DANT HQ has made an inspection visit, and they want to know what can be done. They also repeated the need for improving the potable supply, saying that the likely source of clean water is only 5 minutes' walk away.

A rough scrawl relates to a widow from Wong Mo Ying, Mrs Tang Chau Kang, aged 80. She had already been referred to Kadoorie Agricultural Aid Association for a cow or a sow, not clear which was requested, but had heard nothing as yet.

Tai Po Tsai (70528) (Austin Coates)

Population 56. Surnames: Cheung, Lee (profit). The valley school is situated here, in rented and not very good premises. A new school building has been requested, and the Education Department is being urged for a reply; a site is available. Bridges are required for communication between the various villages and the school, the streams being impassable by children during the rain.

Tai Po Tsai (*Gazetteer* p. 185) (James Hayes)

Over 50 residents are reported. They claim to be all Cheungs, and here for 10 generations they thought. They said they were not very sure, as they had no clan record and didn't know where they came from. They then added that it was at the same time as the Cheungs of Shek Hang.

[There was no mention of the Kan and Tse families listed at Shek Hang as also being at Tai Po Tsai.]

They farm 20 taochung of rice fields and some hill cultivation, all their own, then added that they do rent a little from Tai Mong Tsai people. They have 7 brown cattle.

They are building a well at present and are short of cement. They need another thirty bags. They also requested 50 bags for a footpath. I promised 80 bags for April, when they wanted to start.

There is a school here, which serves the children of this, the Tai Mong Tsai, valley. However, in conversation with the teacher, a Mr Lung, it seems that only 40 attend school, out of 106 of school age. Next term will see a start to two sessions per day, with a second teacher.

Mr Lung has been here for two years and previously taught in Hong Kong. He was born in Canton, but has been in the Colony for twenty years. He is a widower, and says he likes teaching and living here. The Village Representative of Tai Po Tsai is reported to be keen on getting someone to teach English. Mr Lung also seems interested, judging by a few English words written on the classroom blackboard.

The note ends with a remark about the "government" rice that is issued for relief purposes when fields are flooded by the sea, as at Pak Tam Chung recently, or upon request, for villagers to supplement home supplies when they are engaged on their local public works projects, using cement and other materials supplied by the District Office. It was described to me by those present as being "sweepings, and only fit for pigs".

She Tsu *(She Tau)* (707129) (Austin Coates)

Population 29. Surname: Wong.

She Tau (*Gazetteer*, p. 184) (James Hayes)

Population 42, all Wongs and here for 5 generations from Wai Chau. They have no clan record.

They say they farm 8.5 Chinese acres (mu), their own land, and do not rent any from others. There are 10 brown cattle.

They are all Catholics and attend the church in Wong Mo Ying, and in answer to my enquiry confirmed they had no earth god altars in the village.

They would like 60 bags of cement and 600 feet of water pipes to improve the drinking supply, and 80 bags for a footpath. If possible, they would like to start on the dam to improve the water situation as soon after 1 April as possible.

Tam Wat (*Gazetteer* p. 184) (James Hayes)

Over 20 persons, all Wongs from Tam Shui. There are 5 families, farming 10 taochung with 5 cows. There is no clan record.

They had many requests: for water pipes, and cement for a bridge, the footpaths leading to their fields and to the school at Tai Po Tsai, and a dam. They thought they might need 100 bags for the bridge and path. However, they said they needed some assistance in kind to maintain livelihood whilst they were doing the construction work, and had applied for some "government rice", along with the materials.

Tit Kim Hang (705137) (Austin Coates)

Population 21. Surname: Sung.

Tit Kim Hang (*Gazetteer* p. 184) (James Hayes)

There are 24 persons in 3 families named Sung. They say they have been here for a very long time, say several hundred years, and have a clan record. This is presently being copied with updatings in Sai Kung Market.

They farm 20 taochung of rice fields, of which 3 are rented from others.

None of the villagers are abroad or working away from the village. Five of the children have no birth certificates.

They rely on a well for drinking supply, but it runs dry in the dry season, and they would like to dig another. As at Tam Fat, "government rice" was requested here, to facilitate this and other local public works projects.

Shak Hang *(Shek Hang)* (699141) (Austin Coates)

Population 35. Surnames: Tse, Cheung.

Shek Hang (*Gazetteer* p. 184) (James Hayes)

There are 29 persons reportedly resident, and from different clans. There were Cheungs, Kans, and Tses. All the families of these names, currently living at Tai Po Tsai, are said to have lived here first.

The villagers say they farm about 4–5 English acres [why should they use this term?] together with a few hill fields. I noted that there were a lot of abandoned fields in the area. They have 5 brown cattle.

Seven children from the village go to the school at Tai Mong Tsai [Tai Po Tsai?].

I noted that some persons were making rattan baskets.

Several villagers say they want to surface an irrigation channel.

There are two widows, one of whom got rice and the other a cow, from the Kadoorie Agricultural Aid Association.

The villagers carry all their drinking water from a stream. They need about 400 feet of water piping to obtain a purer piped supply, and also want some bags of cement for a footpath.

Ping Tun (707140) (Austin Coates)

Population 31. Surname: Cheung. A high village cultivating rice at a surprisingly high level (as its name suggests). Improvement of water supplied needed (Kadoorie Agricultural Aid Association).

Ping Tun (*Gazetteer* p. 184) (James Hayes)

There are 30 residents, all named Cheung, and settled here for 7 generations. They have relatives at Tai Shui Hang, below Ma On Shan, who have lived there for 6 generations. [This seems odd as Tai Shui Hang is a much bigger village. It implies that the family settled first at Ping Tun, with some members branching out to Tai Shui Hang after one generation.]

The founding family came from Cheong Lok in Kwangtung. There is no clan record at Ping Tun; it is at Tai Shui Hang.

They have over 10 taochung of rice fields and 4 taochung of dry cultivation, with 5 brown cattle.

All the men are living and working at home.

Wong Mo Ying (698135) (Austin Coates)

Population 58. Surname: Tang. Hill village. Army tracks have improved access here from Ngong Wo and the Tai Wan group. Catholic Church, but the villagers are dependent on the valley school. Cement already promised for dams last year, but not yet sent, due to unexpected demands of a more urgent nature elsewhere. This is now on the Kadoorie Agricultural Aid Association list.

Wong Mo Ying (*Gazetteer* p. 184) (James Hayes)

There are reported 61 persons here, all named Tang, from Lung Kong, Tam Shui. They have been here for 5 generations. One of the village men is working in Singapore. Seven children go to school in nearby Tai Po Tsai, and some study in Sai Kung, presumably staying with relatives or friends.

They farm 70 taochung of paddy fields and some dry hill land. They do not rent from others. There are 18 cows and, interestingly, 3 water buffalos.

There is a Roman Catholic Church here, quite a large one. It was built a hundred years ago, they said, and was rebuilt or repaired 20 years back. The villagers have always believed in Jesus, and an Italian priest whose name in Chinese is "Kong Chi-kui" comes to hold services once a month.

There was one request for a "chaser" on a post-registration of birth for a Tang Kam-hei. He had done it through Lee Siu-yam, the Sai Kung Rural Committee Chairman and the Immigration Office. This sparked off a recital of the delays and difficulties being experienced on this; for the villagers, it is very important subject, connected with their plans to go to work in a relative's Chinese restaurant in the United Kingdom or elsewhere. Another young man was having problems of another sort. His birth name was Tang Yau, and his school name was Tang Yung-hei. They wanted the name in "both certificates" to be in the latter form.

I must have enquired whether any widows in the village were in need of help, from the Kadoories or elsewhere, as this has been a standard enquiry on my visits. In this case I had written "Widows?" and ticked it, without comment.

They would like to have 300–400 feet of water pipes to improve the drinking supply. This was promised.

PAK TAM CHUNG GROUP OF VILLAGES

Pak Tam Chung (729132) (Austin Coates)

Population 49, of whom 5 work in Hong Kong. Surnames: Wong, Ng, Cheung. This is the central point of a group of 6 villages. Nearly the entire population of the valley is Catholic. Since 1953 the group's connections with the District Office have been strengthened, and in this we have been helped by Wong Wai Man, nephew of the Village Representative and an employee of Her Majesties Dockyard. This young man speaks fluent English, and has a keen sense of public service; he is one of the up-and-coming men of the Sai Kung region. He has a good deal of technical knowledge, and with his help one of the larger Local Public Works works handled by this Office, a bridge over quite a large river, was completed successfully in 1954. Four small piers were also constructed in 1953 and 1954, serving various points of approach to this geographically complex place. Before the war there were lime kilns in the valley, but the Japanese occupation, and (according to the villagers) the competitive prices offered to the Green Island Cement Co., finally put all the kilns out of business. They have never recovered, and the villagers consider it would be useless now to pursue this question.

Wong Wai Man has been able to get jobs for several of his fellow villagers in the Hong Kong Tramways Co., and since I have known him he has pleaded consistently the urgent need for employment outside the rural area for Pak Tam Chung people. There is a certain amount of remittances, however, and about 18 people work abroad from these 6 villages. A Pak Tam Chung Welfare Association was started in 1954, and a ferry junk operates daily between this valley and Sai Kung, the profits being devoted to village improvements. There is good cultivated land, and the valley produces excellent tangerines. To my palate, they are the best in Southern District; they won 3rd prize in the 1955 Agricultural Show. The Kadoorie pig scheme has been extended here, and it is hoped that the regular ferry service will enable pig-breeding to become a new source of profit to the people.

During last summer's typhoon the sea flooded the river and inundated several fields. Measurements have been taken, and owners will receive rice in proportion to the paddy lost. To prevent further flooding a strong sea-wall is needed just below Pak Tam Chung; this will have to be done this autumn, because the villagers have no time now to undertake so large a job, in view of their other requests. These projects include several dams and wells, and one fairly ambitious irrigation project, including a small tank and system of water courses (Kadoorie Agricultural Aid Association). Because of the inhabitants' experience, it is not necessary here to consult the Irrigation Engineer; the scheme is admirable and will be supervised by my staff.

Forestry licences have not been renewed for a long time. I have forwarded to the Forestry Officer a request for pine seed, to start a communal nursery. Actual lots will be held individually. Workers skilled in the treatment of seedlings are available.

A new school building is urgently required to replace the existing one, which is old and unsatisfactory. The Education Department is being pressed on this.

Altogether Pak Tam Chung, though remote and needy, is a bright spot.

Pak Tam Chung (*Gazetteer* p. 185) (James Hayes)

Pak Tam Chung is one of a group of villages lying beyond the end of the Sai Kung Road. It is more of a central place, giving access to (and from) the sea at the nearby portage village of Wong Keng Tei. There are some houses and shops at Pak Tam Chung, and the newly-built school whose construction and running expenses are subsidized by Government.

Pak Tam Chung has about 54 persons, of mixed names. There are Wongs who have been there for 2–3 generations, and Lams, Chans, and Ngs, also settled there for about the same length of time. Between them, these families own about 3 taochung of paddy, but say they do not rent any. There are over ten brown cows.

The school at Pak Tam Chung serves all the nearby villages. It operates in morning and afternoon sessions, with 54 pupils. Two teachers each teach classes 1–6 at different sessions, and another comes twice a week to teach English. One of the living-in teachers is 59, the other is 46. Not all the children of school age in the villages attend.

The Village Representative made a request for cement and other Local Public Works materials in the coming August to build a badly-needed bridge, and then said that a protective wall is needed at the creek. Salt water had flooded the fields and last year the District Office had provided "relief rice" for the persons who farmed the affected 15 taochung.[7] The request for the wall had been made two years before and the Irrigation Engineer had inspected the site, but nothing has yet been done.

From discussing the relief rice, we got onto the selling prices of local products. Their unhusked rice or kuk is currently worth $39 per picul in the Sai Kung shops. This is cheap, as it is usually $42–43. Dry firewood sells for $3.50 a picul, but the other local fuel, dried grass, is not selling, as no one will take it.[8] Pigs sold for $140 last year (1956), and are now $200. Local tangerines had sold for $120 a picul this year. This is a winter, and especially a lunar new year, crop.

These items are taken to Sai Kung Market by the "Six Villages" Pig Raising Cooperative's own cargo junk. The charge is 20 cents a picul. The Committee had purchased and runs this boat. It had cost $2,000, and they had paid $1,000 for another, smaller one. Five villagers are hired to run the two boats, and each man is paid $90 monthly. The boats need to make $5 a day to meet their running expenses, roughly half this amount each. However, they are not even making $2 each at present, as only about 5 or 6 persons are travelling daily to Sai Kung, paying 30 cents each way, and extra for whatever produce they take with them. The provision of such boats was described as a traditional rural practice in this area, meaning that it was, and still is, a necessary public service.

Pak Tam (724144) (Austin Coates)

Population 58, of whom 5 work in Hong Kong. Surnames: Lam, Maan, Siu, Ho, Tsang.

Pak Tam (*Gazetteer* p. 185) (James Hayes)

This inland village has a reported population of 60 persons, comprising 16 men, 25 children and 19 women, in that order! The Village Representative is Maan Kau, with an Assistant Village Representative, Lam Siu.

The families here are very mixed. They include 5 of Siu, 2 of Lam [one man said 8] and one each of Maan, Tsang and Ho. The Wong families are said to be in the majority. There are said to be no clan records, but this may not be so, as hardly anyone was at home when we visited.

Eight children go to school at Pak Tam Chung, but another 4 of school age do not.

The families of Pak Tam have 80 taochung of paddy fields, and another 20 on dry cultivated hill slopes. Half the crop is said to be exchanged with shops in Sai Kung, but they still have to buy rice for six months. There are 60 pigs including 4 sows, and 40 brown cattle (but another estimate was only 16). They don't buy vegetables, as they grow their own. Nowadays, they cut grass and firewood only for themselves, though they used to take firewood to sell in Sai Kung.

A number of men are working outside the village. One is an Army driver in Sham Shui Po Camp, two are working in restaurants in Kowloon, three are in a paper-making or printing works in Yaumati, another is an assistant in a grocery shop in Sai Kung, and one is working in the phosphate island of Nauru in the Western Pacific.

There were requests in connection with birth certificates. Tsang Yuen-loy and Tsang Shun-lin said they had been processed but had not yet received their papers. Presumably they are planning to go overseas to work.

There is plenty of water for irrigation in the area, and they mentioned a big waterfall near the village. However, they made a repeat request for 280 bags of cement to repair a footpath, and to make a dam, for which they had applied some time ago.

Chik Yu Wu *(Tsak Yue Wu; Tsik U Wu)* (729135) (Austin Coates)

Population 60, of whom 2 work in Hong Kong. Surnames: Lee, Wong.

Tsak Yue Wu (*Gazetteer* p. 185) (James Hayes)

This is the nearest settlement to Pak Tam Chung and almost part of it. There are 28 persons in all. Nine of the families are Wongs, and there are four families of Li. Both originate from Wai Yeung District in Kwangtung, the Wongs coming from Tam Shui. The population, though so small, takes in people living in two other places, Cheng Hang and Shiu A Yiu. At the former, there are two more families of Wongs, but it was not known how long they had been there.

Altogether, the families own 50 taochung of paddy, and something over 1 taochung of dry hill cultivation ("che"). Between them, they own 24 brown cattle for draught purposes. They speak of needing a big sea-wall to protect houses from flooding, and of building irrigation dams.

Several young men request post-registrations of birth. This is a common and frequent request at this time, because many young men wish to go overseas to seek employment, mostly in Chinese restaurants in Britain. The applicants here were Wong Ah-yan, Lee Shek-yung, and Lee Kwun-tai. My note includes another Wong, from Wong Yi Chau, Wong Ting-loy.

Tsam Chuk Wan (714123) (Austin Coates)

Population 53, of whom 5 work in Hong Kong. Surnames: Lee, Lai, Ip, Pang. A coastal village.

Tsam Chuk Wan (*Gazetteer* p. 185) (James Hayes)

Population, 55. The Village Representative's name is Lai Shui-yeung.

Mixed composition, with four families of Li, four of Lai (the Village Representative is a Lai), two of Wong, and one each of Yip and Pang. The Li families claim four generations here. No other details were taken.

Wong King Ti *(Wong Keng Tei)* (723124) (Austin Coates)

Population 57, of whom 4 work in Hong Kong. Surnames: Chan, Yau, Lau, Wong, Tsui, Lee.

Wong Keng Tei (*Gazetteer* p. 185) (James Hayes)

I seem to have no notes for this place, and may have inadvertently omitted it from my visits. Could it be because we landed there, and moved off straight away?

Gazetteer gives population 60, many surnamed Chan.

Wong Nai Chau *(Wong Yi Chau)* (726118) (Austin Coates)

Population 120. Surname: Wong. Church and school situated here. It is possible the Anti-Tuberculosis Association may be establishing a settlement on a promontory near here. There are no local objections, and I consider it may benefit the villagers indirectly. Coastal village, with one of the four piers.

Wong Yi Chau (*Gazetteer* p. 185) (James Hayes)

This is a big village, as local settlements go, with two subsidiary hamlets. The Village Representative (who is also Chairman of the Six Villages) reported a population of 135 inhabitants, and said that the village was 200 years old. The Wongs had come from Tam Shui in Wai Chau, but the clan record was reported as having been lost. They have an ancestral hall. The village also has some Lau households.

They farm around 150 taochung of paddy fields, and exchange the unhusked rice for poorer quality rice at shops in Sai Kung Market, but even so, receive only enough for half the year's needs, and have to buy the rest. They have about 20 brown cattle. Their livelihood also comes from pig-rearing and selling firewood. There are currently 200 pigs in the village of which 15 are sows. Pigs fetch $150–160 a picul at this time. Some remittances come from 5 men working abroad, all employed in Chinese restaurants in Liverpool. Another man works in Hong Kong with the tram company; but no one is currently working in Sai Kung or Kowloon.

Besides the two boats running from Pak Tam Chung, the Village Representative said that he himself runs another boat from this village to Sai Kung. This is a 20-foot sailing sampan, with no motor, carrying between 10–20 persons and charging 30 cents one way, and 10 cents for a basket of rice or vegetables.

One interesting fact about this village is that some of the villagers are Christians. There is a Protestant chapel in the village, said to be already 60 years old. A pastor comes to hold services each Sunday. Some other village people attend the Catholic Church at Pak Tam Chung.

The villagers suffer from insufficient drinking water in the dry season, and wish to have cement and other materials to build a well. They had also applied for cement to build an irrigation dam, four months ago.

Pig-raising is obviously a chancy business. My notebook records that the people at Yau Yue Wan on Junk Bay also reared pigs on a large scale, but on my visit said they had lost a lot of money when 50–60 pigs died just before the Chinese New Year. The disease was reported to the Government Agricultural Station at Sai Kung. Some couldn't repay their loans, and others only managed to repay half, so were unable to get any more. They used to have 600 pigs, but were now down to about 100. They did not know the name of the disease.

Hei Tsz Wan (*Gazetteer* p. 185) (James Hayes)

Situated across a little bay from Wong Yi Chau, this village has its own Village Representative, called Lau Lin-tso. There are 32 inhabitants, including 11 children, five of whom go to school at Pak Tam Chung. They have 30 pigs, including 5 sows.

The Village Representative requested 30 bags of cement for a well to improve drinking supply, and 15 bags for a footpath, to be delivered after April.

Sheung Yiu (*Gazetteer* p. 185) (James Hayes)

Comprises a walled enclosure, with about eight dwelling houses within it, and a number of other old and new structures further along the path. A disused lime kiln is located beside the footpath that leads to the village.

The founding family is Wong, now in its 6th generation here, from Lung Kong, but only two households are now resident. There are also an Ng and a Lam family, here for 20 and 6 years respectively. There are 40 persons in the village population, all told.

The Wongs have 30 taochung of paddy fields and 6–8 taochung of hill cultivation. Only one old man among the Wong families is in the village, the other men are working as earth coolies in Hong Kong. Some whole families are in the United States.

Their drinking water is fetched from the stream.

The village, after its full restoration in 1983, opened as Sheung Yiu Folk Museum. Furnished with farming implements and village-period furniture, the folk museum recreates the atmosphere and environment of a small Hakka village. The original lime kiln has also been restored for public viewing.

EAST SAI KUNG PENINSULA

Nam Fung Wan (*Gazetteer* p. 186) (James Hayes)

I have a vivid memory of looking back at this peaceful hamlet, basking in the winter sun, but do not seem to have a note on our visit. It is only mentioned under Lan Nai Wan, as there is one child from there at school in the village.

Gazetteer groups the population under Lan Nai Wan, with a total population of 205, but does not elaborate.

This village was probably under what is now the High Island Reservoir.

Lan Nai Wan (750116) (Austin Coates)

Population 133. Surnames: Lee, Maan, Chow, Chan. Native place of Li Shiu Yam, Chairman of Sai Kung Rural Committee. Remote and difficult of access; there is one steep track from Pak Tam Chung and the very shallow sea approach. Repairs to the pier and the construction of a small bridge to aid children reaching the school were requested and included in my Kadoorie Agricultural Aid Association list, the only items not concerned with irrigation and wells. For two years they have also asked for a school building; the present school is in a private house and is very small. Employment problems are lessened here by the assistance given by Li Shiu Yam in finding jobs in Sai Kung for his clansmen and fellow villagers.

This village was probably under what is now the High Island Reservoir.

Lan Nai Wan (*Gazetteer* p. 186) (James Hayes)

There are 180 villagers, living in separate settlements named for each of the Li, Chau, and Maan lineages. Each of them has an ancestral hall, but those persons we met, including the Village Representative, were unclear as to whether they had clan records. There is also a Chan lineage.

The Chau and Li families came first, perhaps 300 years age, the other two later, but already for roughly 200 years.

Men working outside the village include 10 working in Hong Kong as labourers, and 3 in Kowloon, with another 5 who are seamen.

17 children attend the village school from these villages, together with some from neighbouring places. The fees are $2.50 per month.[9]

The villagers farm 250 taochung of paddy fields, with a further 80 taochung of hill (dry) cultivation, and have 65 cattle for working the fields. They exchange their rice for cheaper quality rice, in the usual way, and thereby have enough for half the year. Otherwise, they rely on pig-breeding (currently 10 sows and 100 pigs) and the sale of cut wood, together with remittances from the men working outside.

We then got onto the subject of communications. There are two private ferries, sailing boats owned by different people. They are 20 feet long and 8 feet broad. They go to Sai Kung and return within the day, and on average 6–8 persons travel on any one journey. Some other villagers make use of their own boats for going there.

Two widows were brought to my attention, both already reported to the Kadoorie Agricultural Aid Association.

Lee Lin-mui, aged 69, has a son in the United Kingdom, but has had no news of him for ten years. The son's wife is not with her husband, and left her mother-in-law some time ago. There is a grandson of 8 years, and a granddaughter of 6, presumably staying with their grandmother. The widow's house has been affected by wind and flood. Her hardship has been reported to the Kadoorie Agricultural Aid Association.

Another widow, Chau Chung Ying, aged 61, has a son who is a farmer in the village (seen), and there are granddaughters. Their mother may not be present, but the widow does not seem to be as badly off as the first.

The Village Representative said the village had applied to Kadoorie Agricultural Aid Association for materials for repair of two wells, a pier, and a bridge. These had been received, but 100 bags had "gone bad", probably due to poor storage and neglect.

This village was probably under what is now the High Island Reservoir.

Sha Tsui (774099) (Austin Coates)

Population 50. Surnames: Lee, Lau, Tang. Almost entirely Catholic, and including a few families migrated from Long Ke. Combined Catholic Church and school. The main fields of the village run in two narrow strips up valleys heading northeast, of which leads to Saiwan. The village has an ambitious project to construct a dam high up in one of these valleys and connect to the village a direct water supply by cement channel and by pipe. They are of limited intelligence and had not thought out the scheme properly. Later, a Land Bailiff will go out to help them and prepare an estimate.

This village was probably under what is now the High Island Reservoir.

Sha Tsui (*Gazetteer* p. 186) (James Hayes)

Population 57 in ten families. "A sensible and courteous Village Representative, Mr Lau."

There are families of Li, Tang, Lau, Wong and Lai, all coming from Long Ke in 1947–48 and before the war. They used to cultivate the fields here before moving over permanently, and had straw huts for storage or temporary stay-over.

They have 50 taochung of rice fields and 10–15 of vegetable land. They exchange their own rice in Sai Kung, and have rice for ten months of the year, which is very much above average if really the case.

They have eleven brown cattle, and currently eight sows.

They undertake fishing for small fish in the 5th–6th lunar months.

No one is in the United Kingdom or at sea. Two or three men work outside the village, one with the Marine Department in Yaumati, Kowloon. One man is in hospital.

There are 20 children of school age, and all go to the village school, together with a few local boat people's children, the school is big enough, they said.

Their Local Public Works needs include repairs to their small pier, two irrigation dams, and materials to build a wall at their playing field.

Chan So Kiu, a widow of 53, was mentioned. Her husband died three years ago. She looks after a boy of ten and four girls. Surely these are grandchildren?

This village was probably under what is now the High Island Reservoir.

Long Ki *(Long Ke)* (789104) (Austin Coates)

Population 54. Surnames; Lee, Lau, Wong, Tang. All Christians, either Catholic or Protestant. This is the most remote and helpless village in Southern District. Father Joseph Famiglietti, the parish priest for all the Catholic villages in this region, describes it as "hopeless", and indeed it is difficult to see what anyone, short of a multimillionaire of almost unbalanced generosity, could do to improve the lot of these people. Steep hills, of the roughest stone, surround it on three sides; on the fourth side is the sea, with a heavy shore undercurrent. Landing by boat is only possible 5 days a month, and even then it is difficult. Normal approach is from Sha Tsui over the hills. The only permanent inhabitants are two elderly men, women and small children. There are no able-bodied men. The Church does its utmost to find jobs and schooling for the young men, and some of them are at present working at the Catholic cemetery in Kowloon. It is hard for Long Ke boys to find wives, because few women will consent to live here. A former village, Sha Ti, has been abandoned, and many fields have been covered with sand; their outlines are in some cases still visible. Careful and industrious work could enlarge the cultivable area to its former size, but men are needed, and agricultural assistance for a number of years before the new fields are productive. The sand is fine and of high quality. I have recommended to the Administrator, Sand Monopoly, that it should be possible to take sand from here. A wooden aerial pier would have to be built out from the rocks at the side of the bay, with a cement path connecting it with the beach. Junks would wait beneath the pier and be loaded from overhead. The villagers asked for cement for a dam, and assured me they could manage somehow to build it (Kadoorie Agricultural Aid Association). Urgent requests for employment of village sons; several of them returned to the village for my visit. There is a Catholic Church.

Long Ki *(Long Ke)* (*Gazetteer* p. 186) (James Hayes)

Population 45, in nine families, with seven of Wong and two of Lau. The Village Representative is Wong Yuet-yau.

The grandchildren in each lineage are of the fifth generation locally, but the Laus came after the Wongs. They came from Cheong Lok County, north of Wai Chow.

They have 40 taochung of rice fields, and 20 taochung of vegetable land, mostly under sweet potato and peanuts. With rice exchange in Sai Kung, they get enough rice for four months' supply.

There are a further 15 taochung of abandoned fields. At typhoon times, salt water from the sea gets into the fields, despite a sea-wall. They cannot get the full benefit of their fields because of this salt problem. There used also to be some protective bamboos, but say this was pulled out by the Japanese during the occupation, which sounds a bit odd! Generally, the place is too windy, which means that they can get little fruit because the blossoms get blown off their trees.

There are two bays in this vicinity, Tai Lo Wan on the left, and Lo Wan Tsai on the right. Besides some abandoned land, there are 5–6 taochung of rice fields there, owned and worked by two Lau families.

There are currently seven or eight brown cattle, and some 20 porkers in the village, but no sows.

There are seven or eight children of school age who do not go to school. Of the few who do, two attend the Sha Tsui school and board there, and one goes to the Sai Wan school, also staying overnight there.

Mr Gus Borgeest is hoping to resettle some refugee farmers here at Long Ke, as he has done elsewhere in the District, and the people here say they would welcome this, as strengthening their small and isolated community. Mr Borgeest will visit again soon, together with our Assistant Inspector of Works, Chan Chi-nin, as there are a number of Local Public Works projects to inspect. These include repairs to dams and irrigation channels, improvements to the worst sections of the main footpaths, and a second well, as well as how best to prevent salt water coming onto the fields where the two local streams flow into the sea, and in part connected with the tentative resettlement proposals. There is a lot to consider, and Mr Chan plans to spend two days here at the beginning of April.

There is only one well at present, which, though not brackish, is insufficient in the dry season, when they have to fetch water from a stream.

The Long Ke Treatment Centre was established in 1981 to provide drug rehabilitation services.

Sai Wan (784129) (Austin Coates)

Population 72. Surname: Lai. A Catholic village, converted (reminiscent of Glovis) in 1950 when the Village Representative so ordered. With Tai Long and Long Ke, this is one of the most remote villages in the New Territories. Sea conditions are as at Long Ke; it is seldom possible to land provisions from a boat. The normal sure approach is by track, an endurance test for the best pair of shoes, from Sha Tsui. Carrying a load, this takes 2 hours. Allowing (which is never the case) that there is a sampan waiting at Sha Tsui and some wind, it is possible to reach Sai Kung in another 2 hours. Motor-craft can only operate at high tide, because of the shallows, and passage south of High Island is only possible on Wednesdays, Saturdays and Sundays because of the firing range. The District Office launch does it in one hour (30 minutes by rowing boat, 30 minutes by launch from deep water to Sai

Kung). A very healthy village; men of outstanding, uniformly good physique. Catholic church and school combined.

On arrival here I found a supply situation verging on the utmost gravity. The summer typhoons inundated 35 taochungs of paddy land, but the Village Representative, who had been rebuked by me for his negligence, did not report the fact until several months afterwards, after neighbouring Tai Long had received Government assistance and when I thought he might be exaggerating. On top of this the village lost its entire sweet potato crop in the frost this January. When I came to the village, they stated they had enough food for only four weeks. Immediate steps were taken to obtain rice from the Social Welfare Officer, the acreage lost having been measured. Further relief will be needed later in the year. The last 12 months have been the worst within living memory. The spirit in the village is good, and greatly reassured by our visit; it was evident that despair had not been far off. The village has suffered grievously, and will have to be nursed for the next year, until it has properly recovered. A sea-wall must be built to protect fields against future inundations, but the villagers had not worked their scheme out fully. A Land Bailiff will visit later to help them and prepare estimates.

Sai Wan (*Gazetteer* p. 187) (James Hayes)

I do not seem to have a note on this village. *Gazetteer* has "a scattered village with a population of 65, Cantonese, with most villagers surnamed Lai".

TAI LONG GROUP OF VILLAGES

Tai Long (784154) (Austin Coates)

Population 220, of whom 2 work in the United Kingdom, Surnames: Tsam, Cheung, Tai, Chan, Wan, Lee, Lam, Lai, Hui. A Catholic village, with church and school. My predecessor considered this a remarkable village, and I fully agree with him. In some ways it is the most remarkable, for high spirit, endurance and hard work, of all the villages in Southern District. It is one of the three remotest villages, the easiest access being over the hills to Chek Keng (1½ hours with a load), and thence by boat to Taipo. Its trade connections are with Taipo and Sai Kung, the latter predominating. Sea conditions are as at Sai Wan and Long Ke, and with liberal assistance from this office, a pier was constructed in 1954, the most ambitious Local Public Works job ever undertaken in this District. The pier was completed, and then unfortunately followed typhoon Ida, the worst within living memory for Tai Long and Saiwan, the pier was almost entirely destroyed. At the same time, there was an inundation of 95 taochungs of paddy land.

The Village Representative reported within a few days, measurements were taken, and in less than a month a free issue of rice was made, from the Social Welfare Office, compensating a picul of rice for every picul of paddy lost. An approach was then made to Kadoorie Agricultural Aid Association, which granted $4,000 to build another pier, the villagers having stated, to my amusement, that they would like to have a second try if I would help them. The necessary steel and cement are being assembled now, and work will start in the 5th moon, and only time of the year when the sea permits work of this kind. Chief Officer, Port Works, came last year to advise, and I hope this year we shall be successful. A pier is vital for the delivery of supplies from Sai Kung, and for the export of the villagers' rice to Sai Kung. As with most villages, Tai Long sells its own rice, which is of high quality, and buys cheaper rice for consumption at home. The supply boats sometimes get to within 30 feet of the shore without being able to land a bag. The present situation is grave, but in hand. The sweet potato crop was lost in the frost, and relief will have to be sent to this village, as to Sai Wan, for the next year, to nurse it through its convalescence from the most serious disaster within living memory. An additional worry is that because of the drought, the ruined rice fields are still laden with salt, and all efforts to drain off the salt water have been unavailing. I have examined this, and it will be impossible to restore them by artificial means. Only rain will solve the problem. This means that there will be no autumn crop on the 95 ruined taochungs. Arrangements are being made with the Social Welfare Officer, and, failing his assistance, Mr Kadoorie will be approached formally. I have already mentioned the matter to him.

The admirable spirit of this village is exemplified in its former Village Representative, Chung Man Cheong, a man of exceptional vitality, shrewdness, humour and common sense. He was succeeded in 1953 by Tsam Shek, a simpler type of man, but who is shaping well; he is much younger than Cheung, who still plays an important, though discreet part in local affairs. We have good relations with both. A good school teacher has also been an asset. This teacher, at my request, was moved to another part of the New Territories, so that he could take the qualifying course at Taipo, and I note with pleasure that the Education Department has replaced him by another competent and suitable person, a Chan of Tsun Wan. Some minor jobs on wells and dams will be put in hand now (Kadoorie Agricultural Aid Association), and there will be some bolder schemes for later in the year, after the pier is built. I have also asked urgently for the Irrigation Engineer to examine and report on opening up a large area of formerly cultivated hill land, which cannot at present be used because it is waterless. With assistance from Government or Kadoorie Agricultural Aid Association it may go some way towards solving the employment problem, which in Tai Long and Sai Wan is most serious. Special measures will be required to find outside employment for men from these villages, specially during the convalescent period. We have taken on one as a messenger on the District Office staff, and every one counts.

Tai Long (*Gazetteer*, pp. 186–187) (James Hayes)

The valley has a number of scattered villages and hamlets, including Cheung Uk Tsuen, Lam Uk Tsuen (with a Lam Uk Ha Tsuen), Tai Long itself (also called Tai Wai), Lung Mei Tau, Ham Tin with Hoi Ha, and Tung Wan.

I did not seem to have asked about the total population, perhaps because the visit took the form of a walk around the above places, in which various problems and needs were explained. However, the 1960 *Gazetteer* gives 360, and states that they are Cantonese (p. 187). Nor did I ask about lineages or ancestral halls, but from the names of two of the villages it is clear that there were at least Cheungs and Lams living in the area.

Lam Uk Tsuen has 12 families with 43 persons, and 16 cows, with 80 taochung of hill paddy and 10 taochung of vegetable land. It is reported to be the oldest village in the group, settled for about 200 years, but with no specific details.

At Tai Wan, there are two families, with 14 people and 5 cows. They want 30 bags of cement to build a well, instead of having to fetch dirty water from a stream.

To the south of Tai Long Village, there had been a Kwun Yam temple, but we met a man of 50 who couldn't recall it being there when he was a boy. [It is possible that it was abandoned after many of the villagers became Christians when the Catholic Church proselytized the Sai Kung area early in the second half of the nineteenth century.]

My main impression was of the large amount of abandoned land. I was told that the main reasons for this were: (1) the lack of water to irrigate them for rice cultivation; (2) damage from salt sea spray in fields close to the sea; (3) damage to crops caused by wild pigs in some parts; and (4) insufficient manpower. The last seems rather odd, since a population of 360 was large for any group of outlying settlements in those days, and Tai Long was certainly a remote place.

They are all growing rice where they can, but the crops only provide them with a six-month supply, and as is usual, they take most of this to Sai Kung by boat, to exchange for cheap imported rice that will last them all year round.

Despite (or perhaps because of) the amount of land being out of cultivation, these people seem very energetic.[10] This time, they wanted lots of cement and aggregate to deepen and line village wells, make a water tank, dam streams, and surface long lengths of earth channel with cement to bring water securely to the fields, make concrete footpaths, and so on.

They are poor. One man (Cheung Shek-shing) who lives with his wife and child, a baby of 8 months, in a dilapidated storehouse because his house had collapsed, was brought to my attention. He has some rice fields and 2 cows, but the repairs to his house are estimated to cost $1,500, which he simply does not have. The Village Representative asked me if he was eligible for financial help from Mr Horace Kadoorie's Fund; or else from government's Flood Relief vote, as he said the house collapse occurred in a recent storm.

Ham Tin (789148) (Austin Coates)

Population 65. Surname: Wan. A well needs repairing (Kadoorie Agricultural Aid Association). Men from this village, as from the rest, gave free labour for the pier construction, and will do likewise this year.

Lam Uk Wai (Lam Uk Wan) (786158) (Austin Coates)

Population 70, a slightly hillier village. Surnames: Lee, Lam, one family of Cheung. This village may expect to be the principal beneficiary if we can get water to their disused fields.

Notes

1. *Gazetteer* confuses these with the other Ngau Liu in Sai Kung district. This one (as Coates confirms) is clearly the one linked with Tai Lam Wu, although *Gazetteer* has it inhabited by persons of a Lau lineage.

2. A spring of mineral water, crystal clear and heavier than tap water, was discovered by engineering consultants for a proposed hill resort complex in this remote upland area in 1974. See the interesting report in the *Hong Kong Sunday Post Herald* for 17 November 1974.
3. Some years later, I spoke with an old lady in Hok Tsui Village, Cape D'Aguilar, Hong Kong Island, who had been born at Mau Tso Ngam about 1885, and had been married into the Chu family of Hok Tsui. The Chengs had been earlier at Tai Lam Liu near Siu Lek Yuen in the Shatin Valley before going to settle at Mau Tso Ngam, but she did not know when. However, she knew that her father had also been born in Mau Tso Ngam in about 1840–45.
4. *Gazetteer* confuses these with the other Ngau Liu in Sai Kung district. This one (as Coates confirms) is clearly the one linked with Tai Lam Wu, although *Gazetteer* has it inhabited by persons of a Lau lineage.
5. CARE (Committee for American Relief Everywhere) was a relief organization. We made considerable use of this agency to get help for villagers at that time.
6. See the account given in *Beyond the Metropolis: Villages in Hong Kong* (Hong Kong: Joint Publishing Company Limited, 1995), p. 25.
7. "Relief rice" was given to help out in time of difficulty, or to provide food when villagers were working on local public works. It was described to me as being "no better than sweepings" at one or two villages.
8. More people were using kerosene for cooking.
9. Under the government subsidy for building, equipping and staffing village primary schools, this charge, under the name of "tong fai", were levied on parents to provide for extras. In those days, not all families felt able to pay them, although it was usually about $2.50 per month.
10. A few years before, they had put a lot of effort into building a pier with construction materials supplied by the District Office, with strong support from Austin Coates, but it got blown away by a monster storm (Typhoon Ida) not long after completion.

Cheung Chau, 1962. Courtesy of Tim Ko.

A cheerful Hakka woman carrying grass cut from the hillside, c. 1950. Courtesy of Tim Ko.

Fishermen in Cheung Chau, 1955. Courtesy of Tim Ko.

A young girl repairing net in Tai O, 1954. Photograph by Chan Chik.

Junks in Tai O, 1956. Photograph by Chan Chik.

A cargo junk, c. 1950. Courtesy of Tim Ko.

Au Tan, Tseng Lan Shu, 1970. Courtesy of James Hayes.

Chan Houses at Ham Tin, Pui O, 1959. Courtesy of James Hayes.

Fan Pui at Shek Pik, 1958. Courtesy of James Hayes.

Ham Tin, Pui O, South Lantao, November 1994; a large granite slab bridge with modern railings. Courtesy of James Hayes.

James Hayes, 1958. Courtesy of James Hayes.

The famous "Single Cross 5 Cents Double Cross 10 Cents" Drawbridge at Mui Wo, c. 1958. Courtesy of James Hayes.

Interior of a village house in Mui Wo, 1980s. Courtesy of James Hayes.

Shui Hau, Kuk Hang, 1960. Courtesy of James Hayes.

Tseung Kwan O, 1962. Courtesy of James Hayes.

7
The Islands of Port Shelter

IM TIN TSAI *(Yim Tin Tsai)* (707104) (Austin Coates)

An island in Port Shelter. Population 283. A Catholic village with church and school. Surname: Chan. The Village Representative here is dim-witted, but there are some energetic men available, working in Hong Kong, who give advice when needed. A village of some natural beauty, frequented by artists. 40 minutes by motor sampan from Sai Kung. I note with regret that on my first visit, in 1953, the atmosphere was one of broken Government promises. I have tried to put this right. In 1954 we helped with material and technical advice on an ambitious Local Public Works scheme, constructing a causeway connecting Im Tin Tsai with the neighbouring Kau Sai Island, much larger but sparsely inhabited, where the villagers from Im Tin Tsai graze cattle and cultivate some fields. This was slightly damaged in typhoon Ida, and I am arranging for technical advisors to report on the best means of remedying the defect. The trouble here was that the village did half the work before reporting to me and asking for assistance. This was because they had been, they said, let down before by the District Office. As a result, I issued cement, to "comfort" them. But it was political cement, not so enduring from an engineer's point of view. Some years ago they also constructed a sea wall to protect their fields. It has never really worked, and why it does not work and cannot be put right baffles me, and also baffled the Public Works department engineer who inspected it with me last year. I have now asked the Irrigation Engineer to have a shot at suggestions for improvement. These, however, will have to wait till autumn. Another request is for dam improvements on Kau Sai Island. This will have to be referred to the Secretariat in due course. The fields are within the firing range; they have never before been visited by a District Officer, or even by the almost omnipresent demarcators. Thus for five years the British army has fired, while civil and military remained unaware that farmers were cultivating within the range. The place is quite safe, the farmers assured me; in five years only one bomb has fallen there. Am I to grant the cement or withhold it? I would like to grant it, if it can be arranged.

YIM TIN TSAI (*Gazetteer* p. 138) (James Hayes)

Though I remember visiting, I do not seem to have a note on this village. This is a regrettable omission as it was a big and interesting village with an old Roman Catholic church and school. *Gazetteer* gives population 260, with many villagers surnamed Chan.

KAU SAI (730062) (Austin Coates)

The remains of what was once a flourishing village. When the Port Shelter Firing Range was started, in 1950, the land-based members of the village were moved to Pak Sha Wan, on the mainland. The remains consist of two shops, permitted by me on condition the owners do not engage in agriculture or graze cattle on the hills, and a large boat population, gathered round an important fishermen's temple, devoted to Hung Shing, where an annual theatre performance is held around the 13th day of the second moon. The houses of the former land-based villagers were largely dismantled and are in ruins. The village is an attractive place, situated in the narrow strait between Kau Sai Island (Keui Island) and the smaller but uninhabited Jin Island. Miss Barbara Ward stayed there when studying the fisherfolk, and is responsible for the fishermen's almost brotherly affection for officials of the Government. Never can Chinese have been so natural and amenable in the presence of officials. Miss Ward is in fact a possible future deity; "Wah Ku Leung" is mentioned with unaffected admiration and respect. Population figures cannot be given; they vary. The safest way of assessing the fishing population is through the Marine Licensing Offices. The Cooperative and Marketing Department is naturally active at Kau Sai, and the school, run in the west wing of the temple, is their offspring. The fishermen asked if a proper school building could be erected, because not all the children can fit into the present classroom, and I have already addressed the Registrar of Cooperatives and the Director of Education on this. We might do a combined assault, with District Office assistance in material. A water supply is most necessary here. At present all water, in times of drought, has to be carried by boat from Sai Kung. A dam is thus included in my Kadoorie Agricultural Aid Association list. The site is excellent. Two piers need repairing; this can be done in the autumn. A public lavatory was also requested; and, because of the peculiar formation of the village, modesty does rather demand this. They asked for it to be built over the sea, the night soil being unimportant to them as fishermen.

KAU SAI (James Hayes)

At the time of my visit, only two families of 20 persons were living there, one keeping a store, and a pig-breeder. This was because the Hakka village on the island had already been removed five years before, after the establishment of the Port Shelter military and naval field firing range (see under Pak Sha Wan New Village). These families were servicing the 30–40 fishing boats which still used the anchorage, though probably they had been told that they, too, must look elsewhere for an anchorage. There were 16 boat people's families of about 160 persons, I was told, engaged in purse-seine net fishing, and planning to apply to the Cooperative and Marketing Department for help in forming a cooperative society and obtaining a loan, as so many fishermen were doing at that time.

Those present seemed to be ignoring the government's restrictions on the use of the area. The land families were talking about repairs to the abandoned old houses, and about being ready to bring water into the village if we would supply the materials two months from now. The boat people spoke of (repairs to) three existing piers, and the need for a typhoon shelter. The anchorage at Sai Kung was congested, and boats from Leung Shuen Wan and Kau Sai could not get into it in time of need. At present, they said, 100 boats take shelter at an inlet they called Wong Chau Tong [which may be the locality mentioned in (*Gazetteer* p. 138, as being off nearby Yim Tin Tsai) where they said a proper shelter could take two or three hundred boats, or even more.[1]

HIGH ISLAND *(Leung Shuen Wan)* (Austin Coates)

The Chinese name given on maps is Lung Shuen Wan, Provision Boat Bay, a name which is used solely in connection with the general situation of the temple next to Tung A.

Tung A (767072) (Austin Coates)

Population 118, may be described as the capital of High Island. Surnames: Chan, Kwong, Law and one family of Lee. A population part-farmer, part-fisherman. A rocky place, with houses built like European mediaeval fortresses, protecting them from the sea. The temple here is of particular interest, and it is evident from it that High Island was a fishing centre long before it became inhabited by land-based people. The present land inhabitants are in the 6th generation, and the temple was repaired during the reign of Kuang Hsu; but the bell within dates from the 6th year of Chuen Lung (1740). It is a fine bell, and its presence there at that date leads me to suppose that the temple is a good deal older than even the local people imagine. A large well needs repairing and enlarging (Kadoorie Agricultural Aid Association), and later in the year the paths round the sea wall will need cement (Local Public Works). A large project for a sort of reservoir-dam, for which I had previously agreed to give consent but had failed due to unexpected circumstances, is being diverted to Kadoorie Agricultural Aid Association. Tung A has few fields and derives part of its income from the fisherfolk who frequent it in quite large numbers. Small boats only, fishing in Mirs Bay and Port Shelter chiefly, their limits in good weather being Ping Chau and Po Toi. There are two shops, a breaming beach and net-drying grounds. The village fishermen can of course also use these facilities.

Tung A (James Hayes)

The Village Representative is Kong Fuk, who owns a shop. The land population is 101, all Hakka.

There are no paddy fields here, only about 60 taochung of vegetable land. Consequently, there are no cattle. The villagers buy all their rice at Sai Kung, and eat all their own vegetables. They have 20 sows, and currently 200 pigs in all, selling to Sai Kung and Kowloon.

Some villagers use their own boats for transportation. There is also one kaifong (public) boat, rented from others. All the boats in use here are still sailing craft, without engines. The cost of a journey to Sai Kung is reported to be 40 cents (presumably return).

The villagers make their living by trading with the boat people who are here in large numbers. I was told there are 2,700 in the district, which seems a lot, but was unable to verify it because the fishermen's representative was away at Sai Kung on the day of my visit.

In his absence, I asked the school-teacher, another Mr Kong, to let me know, by letter, of any requests they may have.

There is a large Tin Hau Temple here. It is of some age, as the bell is dated in 1740 (Chien Lung 6th year), and points to the presence of many boat people here for at least two centuries.

None of the villagers are overseas, nor any working in Sai Kung or Kowloon. However, about 25 are reported to be working in Shaukeiwan on Hong Kong Island. They are said to be hawkers and labourers, though one has prospered and runs an engine and machinery repair shop.

There are 40 children in school, but only 10 of these come from the land-dwellers, the rest from the boat people. This is rather similar to the situation reported at Po Toi O.

Pak Lap (*Pak Sik; White Wax*) (773076) (Austin Coates)

Population 85. Surname: Lau. A fine village, entirely agricultural, over a steep hill from Tung A. I was informed it was the first visit of a District Officer for many years. A healthy village; men with exceptionally good physique, like Sai Wan people. A large area under cultivation, much larger than appears from the 1:20,000 maps. The village calls itself Cantonese, but due to a tradition of marrying Hakka women it is actually of mixed race. The people are bilingual, their Hakka being rather better (and more often used) than their Cantonese. It is evident that the cartographer, on reaching this village, found it necessary to make a note to the effect that the Pak in its name meant white and not north (as in the name of the neighbouring village, Pak A). He apparently noted the name therefore as Pak Sik Lap, but subsequently forgot to cross out the Sik, which is still found, erroneously, on maps.

The village also cultivates fields on the northeast coast of the island, in the small areas of flat land known as Kon Mun, Mong Tin Kok and Nam Sha Tsai. No permanent dwellings are erected in these places, but at Kon Mun there are ruins of former small stone houses, used before the war, the villagers informed me, as watch houses. The irrigation throughout all the area cultivated by Pak Lap is inadequate but easily improvable. Attempts have been made in former years to improve irrigation works, but only small amounts of cement were used and many of the works collapsed after only one summer. There is no evidence of any Government assistance since the war, and the villagers appeared grateful for help and enthusiastic to do the best job possible. An immediate provision was therefore arranged from Kadoorie Agricultural Aid Association for building various dams in the neighbourhood of the village, and a number of other necessary works at Kun Mun, Mong Tin Kok and Nam Sha Tsai will have to be carried out in the autumn. In these latter areas all water supplies have dried up with the exception of a small spring at Hong Tin Kok.

The fields are not good enough, and are not sufficiently well fertilised, to grow the highest grade of rice, and as a result when bartering their paddy at Sai Kung the rice received in exchange varies, according to the prevailing price, from 70 to 90 catties per picul. The public works here will require a minimum of 335 bags of cement.

Pak Lap (James Hayes)

The Village Representative is Lau Chuen. He has a son and grandson working in England. The clan record is lost, and they do not know where they came from. He himself is of the 10th generation.

Population 94: but said to have been twice as many people here before the war. The families were originally Punti speakers, but through local intermarriage with Hakkas say they are now half and half.

There are 40 children of school age, but only 15 attend school, at Tung A. Whatever the reason, it is not a matter of accommodation, as the school can take up to 90 pupils.

There are five men abroad, and three working as apprentices in Wanchai, Hong Kong. Another man is with the Army as a muleteer in a Royal Army Service Corps pack squadron, and two others are also soldiers.

They have 80 taochung of paddy fields, and 40 of vegetables, growing mostly sweet potato. They get three months' supply from each rice crop, and exchange it in the Sai Kung shops. The vegetables are for household use only.

At present they have 50–60 pigs, including three sows. They had 17 cows, but four big ones died in the first half of the 12th moon, though none are sick now. They say they keep them indoors at night, all the year round. The Village Representative's nephew lost his only cow in the epidemic, making it difficult to prepare fields for planting out.

The Village Representative said he had applied to the Kadoorie Agricultural Aid Association for a $200 loan for each family through the Agricultural Department's Sai Kung office, but have heard nothing as yet.

In answer to a query whether any land was available locally for new settlers, they said that another 40 taochung could be opened for dry cultivation, but added that newcomers would not be welcome.

Local Public Works: one water-channel needs repair. I said that if they would clean it out now, materials could be given from the current year's allocation, i.e. by 31 March 1958. Repairs to a little path to the village were also mentioned. A supply of DDT is requested.

Post-registration of births can be done at Sai Kung. Apparently a team from the Registration of Persons office visits there twice a month, but there has been a problem, because the doctor in the Sai Kung clinic refuses to provide attestation for the births. This is hardly surprising when all the births in question are likely to be pre-war.

There is an interesting Yuen Tan Temple here. It was built to protect the village from the white tiger hill, the temple being dedicated, they say, to a celebrated killer of tigers! Forty-three years ago [about 1918] the building was moved slightly southwest, because the tiger had made all the village women barren. Hence the information provided during the visit: that the population is now twice what it had been previously.

Pak A (761076) (Austin Coates)

Population 209. An inhabitant aged 66 claims to be the fifth generation living here. It would appear that the village was found shortly after the present inhabitants arrived in Tung Ah, which is opposite Pak A, across the bay of Lung Shuen Wan. Surnames: Wong, Ip, Chung, Ng, Chan. Like the Tung Ah people, Pak A have their own boats (7 small ones) and are part-fisherman and part-farmer. They have very much less profit from visiting fisherfolk, however, and as a result are obliged to take agriculture more seriously. One family keeps pigs. The village cultivate a total of 70 taochungs, some of which is on the northeast coast of the island, at Nam Sha, and at Nam Sha Tsai, which area is divided among holdings owned by Pak A and Pak Lap.

The maps of this area are inadequate. The grid references of the four cultivated strips on this coast are: Nam Sha (768093), Nam Sha Tsai (771092), Mong Tin Kok (774091), Kau Mun (782088). As with Pak Lap, paddy produced is not of first quality. At the first harvest (6th moon) they seldom get more than 70 catties of rice per picul of paddy; but at the second harvest (10th moon) they usually get around 90 catties.

As opposed to the fisherfolk, who sell their catch through the Government Fish Marketing Organisation, the village fishing folk sell to fish dealers from Sai Kung, and at Tung A at any rate there is opposition to the Fish Marketing Organisation. I suspect, however, that the full story is that were the fisherfolk not under the wing of the Fish

Marketing Organisation; the villagers of Tung A and Pak A would be their middlemen. The case of Pak A, nevertheless, deserves investigation by the Director of Marketing; for Tung A I feel more certain in the view given above than in the case of Pak A.

The village has one shop. There are 20 able-bodied men under-employed, by which is meant that they spend the year fishing. It would, however, be a help if labour in town could be found for up to 10 here.

By dint of persuasion of this Office and the Education Department the three villages, Tung A, Pak A and Pak Lap, have at last agreed on a site for a High Island school. The site is crown land situated on a fine hill between Tung A and Pak A; I hope it will be built this year. An important public work in connection with the school is the building of a bridge over the large stream which in the wet season cuts off Pak A from the school site. This will require cement and steel, and I intend to take as the plan the bridge being built by the villagers of Shek Pik, Lantao Island, in collaboration with the army. At least 140 bags of cement are needed for works at Nam Sha.

Pak A (*Gazetteer* p. 132) (James Hayes)

The Village Representative here is Chung Hei, who is also a shopkeeper.

The population, all landsmen, is 196. There are five clans here, the majority Wongs. Also Ng, Yip, Chan, and Chung. The Village Representative's clan (Chungs) has a genealogy, but all the rest are said to have lost theirs.

There are 60 children aged between 6–16, of whom 28 go to school. The rest can't attend, because their families are reported to be too poor.

They farm 110 taochung of paddy fields, and another 60 of hill paddy and vegetables, but many families have no fields of their own. The average yield is only enough for two months, and they exchange all their home-grown rice for a larger amount of poorer quality imported rice at Sai Kung.

Although many vegetables are grown, only a little is sold, and not to the local boat people. No firewood is sold either, and only a little (dried) grass, presumably to boat people for breaming their craft. One would have thought that the local boat people would be the ready market for all three commodities.

There are at least 35 cattle, and around 400 pigs, including 20 sows. They go in their own boats to Sai Kung with their pigs, and hire lorries to take them directly from there to the pig dealers in Kowloon.

The water supply comes from three wells, but is insufficient in the dry season. The fishermen were said to have a free water supply, presumably from their own traditional sources.

Seaweed is used as pig food. It is collected, dried, riddled, boiled, and mixed with vegetables.

No-one is working in Hong Kong, but there are 10 in Kowloon, including 2–3 in a cotton mill, one in a weaving factory, and five in a rattan works.

Ten men are working in the United Kingdom, the majority of them in Chinese restaurants in Liverpool.

A tuberculosis sufferer, Ng Chiu from this village had recently been sent to Kowloon Hospital for an X-ray. He produced Form AB 85216 (reason entered as pulmonary TB) dated 9 October (or December) 1957, but said he had not yet heard from the hospital. He had suffered from coughing with blood for a long time. I said I would follow up.

As regards Local Public Works, our Assistant Inspector of Works must pay an early visit, as there is a lot to do here and in Pak A. The people are willing, and want to do the work, but they say it must be before the onset of the wet season.

The following note may relate to Tung A and Pak A.

A widow, Kong Kwun-mui asked for help. Aged 52, with 6 taochung of rice fields, she has a son aged 30 working as a labourer in Kowloon. She lives with his wife, son and daughter.

There is another woman, Wong Kang aged 38, with a husband who is also working as a labourer in Kowloon. They have a son, "a nice little chap of about 12", and two daughters, and have four taochung of paddy fields.

The widow said there was no well in the village, and she wanted one for herself, to be used for irrigation too. This was an odd request. I didn't record what the other had said. Neither seems to warrant help, with land of their own and a family member working outside.

She Wan *(Tai She Wan)* (763079) (Austin Coates)

Population 8. Surname: Cheung. A man of 27 is the 4th generation settled here, the clan having come from Tamsui District. The family has 10 taochungs of hilly cultivable land, but its major means of livelihood is from fisherfolk, who come here for cleaning their nets, a process which they state has to be done on the average 10 times a year. The Cheungs charge fees for their cleaning vats. There is also an inferior breaming beach with walls suitable for use as net drying grounds. No fees are charged for grass for the vat fires provided the fishermen cut and carry their own grass, which in effect means that fishermen are charged \$8 per picul of grass. The family asked for assistance in constructing a pier, but I do not favour this. Since the village is virtually a shop for fishermen, the request can be compared to that of a shopkeeper asking the Government to repair his front door.

Tai She Wan (*Gazetteer* p. 132) (James Hayes)

Only two families of Cheung left of the five or six who used to live here. They had been here for several generations. A woman of around 50 is of the third generation, or rather her husband was. The other families have died out, they said.

The remaining two families do not live here now, but their members come over from Pak A (where they now stay) to work on their fields during the day. A woman, her son, and a brother's son do this.

They have 20 taochung of rice fields, and ten taochung of vegetable land. The rice fields are all under cultivation, together with part of the vegetable land. They have nine brown cattle, and currently two porkers.

When asked why they do not stay in their houses here, they said there were no ghosts (of their dead relatives) but being so few in number they were afraid of robbers.

Two men work in Kowloon.

Nam Fung Chau (S. H. Peplow)

Nam Fung Chau, an island near Leung Shuen Wan (High Island). Nam — south, Fung — wind, Chau — an island. An island open to the southern wind and protected from the colder winds from the north. An ideal spot.

TOWN ISLAND *(Fu Tau Fan Chau)* (Austin Coates)

Is now uninhabited. Fishermen anchor regularly off the former village and use the ruins of the old houses as places of storage for tackle and gear.

NINEPINS *(Kwo Chau Kwan To)* (Austin Coates)

Are uninhabited. Fisherfolk dry nets and fish on the grassy slopes of these islands.

SHELTER ISLAND *(Ngau Mei Chau)* (Austin Coates)

Is uninhabited, and is unusable, even temporarily, due to its being in the middle of the Port Shelter Range.

Note

1. In 1964, I was fortunate to meet Mr Fan Leung, born in 1880 in Nam Tau, Po On, who had lived at Kau Sai with his father and uncle, who each had their own salt fish businesses there, from 1883 until he went to work as a seaman about 1895. There were four or five such concerns, each employing 20 men and with a large boat. Kau Sai was more prosperous then, with more boats based there and more visiting craft than later on. He attended school for three years in the Hung Shing Temple, between 1888–91, where his fellow pupils were from both the land and sea population, and their teacher was a Wai from Sha Kok Mei. The temple was repaired about that time, and he recalls the puppet opera performed at its re-opening. His mother, a Nam Tau person, had not moved to Kau Sai but resided permanently at Nam Tau.

8
The Hang Hau (Clear Water Bay) Peninsula

HANG HAU PENINSULA (Walter Schofield)

One of the quietest parts of the District was the area of the Lyemun and Hang Hau peninsulas, where the traditional ways of life were kept going, and people rarely dealt in land, or brought their disputes to me. Hang Hau peninsula was served by only two good lines of communication: the Hang Hau ferry from Shaukiwan, connecting with a launch that ran from the east side of the Hang Hau isthmus to Saikung, and a solidly built Chinese paved road running along the ridge north and south down the peninsula. On Nam Tong, by the Fat Tan Mun stands a fort with a gun on the south rampart for light artillery; this was said to have been a pirate stronghold originally. West of this fort lay some old deserted fields, which at the time of my visit were being tilled by a squatter. I suggested to him that he might become a regular landowner and start paying Crown rent, but apparently the rent suggestion frightened him off, for next year the land was deserted.

HANG HAU PENINSULA (Paul Tsui, written after he retired)

1950 was the year following closely after the changeover to Communist rule on Mainland China. People in High Command were nervous about the uncertain situation. The garrison in Hong Kong was greatly expanded, and much space had to be found not only to accommodate the Armed Forces but also to store their arms and ammunition and for training grounds for the troops. For storage of ammunition, mostly artillery shells, the whole length of the then Clear Water Bay Road from Tai Po Tsai right down to the far end of the road was gazetted as a Protected Area. Stacks of shells were placed at short intervals in open air, with only flimsy shades to cover them from the wind and rain. To gain access to that part of the New Territories everyone including local villagers had to show a pass issued by the Military Authority on the recommendation of the District Officer who would vouch for their bona fides. As a District Officer I personally found it a frightening experience being driven through a gauntlet of stacks of artillery shells. We could never be sure that some of the shells would not be blown up by accident or by the intense heat of the summer sun.

To the east of Clear Water Bay Road was Port Shelter which the Royal Navy gazetted as an area for target for firing practice. Again, it was the District Officer's duty to see to it that a warning system, understood by unsophisticated villagers, was devised to minimise the danger to them.

THE HANG HAU PENINSULA (Austin Coates)

The Hang Hau Peninsula is more uniformly prosperous than the Sai Kung region. It is predominantly Hakka, with the exception of the Cantonese village of Taipo Tsai, founded in 1629 and one of the older villages of the District, and Rennie's Mill, the refugee settlement set up in 1949 to cope with the influx of disabled soldiers from the Nationalist Army and which has subsequently attracted refugees of many kinds. The latter is at present administered by the Social Welfare Officer. Apart from Rennie's Mill there are few refugees in the region. The only village of mixed Hakka and Cantonese is Po Toi O. The region contains three of the most prosperous villages in the Southern District, Tseng Lan Shu, Taipo Tsai and Mang Kung Uk.

The present population of the region, excluding military personnel and Rennie's Mill refugees, is 3,800, of whom 60 are refugees. The population of Rennie's Mill is approximately 7,900.

For convenience the region may be divided into three sections. The first reaches from the borders of New Kowloon as far as the Closed Area and includes Tseng Lan Shu and Ma Yau Tong, villages close enough to the urban area to find grass-cutting a main source of livelihood, and Taipo Tsai, which relies largely on remittances and good pig-breeding. The second section consists of the Closed Area villages lying within easy reach of Clear Water Bay Road; most of these villages are comparatively well-off, with the exception of Sheung Sze Wan and Tai Hang Hau. The third and smallest section consists of the villages away from the main road and whose natural connections are more with Shaukiwan than with Kowloon; these include Tin Ha Wan, Fu Tau Chau and Po Toi O.

The oldest villages are Taipo Tsai, Mang Kung Uk and Tai Wan Tau, but the region also includes the Tai Miu, known in the European tradition as the Joss House, from which Joss House Bay takes its name. This is the oldest traceable religious foundation in the Southern District. Buried in thick undergrowth behind the temple is a stone tablet, the wording of which is being examined by experts, but which suggests that there has been a temple building on this site since the reign of Hsien Shun (1265–75) of the Southern Sung dynasty. Prior to that date there was evidently yet an older temple, a forerunner to this founded in 1013.

The Military Closed Area, established in 1949, is not so much of a nuisance to the villagers as might be imagined. But it would be of advantage to the people if the Army would employ more men locally.

One of the chief problems is education. Most of the villages have small schools, but it is clearly desirable to have large schools, attracting better teachers, serving several villages at once. The Education Department and this Office obtained agreement among all the Closed Area Villages in 1954 for the building of a school serving Hang Hau, Mang Kung Uk, Pan Long Wan, Sheung Yeung, Ha Yeung, Sheung Sze Wan and Tai Hang Hau, but further investigation shows that the latter villages are not satisfied and would like a school nearer their villages. It seems that two schools will be required instead of one, and the Education Department has been addressed on this. It is hoped also to have one school at Au Tau serving that village, Ma Yau Tong, Cheung Kwan O, and if possible Yau Yu Wan.

It would also be of advantage to all the Closed Area villages to have a pig food cooperative, and the Registrar has been approached with this suggestion.

HANG HAU PENINSULA (James Hayes)

The population of the sub-district is largely Hakka, with Cantonese villages only at Tseung Kwan O and Tai Po Tsai. Even Hang Hau town is mostly Hakka in origin.

HANG HAU (673032) (Austin Coates)

On the coast of Junk Bay was formerly the principal entrepot for all the most profitable local trade passing from the Sai Kung region and the Hang Hau Peninsula on the one hand, to Shaukiwan and Kowloon on the other. Tradition asserts that it was founded prior to the British annexation of Hong Kong, and that there was a time when it was bigger and more prosperous than Hong Kong. The founder families were the Yu clan of Mang Kung Uk, who are Hakka, and a Hakka family of Wong who came from Namhoi. A man of Wong aged 51 is, however, only of the 4th generation, and it seems therefore that if Hang Hau was founded before Hong Kong, it only preceded it by a few years. There is a suggestion of confirmation in this from the fact that the oldest commemorative tablet in T'ien Ho Temple dates only from the 20th year of Tao Kuang (1840). It seems from the evidence immediately available that families from Mang Kung Uk may have been using Hang Hau as a port in a small way for many years before 1841, and possibly having (as they have to this day) subsidiary houses there, but that actual permanent residence of families with no connection with one or other of the peninsula villages only started about 1830, when British shipping started using Hong Kong harbour regularly as an anchorage.

The foundation of Hong Kong as a British Colony gave an appreciable impetus to the growth of Hang Hau as a port of supply, and the place continued to flourish as an entrepot and market town until the construction of Clear Water Bay Road in 1935, as a military measure. This road gave direct access to Kowloon to many of the villages which had hitherto used the facilities provided by Hang Hau, and since that date the market town has declined in numbers and prosperity, until today it seems to be creeping slowly towards a final decrepitude. Nothing, in my opinion, can save it. Its cultivable hinterland is small, its water supply inadequate, and the sea in front of it is very shallow. A branch road from Clear Water Bay Road could be built, though for so small a population it would be more a work of charity than of development. The rise and fall of Hang Hau is a good example of the drastic changes which the establishment and growth of Hong Kong has had upon the surrounding villages and islands of Kwangtung.

There is a more or less floating population of 201, chiefly Tanka, owning single masted small boats and sampans. Ashore there are 30 families, with a total population of approximately 130, of whom 20 are Cantonese, the rest Hakka. There are also a number of houses owned by villagers from Mang Kung Uk and Pak Long Wan, but these are only subsidiary houses, and many empty spaces where there were once houses that have since entirely collapsed. The surnames are Yau, Yu, Shing, Wong, Chan, Cheung. The people themselves are concerned about the decline of the town and have suggested that, to aid recovery, the military barrier of the Closed Area should be moved back from its present site to Sheung Yeung, and that industrial concerns should be encouraged to move here. They also requested a ferry service between

Hang Hau and Shaukiwan. It would be possible to move the barrier back to Sheung Yeung if the Army would make more economical use of space for their ammunition dumps, but the move would certainly meet with strong Army opposition. Industries could move to Hang Hau if the water situation could be improved, and this will be examined; but at best there would only be enough buildable land for a few factories. Still, it might make a difference.

There has already been an application for a ferry service between Shaukiwan, Rennie's Mill and Hang Hau, which this Office has strongly supported. The Cooperative Department has recently set up a school near Hang Hau for fishermen's children, and the Rural Committee asked that a road be built between the town and the school. They also asked for the pier to be repaired; it was apparently last repaired around 1935. The Tin Hau Temple has recently been completely renovated out of the proceeds of the Chinese Temples Fund, and repairs to a bridge between town and temple were requested. As this bridge serves no really useful purpose, it is difficult to make provision for this. The existing wells are unsatisfactory, and cement will be supplied for their improvement. At the same time it is proposed to ask the Chief Engineer, Water Works, whether a supply of potable, or at least washing, water could be obtained from the large stream flowing down from Mang Kung Uk. This water is unfortunately dirtied by the upper village, and may be useless for Hang Hau. There are 5 shops and a tea-house, the owners of which, most unsatisfactorily, completely control the town committee. There is a breaming beach run by Mr Lau Yu Fuk, a shipyard owner in Hong Kong, whose local organisation sells grass and water to fisherfolk, but who himself appears to be doing nothing of any value to help Hang Hau in its present sad state.

Pig-breeding is the principal local source of livelihood. Agents from Kowloon purchase the pigs in the village and transport them to the Kowloon slaughterhouse by the lorry owned by Sheung Yeung and which serves as the main means of provision transport between the peninsula and Kowloon.

There are 30 taochungs of rice land. Vegetables grown are sold at Shaukiwan direct to stalls in Shaukiwan market. They are taken into town by Tanka boats, the boatmen charging 10% of proceeds of sale as a transport charge. Seaweed is used as pig food, in addition to other kinds of fodder and is collected from Nan Tau Sha, chiefly by women.

There is an annual theatre performance financed by land and sea people, and with free admission, on Tin Hau's birthday, the 23rd day of the 3rd moon.

HANG HAU (*Gazetteer* p. 129) (James Hayes)

I paid an early visit (24 October 1957) to the Hang Hau Rural Committee, which represents the township, together with the villages of Junk Bay and the Clear Water Bay peninsula.

Hang Hau is a portage point for the area, with ferry links to Shaukeiwan on Hong Kong Island and the post-war settlement of ex-KMT soldiers and other refugees at Rennie's Mill. The township is linked by road to Kowloon, with connecting bus services.

Judging by its several streets of old and quite large nineteenth-century two-storey shop-houses, with their decorated under-eaves, the township was more prosperous in the past than at present. Its population is reported as being around 300, with many named Shing, as at the nearby large village of Mang Kung Uk. There is a fairly large village temple here, dedicated to Tin Hau.

The pier, used by the ferries and cargo junks, needs attention. A Public Works engineer came to inspect it with my predecessor fairly recently, they said, and asked me to check on the position. In regard to materials for local public works, which the Committee can arrange to do, they requested 2-inch pipes to bring a better water supply to different parts of the town. They will take measurements, and let us know the requirement, suggesting that I ask the Water Supplies Department if they have any second-hand piping available. The Committee also requested materials to repair drainage channels.

It seems that a government scavenging coolie visits once a week, which (not surprisingly) they said was not sufficient. Apparently there is a shortage of scavenging staff at present. The sanitary and scavenging staff are controlled by the Principal Medical Officer of Health, New Territories. Pre-war, they said, the coolies came under the District Office, which paid their wages.

There were complaints about the British Army's Closed Area, and the difficulty created by its not permitting entry by private lorries to pick up locally-grown vegetables for sale in Kowloon. There is a problem because, as yet, there is no Vegetable Marketing Organization cooperative society or collection point at Hang Hau. There was also a request for the boundaries of the military camps in the area to be reduced.

We then discussed various forms of assistance, including loans and grants from the Agricultural and Fisheries Department and the Kadoorie Agricultural Aid Association, but I did not record these in detail. The Hang Hau area had been hard hit in recent rainstorms, they said, and some families still needed help.

MA YAU TONG (644032) (Austin Coates)

Population 194. Surname: Lee (11 families), Tsang (10 families), Chu (4 families). The village is in three distinct family sections. The families came together, from Sha Tau Kok; a man of 51 is of the 7th generation, giving a foundation date c. 1775.

Repairs to wells will be made under Local Public Works as soon as possible. The best access to the village is via the small road through number 7 cemetery; this road has apparently been damaged recently when a new water installation was erected by the Government. The matter will be investigated, and if necessary the Public Works Department will be asked to repair the road. Complaints were received here about refugee grass-cutters from Rennie's Mill. 5 men are working at Kai Tak.

MA YAU TONG (*Gazetteer* p. 128) (James Hayes)

Gazetteer has "Population 125, with the surnames Chu, Cheung, and Lei predominant".

The village consists of several groups of houses, with the Tsang and Chu families living in the lower, and the Li families in the upper. Each of the lineages has an ancestral hall. The Village Representative, Li Tak, who was in his early sixties in 1957–58, says that his family came here from Wo Hang near Sha Tau Kok between two to three hundred years ago, and arrived here before the Chu and Tsang founding ancestors. [A later check with the Wo Hang people did not confirm the connection, but this may be because the removal had been so long before.]

MAU WU TSAI (654031) (Austin Coates)

Population surnames: Yuen, Ng, Tsang. A remote village in the hills not far from Ma Yau Tong. Occupations similar.

MAU WU TSAI (*Gazetteer* p. 128) (James Hayes)

Gazetteer has "a very poor Hakka village. Population 65, with the surname Yuen predominant".

The village is small, with two separate groups of old houses set at some distance above and from Anderson's Road, which leads into this area, with the Haven of Hope Sanatorium and Rennie's Mill, from Clear Water Bay Road. The houses all look at least a hundred years old. In between the two groups are pigsties made of rubble stone, with five or six large lichi or lungan trees.

There are two lineages in the village, the Yuens and Tsangs, the former being in the majority. Each has an ancestral hall.

RENNIE'S MILL (651015) (Austin Coates)

Population 7,900. The place acquired its English name from a European who, in the first decade of the century, set up a flour mill here, the ruin of which still stands. His business failed and he committed suicide in one of the buildings. The map name given is Tiu Keng Wan (Hang-by-the-Throat Bay), which in 1949 was changed to Tiu King Ling (Harmonious Hill View). The area has been used since that date as a settlement for refugees from China. The first refugees were disabled Nationalist soldiers, and the sentiment of the place has remained overwhelmingly Nationalist.

The District Administration has had little to do with the settlement, which is administered by the Social Welfare Office. At present there are about 1,160 former soldiers, including families, 740 destitutes and nearly 6,000 refugees who somehow fend for themselves without Government assistance. The place is managed fairly efficiently under what calls itself a self-governing committee which settles disputes and looks after sanitation, with the help of unpaid volunteers. There is a market at which food is sold after being carried over the hills from Cha Kwo Ling. Recently a fairly good pedestrian road has been constructed connecting the two places. There are three small ferries operating between Rennie's Mill and Shaukiwan. The inhabitants earn their living as rattan workers, casual labourers in Kowloon, professional beggars, magazine sellers, matchbox makers and embroiderers. Several churches and missions operate in the area.

RENNIE'S MILL (*Gazetteer* p. 113) (James Hayes)

Not included in my notes, being a post-war settlement of a rather special nature, the main body of settlers being ex-Kuomintang soldiers and their families.

YAU YU WAN (*Yau U Wan*) (662039) (Austin Coates)

Population 119. Surnames: Yu and Chung. The Tu clan, an offshoot of the Yus of Mang Lang Uk, were the first settlers, arriving about 1850, but the size of the clan is shrinking, until now they are greatly outnumbered by the Chungs. There are 16 families of Chung and only 6 of Yu. The village, situated on the short of Junk Bay, with few fields and poor water supply, is a rather miserable place. Every man of any intelligence works outside, the village being left almost entirely in the care of the women. 5 men of Yu are working in Hong Kong or at the Jubilee Reservoir; 10 men of Chung work in the urban area as mechanics, electricians, drivers, etc. There are about 10 under-employed youths in the village. The principal local occupation is grass-cutting.

The village has three fairly large sampans used for transporting grass for sale at Shaukiwan, and 7 smaller sampans used for fishing. Fish is also sold at Shaukiwan. There are some pigs, which are sold to Kowloon; the villagers carry them up the steep hill to Clear Water Bay Road whence they are collected by lorries. The path to Yau Yu Wan leaves the road about ¾ mile south of Taipo Tsai. Women cut grass and row the sampans to Hong Kong, a job which requires 5 or 6 women per boat. The fishing sampans (crew of 3) are operated by men.

If some pipes can be obtained, it should be possible to make an inexpensive but good improvement to the water supply of the village. Various schemes for irrigation were suggested, but as all the senior members of the clans are absentees the projects had not been properly thought out, and no action can be taken until they are.

During the war there was trouble between this village and Cheung Kwan O, further round the coast. Apparently at that time there were no men at all in Yau Yu Wan, which was then raided by Cheung Kwan O people and looted of pigs, chickens, etc. The Cheung Kwan O people claim that in raiding Yau Yu Wan they were acting on Japanese orders, which may or may not be true. In any case, the Yau Yu Wan people will not send their children to the Cheung Kwan O school, nor to Taipo Tsai (with the exception of 4 children). They have adopted a sullen attitude in the matter: unless they have a school in their village, they will be illiterate. As the village is never likely to be large enough to have a school of its own, the villagers' attitude is unsatisfactory, and the Education Department will be asked to advise.

YAU YU WAN (*Gazetteer* p. 128) (James Hayes)

Population 138. There are 19 families of Chung, and 6 of Yue. The Yues came first, 9 generations ago. The Chungs have been settled for 8 generations. The Yues are from Cheong Lok in northeast Kwangtung, and the Chungs came direct from Fukien. Both have ancestral halls in the village.

My general comment was: "Quite a prosperous village, one or two new houses [a rarity in most villages in those days] and intelligent-looking children."

Of the children of school age, 21 attend the school at Tseung Kwan O, but over 10 do not. This is mainly on account of the bad footpath connecting the two villages. There is no bridge over the stream, and the children have to climb up the hillside and down again to get across it higher up. A bridge would take 10 minutes off the journey and make it safer for the smaller children, especially in poor weather. They had cut the steps in the hillside to assist the children, but this is not a long-term solution.

They said that over 10 taochung of paddy fields are owned by the Yues. [The Chungs are not mentioned in my note.] There are 40 taochung of vegetable land. Also 6 brown cattle. There are 15 men working in Kowloon or Hong Kong, but no one is abroad or at sea. The villagers catch fish all year round. There are 10 sampans, big and small, and there are stake-nets too.

The villagers rear pigs, but say they lost a lot of money when 50–60 pigs died just before Chinese New Year. The disease was reported to the Government Agricultural Station at Sai Kung. Some couldn't repay their loans, and others only managed to repay half, so they couldn't get any more cash advances. They used to have 600 pigs, but were now down to about 100. They do not know the name of the disease.

Local Public Works featured prominently in our conversation. Apart from the bridge and improving the path to Tseung Kwan O, they want materials to continue work on a seawall, on which they say they have spent $700 already. There have been recent improvements in their drinking supply. Every family now has a water tap in its house, but cement is needed for a storage tank at the new dam site.

Some of the men would like to work with Gammons, the construction company for a government project, and I said I would ask on their behalf.

TSEUNG KWAN O (S. H. Peplow)

Tseung Kwan O. Tseung Kwan — an officer, O — a cove. The shape of the rocks forming the cove has the appearance of an officer, hence the name.

CHEUNG KWAN O *(Tseung Kwan O)* (649042) (Austin Coates)

Population 237 (45 families). Surnames: Chan and Ng. At the end of the war about 20 people from here went to Malaya, and there is probably a small amount of remittances from this source. The village began as an offshoot of Nga Tsin Wai, near Kowloon City. The Ng family came first, followed after a year or so by the Chans. A man of Chan aged 42 is of the 7th generation, which gives a foundation date c. 1785. The reason given for migration was that at the time there was a surplus of population around Kowloon.

This is a needy village; there are 240 taochungs of rice land and 45 of vegetable land but the place seems to have less outside connections than other places in the peninsula. There are 24 under-employed youths, and 31 men over 40 who would be fit for city jobs of one kind or another, if they could find them. A number of them have been bar waiters, club servants, etc. Means of livelihood are rice, sweet potatoes (used for human consumption), and pigs. The pig industry should be improved, and the Registrar of Cooperatives is being approached about a pig cooperative here.

The village owns no boats or sampans, and in spite of being situated on the sea the people do not fish. They complain that before the war they did fish, but that later on dynamite fishers spoiled their chances, and the bay now holds less fish than it used to. Pigs are sold at Kowloon, and also grass, cut by women and transported on foot by them.

It is proposed to erect a new school to serve Cheung Kwan O, Ma Yau Tong, etc. After a good deal of careful study of the problems, the Education Department is recommending Au Tau (in the Tseng Lan Shu group) as a suitable site. This will be examined in due course. There is still the problem of Yau Yu Wan to be considered. This problem, and Cheung Kwan O's dispute with that village, are described above.

Requests were received to build another bridge over the big stream east of the village; in 1954 the villagers did an excellent job with a bridge above the village, without assistance from the Government. A well needs improving.

A complaint was received about the nuisance caused by refugees from Rennie's Mill cutting grass and trees in the Cheung Kwan O area, and assistance from the Police and Forestry Officer was requested to put a stop to this. Sick people here depend on the travelling dispensary passing along Clear Water Bay Road. A noticeable feature of Cheung Kwan O is that while the majority are poor, one or two are obviously quite well-to-do; the villagers admit that a number of them rent their fields from members of their own clans, and it seems likely that during the war one or two landlords profited by their clansmen's misfortunes to enrich themselves by buying up their land.

TSEUNG KWAN O (*Gazetteer* p. 129) (James Hayes)

A big village: 47 families with 250 people, according to the Village Representative. All are Cantonese. There are 214 taochung of paddy fields and 40 taochung of vegetable land, with 60 brown cattle and 200 pigs.

Many of the men work outside. They include seamen, clerks, policemen, and brick workers (at the South China Brick Works at Kai Tak), but there are still 15 unemployed youths.

There are 90 schoolchildren in the one-roomed village school. They applied for an extension last July, but it has not yet been approved. There are morning and afternoon sessions for primary classes 1–6, with 3 teachers. The teachers live outside, and travel to and from home every day. The village raises $500 towards recurrent costs. They need some cement to repair the drains at the school. The Education Department gave a grant recently for a balcony to keep off the rain.

There are a few points about Local Public Works. The villagers don't want any help with wages or rice assistance, etc. The Government Stores barge can bring the materials to the shore. We have to pay an extra 10% on the purchase price per bag to pay for coolies to load the cement.

The village has a Forestry Permit of 37,000 square feet.

Some post-registrations of birth are requested, and I received queries about how to replace lost certificates of title to land.

They have a major complaint against both the International Pig Farm and the Tseng Lan Shue villagers living alongside the two upper branches of the stream that provides their drinking and irrigation supplies.

The International Pig Farm is dumping dirt and rubbish into the stream, and some Tseng Lan Shue people are throwing dead birds and animals (pigs, ducks and geese were mentioned) into it. The Village

Representative says it is no use his speaking to the Tseng Lan Shue Village Representative. He cannot do anything about it, as he has no authority. I said I would ask a health inspector to visit.

The Village Representative mentioned that his ancestral trust ("a tso") had had to surrender agricultural land for a public purpose, and that the managers (himself included) had requested land in exchange, instead of the usual cash compensation. He added that land of building status upon payment of premium had been tentatively promised, but nothing had been done for two years. Would I look into this?

MANG KUNG UK (*Man Kung Uk*) (680027) (Austin Coates)

Resident population 580. A further people from this very prosperous village are working abroad or in Hong Kong and Kowloon. There are 111 resident families, of 5 surnames, the village being roughly divided into 5 sections. The clans are: Shing, 60 families, 310 people; Yu, 30 families, 180 people; Chan, 10 families, 36 people; Lau, 6 families, 26 people; Hung, 5 families, 28 people. Most of the village work is done by women, and there are 53 able-bodied men between the ages of 20 and 40 who are under-employed.

There are 260 taochungs of rice land, giving a population of less than 1 mouth per taochung, which means that Mang Kung Uk is the most densely populated rural area in the District. The oldest family at Mang Kung Uk is the Yu, of which a 45-year-old man is of the 13th generation. This gives a probable date of settlement c. 1635. The Shing clan (12th generation in the first decade) apparently arrived c. 1700. The Lau (45-year-old in the 9th generation) followed c. 1730; and the Hung and the Chan, step-brothers born of the same mother, came around 1775. The Shing clan owns land at Nam Wai, where they have relatives to whom they rent fields, and at Taipo Tsai, where they also rent a few fields. The later arrived clans rent fields from the older inhabitants, but also have fields of their own. For example, the Lau clan tills 28 taochungs, of which they own 20. This clan is related to the Laus of Pak Wai.

The village's chief source of income is from remittances, one of the biggest of which comes to the Shing family from its clansman who owns the Star Taxi Company in Kowloon. Pig-breeding and rice-growing come next. They cut wood and grass, but use it entirely for their own purposes, for fuel and for preparing pig food. There is one shop and one rice shop on Clear Water Bay Road. The chief request was for employment as ammunition carriers; no men from Mang Kung Uk are employed by the Army. Wells and dams need improvement, and I propose that the Chief Engineer, Water Works, should send someone to advise on these at the same time as he investigates the Hang Hau water problem. 100 bags of cement are needed for repairs to paths. The Director of Agriculture will be requested to supply a boar for service, and for the sale of sows from Agricultural Stations at normal price.

MANG KUNG UK (*Gazetteer* p. 109) (James Hayes)

This is a large Hakka village, of around 700 inhabitants in and around the area. The several parts include Lo Uk, Chan Uk, Hung Uk, Nga Yiu Ha, Pun King, Wo Tong Kong, Pak Kung Pa, and Hang Mei Ting. Wo Tong Kong is described in *Gazetteer* p. 109, as "a subsidiary village of Mang Kung Uk". The rest are not mentioned.

They have 250 taochung of paddy fields, and 130 taochung of other cultivated land, mostly on slopes, on which they grow vegetables and sweet potatoes. There are 130 brown cattle in the village.

The rice crops produce only three months' supply, which they exchange variously, in shops at Tai Po Tsai (Clear Water Bay Road), in Kowloon or here (there are four little shops in the village). There are some hill tea bushes, but not many, and we were not given any hill tea to drink here.

Some employment matters were raised. Until recently, 14 men were employed in the British Army's ammunition depot at Clear Water Bay Road. 12 of them lost their jobs. They said that of the six watchmen employed here by the Army, only one is a local person, and requested me to ask the military authorities if they would consider taking more locals. Five or six men wanted to be sanitary coolies with the government.

Postal services came up for discussion. Despite its size, the village has no mail deliveries, nor even a letter-box in which outgoing letters can be placed. All letters have to be collected under local arrangements from Tai Po Tsai Village on Clear Water Bay Road.

Schooling for the many children here is a major issue. Only 85 children aged between 6–16 are reported to be at school, out of a claimed total of 275. The old school is too small, and not even operating bi-sessionally, but construction of new, much larger, premises by the government is being delayed by a necessary revision to the building plans. Perturbed at the situation, I had made a note to push through the revised plans for the new school as fast as possible.[1]

They pressed for an electricity supply, and I agreed to speak with the China Light and Power Company, whose remit covered Kowloon and the New Territories.

Local Public Works needs included repairs and improvements to 6–7 wells, from which they obtain their potable water. Assistant Inspector of Works Chan will visit to see and discuss.

No widows had yet been registered for assistance from Kadoorie Agricultural Aid Association, and they were asked to send in their list.

PAN LONG WAN (*Pan Lung Wan*) (690027) (Austin Coates)

Population 153 (33 families), of whom 16 are working in Hong Kong as mechanics, drivers, fitters and painters. 11 under-employed youths. Surnames: Lau (10th generation in the first decade) settled around 1755. The clan came from Wong Chuk Shan, in Taipo District, and before that from Shumchun. Before the war about 30 members migrated to Sandakan, North Borneo. Rice, the principal crop, is bartered for broken rice at Taipo Tsai, 85 piculs of paddy for 100 piculs of rice for the first crop, 90 for the second crop. Pig-breeding is the second main source of income, and the village has applied to start a Cooperative Society for the purchase of pig food.

Repairs are needed to two bridges, two new wells and a bridge are required. The construction of Clear Water Bay Road has interfered with various watercourses, and the Roads Office will have to be approached with a request for channels to be constructed diverting flood water from paddy fields which every year are liable to flooding. A small road from the main road down to Nam Tau Sha has been constructed in the last few years by the Military Authorities, using a former dirt track made by the villagers. The latter are now forbidden by the Army from using the new road for transport lorries, and an attempt will have to be made to get special permission to have this order withdrawn. Pan Long Wan is subdivided into three small groups, Lo Uk, the principal section, Sun Uk, and Tung Sam Lung.

PAN LONG WAN (*Gazetteer* p. 109) (James Hayes)

Population around 170, all belonging to the Lau lineage. They said that 30 children attend school. There is enough room in the village school for more to attend, but it appears that some families are too poor to send their children to class.

There are 50 taochung of paddy fields, and 40–50 taochung of vegetable land. Sweet potatoes and water hyacinth are the main vegetable crops. They exchange their rice in shops in Kowloon, and get enough for a six-month supply.

They have 80 cows. These people bring their cows in at night, in all seasons of the year, and so (they said) do persons in all the local villages. They have 20 sows and 100 pigs at the moment.

Their Local Public Works needs include a footpath, and two footbridges over small streams. Their potable water supply is obtained from two wells, but each of these needs improving.

Over ten men are unemployed, and are seeking work.

I asked about village shrines, and was told that there is one Tai Wong shrine, and two earth god ("pak kung") ones.

SHEUNG YEUNG (689023) (Austin Coates)

Population 195, of whom 50 are working in Hong Kong as mechanics, etc. Surname: Lau. Founded at the same time as Pan Long Wan by a member of the same clan. The tradition is that the three villages of Pan Long Wan, Sheung Yeung and Ha Yeung were founded by three brothers. 24 generations ago (Yuan dynasty) the Lau clan lived in Ng Wah District, moving thence to Shumchun, Wong Chuk Shan and here. 14 under-employed youths.

This village is situated just below Clear Water Bay Road; one of its members runs the lorries used by all the villages of the peninsula for transport. This is the most prosperous of the three Lau villages. Repairs are needed to a bridge and two wells. The construction of Clear Water Bay Road has interfered with the flow of flood water, and the Roads Office will have to be asked to investigate and divert the water into channels safeguarding the fields.

Remittances, rice and pigs, in that order are the main sources of income. As with other villages in the immediate neighbourhood (Mang Kung Uk and Pan Long Wan), grass and wood cut by the villagers is used only for village purposes.

SHEUNG YEUNG (*Gazetteer* p. 109) (James Hayes)

Population 160. Mostly families of Lau, and Hakka. Some men work in Kowloon, and there are 3 in Borneo, who went there to work three years ago, and one man is at sea.

There is a village school, with 40 children in Classes 1–3. No provision for classes 4–6. The Village Representative asked for a school for "the Five Villages".[2]

They have 80 taochung of paddy fields, and (what looks like) 14 taochung of vegetable fields. Some people exchange their own rice for imported rice in the shops at Tai Po Tsai Village, others in Kowloon. With exchanges, they have enough for six months' supply.

There are 39 brown cattle, 30 sows, and 100 pigs at present.

Their Local Public Works needs concern footpaths, and Mr Chan, the Assistant Inspector of Works, will make a visit to discuss the materials required. They say they have enough potable water for their needs.

Asked about widows, and whether any needed help, the Village Representative said they would write in.

Buses run along Clear Water Bay Road, and there is a bus stop at their village. Asked if they had any problems, they said that it was OK in winter, but that at weekends and certain times of day in the summer, swimmers using the beaches near the end of the peninsula got on at the terminus for their return journey to Kowloon, which meant that local people were unable to board the buses along the route.

HA YEUNG (693015) (Austin Coates)

Population 120, of which 2 families (6 people) and 2 men live in Hong Kong; of the two men one works in Her Majesties Dockyard, the other is unemployed. Surname: Lau. Founded at the same time as Pan Long Wan and Sheung Yeung, this is the poorest of the three Lau villages, receiving far less in remittances from outside than the other two. 20 under-employed youths, from Ha Yeung and its subsidiary hamlets (see below).

Rice-growing and pig-breeding are the principal occupations, and it seems to me that a pig cooperative is needed here. The Registrar is being advised accordingly. There are 140 taochungs of rice land and 70 taochungs of hill land used for sweet potatoes. Vegetables and grass are used for home consumption. 7 out of the 28 resident families in Ha Yeung and its hamlets have sows and breed pigs. Assistance is deserved to promote this change from keeping to breeding.

The village is in many ways typical of the Southern District, presenting most of the problems which are found in greater or less throughout the District. It would be a suitable subject for a detailed social and economic survey, being sufficiently remote but at the same time within easy reach of investigators coming out by car from

Kowloon. Requests were received for repairs to two wells, 2 bridges, and irrigation channels, all autumn jobs.

HA YEUNG (*Gazetteer* p. 109) (James Hayes)

Population around 200. The Village Representative is Lau Kin-fuk, whose father had also been Village Representative. The Laus say they are in their 13th generation here. There is a Lau ancestral hall.

They have 150 taochung of paddy fields, and another 50 of hill paddy, but have very little vegetable land. With exchange, they have enough for five months' supply of rice.

We had hill tea offered to us to drink here. The villagers say they have 100 trees, not wild but planted and tended by themselves. They are old bushes, which have been there a long time. In fact, we had hill tea to drink at all the local settlements of "the Five Villages".

There are 30 cows: five were born last year.

They have currently 100 pigs and 10 sows. They sell their pigs to a pig dealer at the Ma Tau Kok slaughterhouse in Kowloon.

20 children presently go to school. Another 15 of school age don't go at present. They are asking the Education Department (and the District Office) for a new and separate school for "the Five Villages".

Some used to go to the old school at Sheung Sz Wan, but after it collapsed, they now go to the one at Sheung Yeung, which has two classrooms.

I am requested to write to the Education Department, asking them to transfer a boy who is studying in Form 4 of the Kowloon Chamber of Commerce Primary School in Kowloon to a government school there.

Eight men work in the British Army's ordnance depot at Clear Water Bay. They are on a weekly wage. Ten men work in Kowloon. Two men are at sea, one of them working as a steward.

MAU PO (699012) (Austin Coates)

Population 20, surname Lau. A subsidiary hamlet of Ha Yeung, under the same Village Representative.

TAI YUEN (699015) (Austin Coates)

Population 35. Surname: Lau. A subsidiary hamlet of Ha Yeung, under the same Village Representative. It is incorrectly described on maps as Shui Hang Hau. The acreage of both these hamlets is included with that given for Ha Yeung, and it can be seen that the area is heavily populated, 140 taochungs of rice land supporting 167 people. Tai Yuen, being situated near the sea, owns one sampan and makes some money collecting fish from visiting fishing boats and selling it locally (chiefly to Ha Yeung and Sheung Yeung).

TAI YUEN (James Hayes)

This hamlet is next to (south of) Sheung Sz Wan. It comprises six houses, and the people came 50 years ago from Ha Yeung, under which they are listed in *Gazetteer*.

This means that they probably came here in the first decade after the Lease of the New Territories in 1898.

They are Hakkas, like those in the parent village.

SHEUNG SZE WAN (*Sheung Si Wan*) (687017) (Austin Coates)

Population 85, of whom 6 are working in Hong Kong. Surnames: Lam, Lok, Wong, Lau. The oldest families here are the Lams and the Loks, who came together. A man of Lam aged 58 is of the 7th generation, giving a probable date of settlement c. 1775, about 20 years after the three Lau villages nearer town. The Lam clan are from the Po Toi District of Fukien. The Loks are a branch of the same clan at Pak Kong. 8 under-employed youths.

Rice growing, pig-keeping and fishing are the chief occupations. Assistance is needed in forming a pig cooperative. Paddy is bartered at Taipo Tsai. The village owns 3 small boats; angling is the method of catch. The boats are also used for collecting fish from visiting fisher boats and selling locally at Ha Yeung and Sheung Yeung.

There is an unsubsidised school, with a good teacher, Lau Kai Lui, a graduate of Kwok Man University, Canton. [The school problem here is dealt with in the introduction to this section.]

Irrigation channels need improving, and we were unable to supply cement last year when asked. This must come up in the autumn in the Kadoorie Agricultural Aid Association list.

SHEUNG SZ WAN (*Gazetteer* p. 109) (James Hayes)

Population around 90 persons. The Village Representative is Lam To-shing.

The Lams are in their 24th generation, from Fujian Province, but clearly settled somewhere else, before coming here. They say they are Punti (Cantonese) but having lived among Hakkas for so long they have absorbed their speech and ways, and are now "half and half", as they put it. There is a Lam ancestral hall.

An interesting point came up about the name of this village. Fifty years ago, they say (that is, about 1907) it was changed to the present one from Chap Shu Wan. I should have asked why. I imagine it has to do with finding a more harmonious sounding name.

They have 60 taochung of paddy fields, and about 50 taochung of hill paddy. They called it shan che, meaning sloping hill land. With exchange, they have a six-month supply of rice. They eat their own vegetables.

They have currently 15 sows and 100 porkers (young castrated male pigs, being fattened for slaughter; I had to consult the dictionary for this one!).

They sell to pig-dealers at Ma Tau Kok in Kowloon, where there is a large government slaughterhouse.

25 of their children attend the school at Sheung Yeung, along with others from the adjoining settlements of Tai Hang Hau and Mao Po.

There was an old school in their own village which has now collapsed. It used to take 40 children.[3]

One man is working locally for the British Army's Ordnance (ammunition storage) Depot along Clear Water Bay Road. Three are seamen, working for the Blue Funnel Line, and three work in Kowloon.

Local Public Works needs included materials for making the path between this village and the school at Sheung Yeung into an all-weather one, to benefit the children attending there. There was also discussion about repairs for a ruined seawall, which had protected their rice fields.

TAI HANG HAU (701014) (Austin Coates)

Population 108, of whom 5 work abroad and 5 in Kowloon. Surname: Leung. A man of 80 is of the 5th generation, giving a probable date of settlement c. 1800. The family came from Tungkun District, to Ma Ku Nam, in Taipo District (Northern Sai Kung region), and thence here. This is a healthy village with young men of good physique, and the most urgent request was for employment as ammunition carriers in the peninsula. Few of the under-employed, of whom there are 16, wish to go abroad or to Kowloon.

74 taochungs of rice land and 43 taochungs of hill land growing sweet potatoes. This is a needy village of good spirit, and if work is available from the Army it should have priority over other peninsula villages. A pig cooperative is urgently needed here, and the Registrar of Cooperative Societies will be advised of this.

Cement required for urgent repairs to two wells was granted immediately by Mr Kadoorie on request, and sent within a few days of my visit. Road repairs and irrigation channel improvements will be needed in the autumn.

TAI HANG HAU (*Gazetteer* p. 109) (James Hayes)

There are 120 people here, all Hakkas, and, they say, more than before the war. The Village Representative is Leung Shing.

According to one of my later informants, his (the Leung) lineage was originally from Tung Kwun county, and came first to Pak Kong, and then branched out to this village. Born in 1877, he was of the 5th generation to live here. They may originally have been Cantonese.

They have 70 taochung of paddy fields, and 15 taochung of vegetable land. They exchange their rice in Kowloon, and get enough for 4–5 months' supply. Though they grow their own vegetables, they still have to buy a small amount. There is a stream on the eastern side of the village, and the vegetable fields are on the other.

They have over 10 sows and 140 pigs at present, selling them all the latter at the Ma Tau Kok slaughterhouse.

They used to fish in local waters at certain times of the year, from sampans, but since the Port Shelter Field Firing Range was opened in 1950, can no longer do so, and nowadays there are no boats left in working condition.[4]

Only six children from this village go to school, at Sheung Yeung. Prompted by my query, as to why this number is so few, in what is quite a large settlement, they said that 10 children used to attend the school at Sheung Sz Wan before the old school building collapsed. Currently, over 10 children have no schooling.

As many as 10 men had worked for the British Army in this area, but they are now being retrenched. The Village Representative said that a letter had been sent to the District Office about this. Another 4–5 are working as labourers in Kowloon. There is no one at sea.

They requested Local Public Works materials to improve irrigation channels.

TAI WAN TAU (701999) (Austin Coates)

Population 158. Surname: Lau. The last village within easy reach of Clear Water Bay Road, from which it is situated 10 minutes walk down to sea level. This clan came from Ng Wah district and is not related to the Laus of Sheung Yeung, etc., 14th generation in the first decade, giving a foundation date c. 1650. One of the older villages of the peninsula. The reason given for migration was fear of bandits in Ng Wah. 4 men working in the urban area as mechanics, etc., 3 in North Borneo; 1 in Shanghai. 16 under-employed youths.

There are not many rice fields, but the villagers do well with hillside cultivation. Pig-keeping has been the chief source of livelihood, and the village has recently gone over to breeding. Due to the fact that they buy pig food at high rates of credit in Kowloon, a cooperative is required here, and the Registrar will be asked to send someone to advise. Pigs are sold to Kowloon, being collected at the top of the road.

The village also makes money from trippers holiday-making in Clear Water Bay. The village asked for an extension of the main road down to Tai Wan Tau, stating that if they could be subsidised at the rate of $2 per day for men and $1 per day for women they could carry out the making of a dirt road themselves.

TAI WAN TAU (*Gazetteer* p. 110) (James Hayes)

Population given as 165, inclusive of the smaller settlements located on the track which leads up to the terminus for the bus services to Kowloon.

Many urban people come here in summer, to swim at the nearby beaches.

In 1957, at the time of my visit, a small military camp with Nissan huts was located here.

Gazetteer gives 160 persons, with the surname Lau predominant.[5]

TAI O MUM (699004) (Austin Coates)

Incorrectly described on maps as Tai Au Nam, a subsidiary of Tai Wan Tau, situated close to Clear Water Bay Road. Population included in that of Tai Wan Tau. Surname: Lau.

TAI O MUN (James Hayes)

Is located near the terminus. It has 7–8 houses, and is said to be over 100 years old. There are 23 people, and three children attend school here, together with 2 from Tsiu Wo, another little hamlet further down the track. One more hamlet is located at nearby Ngau Lan Wo.

The main village has 50 taochung of paddy, and another 20 taochung of hill paddy. There are 25 brown cattle. They exchange their rice in Kowloon, and have enough for four months. They grow their own vegetables, and have no need to buy any.

Pig-breeding is the major occupation. The villagers say they lost a lot of money when many animals died just before Chinese New Year. The disease was reported to the Government Agricultural Station at Sai Kung. Some couldn't repay their loans, and others only managed to repay half, so they couldn't get any more cash advances. They used to have 40 sows and 600 porkers, but were now down to about 100 pigs all told. They do not know the name of the disease. Porkers can be sold for $140 a picul at present to the pig dealers at the Ma Tau Kok slaughterhouse.

There are no village boats or local fishing at the present time.[6]

There are 27 children in the village school, from the village and its hamlets, but there are more than 10 others who should be attending in the first year of primary education, but are not in school. Yet the school can accommodate 45 children, they say.

When I mentioned the seemingly overall shortage of school places in the Clear Water Bay area, they commented that they did not think children from other places would come. This was probably due in part to distance and terrain, but perhaps also to inability, or unwillingness, to pay the small monthly charges.

Their Local Public Works needs were for materials to surface a little path to the school, a "car park" at the head of the track, near the bus terminus (for use by the lorries taking pigs to market), and for some drainage channels.

CHIU WO (697003) (Austin Coates)

Not marked on maps, halfway down the hill, more or less between Tai Wan Tau and Tai O Mun. Population included in Tai Wan Tau. The surname is a comparatively rare form of Chow, and the village was founded about the same time as the other two. The story is that one of the Lau founders married a girl whose husband had died shortly after their marriage and who was already with child by him. The child born was a son, and was brought up in the Lau house; but he and his descendants who inhabit the village have always borne their blood name of Chow, although included in the Lau clan arrangements, and being entitled to a one-third share in the Lau clan property.

PO TOI O (*Po To Au*) (703981) (Austin Coates)

Incorrectly described on maps as Po To Au, a multi-clan village, partly Hakka and partly Cantonese. Surnames: Tang, Cheung, Fong, Lee, Lai, Chan, Leung, Kwok, Shek, Ko, Po, Wong, Chung, Lau. Population 447, of whom 76 are land people and 371 boat people. The first-known inhabitants were Chim, who have now died out; they were followed about 30 years later by Tang (Cantonese from Po On District but not related to the Tangs of Yuen Long). They evidently came about 1885, which gives a probable date of settlement of 1885.

Po Toi O has a splendid natural harbour for small boats, and it earns its livelihood in the same way as most of the islands do, by the peculiar relationship of extortion which exists between settled fishing communities and the land people who serve their needs. The fishermen here, operating small single masted boats, use only local waters, and have no seasonal movements.

There are less than 5 taochungs of fields in this rocky place, and therefore the most has to be made out of the fishermen. The villagers deny that they sell grass and water to the fishermen, but subsequent enquiries revealed that they sell both, and also rent vats for boiling nets. Grass is sold at $8 per picul, about double its market value, and the village controls the beaches both in the harbour of Po Toi O and over the other side of the low hill separating them from Joss House Bay, in front of the Tai Miu (the Joss House). The Lau family control the beach at Po Toi O, the Chans at Joss House Bay. They propose to spend their $800 reward money, which is now being obtained for them, for war services, on the improvement of water supplies, probably, I fear, to improve water sales.

Pig-keeping is the next source of livelihood. Because the slaughterhouse at West Point is considered too far from the village, pigs are not taken, as one would suppose they should be, by boat from Joss House Bay to Hong Kong. They are shipped by sampan to Tai Wan Tau, where the Po Toi O people pay the Tai Wan Tau people $2 per head to carry the pigs up to Clear Water Bay Road for transport by lorry to Kowloon.

It takes a villager at least 2½ hours to go from Po Toi O to the District Office.

The school is in very bad condition, being over 60 years old. A small school might be built here, and the Education Department will be approached about this; the present fees, $3.50 per month, are however considered by the teacher to be too high if the fisherfolk are to be encouraged to join the school. Without fisherfolk children, however, a new school would not be warranted; the fees in this case might be lowered to $2.

Rifles and medical facilities were requested. The village received a commendation from the Commissioner of Police in 1954 for the help given in arresting several tens of returning deportees.

Po Toi O has a small Hung Shing temple with what appears to be a very old wall; the date is obliterated, but it suggests that the present land inhabitants are by no means the earliest residents. A small theatre show is held annually in the first week of the 5th moon.

PO TOI O (*Gazetteer* p. 110) (James Hayes)

A particularly interesting place. The village representative (a landsman) said that the greater part of the population were long-established boat people using the anchorage as their home base. There are approximately 500 of them, compared with the 50 local Hakka villagers.

In response to questions about the local fishing, I was told that between March to July, about 100 boats fish at the Ninepins and off Waglan (where the lighthouse of that name is located). Some bigger boats go further out to sea, large vessels over 60 feet long with three masts. The implication is that these are all boats based on this place.

The local temple is dedicated to Hung Shing, a popular deity among boat people. His annual birthday is marked by the customary opera performances, and the committee responsible for making all the arrangements comprises two boat people and one villager.

Tenders are currently being called for a new government-subsidized primary school, to replace the existing school premises in one of the side-rooms of the Hung Shing Temple. Due to its small size, the school has only 20 pupils, of whom most come from the boat population.

Two facts stand out from my notes. Firstly, that in this more or less shared settlement of land and sea dwellers, the floating population has a part to play in temple business, including the arrangements to celebrate the annual festival. Secondly, that some of the boat families, albeit not many, are sending their children to school.

TIN HA WAN (*Sheung Lau Wan* [686987] and *Ha Lau Wan* [683984]) (Austin Coates)

Population 96 (45 & 51). The former is incorrectly marked on maps as Sheung Lau Wan, and the latter as Tin Ha. Surnames: Ip, Sit, Sheung, Chan, Lee, Fung, Tsang, Chu, Wong, Lin. A number of boat people anchor here regularly, but are included in the floating population of Hang Hau.

Tin Ha Wan is situated on the coast of Junk Bay, in a fairly deep harbour conveniently protected by the small island of Fu Tau Chau. A small ferry sampan operates between here and Shaukiwan; a hill track (45 minutes with a load) connects with Clear Water Bay Road. Ha Lau Wan is the older of the two villages; it was founded by the Sit family, of which a 40-year-old man is of the 4th generation, giving a foundation date c. 1865. Sheung Lau Wan was founded shortly afterwards by a branch of the Lau clan of Pan Long Wan, which still owns three houses and some land in this village. The villagers asked that the Police allow them to carry more passengers on their ferry, a request common to all villages with a ferry service of this kind.

Sources of livelihood are grass-cutting (sold at Shaukiwan), pigs and ducks. Tai Hang Hau people cut grass on the hills near the villages and sell to them.

There is no school. Some of the children with relatives at Pan Long Wan live there and attend its school, but it is not possible for others to go there daily; the whole walk takes 1½ hours. A possible solution might be for a school serving the combined needs of Tin Ha Wan and Fu Tau Chau, but the latter place is not very interested. The Education Department will be asked to investigate.

TIN HA WAN (*Gazetteer* p. 110) (James Hayes)

Tin Ha comprises Upper and Lower Lau Wan, but whilst the bay is divided the settlements are under one name. Population is around 150. The Village Representative is Sit Ping, and the Assistant Village Representative is called Yip Kan.

There are 30 taochung of rice fields, and some ten taochung of hill paddy, with ten brown cattle. Only two families have rice fields. The others gain a livelihood by breeding pigs, raising ducks, and grass-cutting. There are currently 25 sows and 100 porkers.

They sell all their produce and livestock, and exhange their rice, in Shaukeiwan. Since no ferries of any kind call here, or at Fu Tau Chau, they operate a hired boat which is 37 feet long and 12–13 feet wide to transport their goods. It can carry up to 250 piculs.

Children of school age number around 40, with some others on nearby Fu Tau Chau. At present some bigger children walk to Hang Hau/Mang Kung Uk for their schooling, and some are taken to Shaukeiwan. This, of course, has been a most unsatisfactory situation, but a school has been approved for the village, a site has been found, and construction should begin soon.

They would like to have Local Public Works materials to open an all-weather footpath to the new school, and want to build a small pier. The Village Representative says they will send us a letter about these projects, and also about their proposal to buy a motorized junk.

Finally, they complained that their applications for land and houses on Crown land permit had not been actioned so far by the District Office. This is, I think, part of the backlog which has been building up in the past few years, owing to increased demand in some parts of the District (notably in Tsuen Wan) and inadequate staffing.

TAI CHIK SHA (683001) (Austin Coates)

Not marked on maps. 2 Hakka refugee families have settled here since 1953. The population is probably not more than 6.

FAT TAU CHAU *(Fat Tong Chau, Fo Tau Chau, Fu Tan Chau, Junk Island)* (S. H. Peplow)

Fat Tau Chau. A small Island between Lyemun and Sai Kung. Fat Tau — a Buddhist's head, Chau — an island. The name is so given because the island, according to the Chinese, is similar in shape to the head of a Buddhist priest.

Ku Shan Wan. Ku — ancient, Shan — a hill, Wan — a bay. Ancient or sacred hill bay. There is a very old temple here.

Fat Tong Chau has now been joined to the mainland by reclamation in connection with the Tseung Kwan O landfill and industrial estate.

FU TAU CHAU (675986) (Austin Coates)

Population 23. Surname: Ip. The clan comes from Tamsui District, a man of 46 being of the 3rd generation, which gives a foundation date c. 1885.

The village, which is on a small island, as the name suggests, has one sampan operating irregularly as a ferry between here and Shaukiwan. Sources of livelihood are grass-cutting, pigs, ducks and chickens. The ducks are excellent. Shaukiwan is the principal market.

The Ip clan is intermarried with the Laus of Pan Long Wan, and most children stay at Pan Long Wan with their in-laws while attending school there. There is no great interest in a school at Tin Ha Wan, though the villages did ask for the services of a teacher. This might be examined, but it is unlikely that the Government would agree to providing a teacher for so small a community.

The Ips have no one working abroad, and only one man working as a driver in Hong Kong. Nevertheless, this is a comparatively prosperous place, with well-built houses.

FU TAU CHAU (*Gazetteer* p. 110) (James Hayes)

This has to be Fat Tau Chau, whereas the inhabitants use Fu (axe) instead of Fat (Buddha).

"This is a nice little island, well-wooded in places," I wrote in my note. But unfortunately there is not sufficient water to irrigate land for rice farming, and in consequence they only have vegetable fields, amounting to some 20 taochung. All are currently under a sweet potato crop.

Population 34. The Village Representative is Yip Sin. We met him and his younger brother.

They gave us hill tea to drink, harvested from their own bushes on the island.

Besides being farmers, they are also pig breeders, with currently 11 sows and 80 porkers. They had also kept ducks, but they had all died, and so discontinued this line of livestock breeding.

One man is a lorry driver in Shaukeiwan. No one is working at sea. The children of school age are not being sent to school for the present. Travel to and from the island has to be done by sampan, and they would like the Hongkong and Yaumati Ferry Company to provide a service, but there is no ferry pier and the demand is probably insufficient to warrant one, even if combined with a service to Tin Ha Wan. They have, therefore, to hire a private junk or motorized sampan, at need.

A stream embankment was breached in recent floods. They want materials to repair it, and also to surface a footpath.

Notes

1. Another large village, Tseng Lan Shue, along the road from Kowloon, was in a similar plight, with only 90 children out of 132 aged between 6–16 attending an old school, in morning and afternoon sessions. See under Tseng Lan Shue.
2. These are Sheung Yeung, Ha Yeung, Tai Yuen, Sheung Sz Wan and Tai Hang Hau. Tai Wan Tau and its hamlets have a school of their own. As readers will see, the provision of school places was grossly inadequate, and there was an urgent need to improve the situation.
3. This must have been a very old school. One of my oldest informants, interviewed in 1964, who was born in nearby Tai Hang Hau Village in 1877, attended this school beginning in the late 1880s, between the ages of 13–16 years by Chinese reckoning.
4. When my informant was a boy, in the 1880s, the village was heavily engaged in local fishing, from small boats, and by stake-nets from the shore.
5. In 1964, I interviewed men from the 7th, 8th, and 9th generation of the Lau lineage living at Tai Wan Tau. The oldest was 80, born in 1885. The lineage had come from Cheong Lok County (near Foochow) in Fujian Province, and came directly to Tai Wan Tau.
6. But my elderly informants would tell me in 1964, that they had fished from boats and by stake-net up to the Japanese occupation, when, they claimed, their boats were confiscated and taken away.

9
Author Biographies

WALTER SCHOFIELD

Walter Schofield (1888–1968) was a cadet officer in the Hong Kong civil service from 1911 to 1938.

After serving in various posts, including that of district officer, South, he was first police magistrate in 1931–33, and again in 1934–37, it being customary for such posts to be filled by cadets up to 1941.

He was a keen amateur archaeologist who conducted digs at Shek Pik on Lantao Island in 1937 where he found six burials under a cultural layer which yielded bronze artifacts. A gulley adjacent to the site contained the remains of six sandstone moulds for casting socketed axes.

He was active in identifying and mapping prehistoric sites and as an expert in geology he suggested that the change in distribution of bronze age against earlier sites reflected an increase in population and adoption of rice agriculture.

He was particularly interested in the sequence of sites in raised beaches noting the stratigraphical distinction between so-called soft geometric pottery with stone artifacts and the later hard fired pottery with bronze remains.

Investigation of the Lung Kwu Chau (Tung Kwu Island, north of Tai O) Archaeological Site was conducted by him from 1925 and others later focused on the low southern isthmus. They found archaeological remains including coarse corded pottery, polished adzes, polishing stones, soft and hard geometric pottery, plain and incised chalky pottery, and human burials.

The deposits on the site belong to several major cultural periods of Hong Kong history, including geometric pottery of the Bronze Age, celadon of the Tang and Song dynasties, and blue-and-white porcelain wares of the Qing dynasty, and glazed Han/Six dynasties pottery (*JRASHKB*, volume 9).

Schofield was also interested in the later history of the area, and in particular contributed to our knowledge of the Kowloon Peninsula, especially in the area of the Sung Wong Toi and the Kowloon Walled City. He copied historical inscriptions and noted other finds, as recorded in his notebooks. These were left to James Hayes, and were passed to the Hong Kong Museum of History some years after his death. They may now be held by the Antiquities and Monuments Office. Schofield and Hayes had frequent correspondence, and Schofield's letters are deposited in the Hong Kong Public Records Office.

All Schofield's work was done pre-war, during his government service. He retired to England in 1938.

The following obituary was printed in the 1969 *Journal of the Branch of the Royal Asiatic Society*.

> With deep regret we record the death of Mr Walter Schofield, one of the pioneer figures of Hong Kong archaeology. He died on 8th December 1968 at his home in England at the age of 80. His opportunities in this field stemmed from 27 years' service in the Hong Kong government (1911 to 1938) during which, as a Cadet Officer, he held important posts in the administration and the Judiciary.
>
> For the last ten years of his service he was, along with Professor Shellshear, Dr Heanly and Father Finn, one of the active band of capable enthusiasts who ranged far and wide in the Colony charting well over a hundred hitherto unknown sites of neolithic man and establishing a corpus of written work for their successors to build upon. Walter Schofield's work in the field and at his desk was of the highest order and is typified by the careful account of the excavations at Shek Pik which he conducted with Professor J G Anderson in 1938 (see W Schofield, 'The Proto-Historic Site of the Hong Kong Culture at Shek Pik, Lantau, Hong Kong', *Proceedings of the Third Congress of Prehistorians of the Far East*, Singapore 1940).
>
> Mr James Hayes, who knew him well and corresponded frequently with him in the last ten years of his life, writes: "Walter had a lifelong interest in Hong Kong ethnography and archaeology which was as keen at the time of his death as at any time in his life. When he died he had just finished writing up from his notes an article on his pre-war excavations at Tung Kwu, which displays anew his passion for accuracy and keen powers of observation.
>
> He applied to Hong Kong archaeology a profound knowledge of geology, which he also practiced widely in the Colony and indeed, as his notebooks testify, anywhere he had the chance, and this added to his stature and reputation as an archaeologist. He was also a sinologue and Chinese scholar, and much else besides, such was the range of his interests, however, knowledge and abilities. No one could fail to recognise in Walter Schofield the gentle but firm wisdom of the true scholar, and one, too, whose knowledge and capacity increased rather

than diminished with age. It was truly a privilege to know him and share his friendship."

Dr S M Bard writes: "W Schofield for me will always be associated with what is best in the British Colonial Administration. He combined a deep interest and concern for the welfare of ordinary people with an enquiring mind and practice of scientific methods. As one of the pioneers of archaeological research in Hong Kong region before the war, he stands out prominently for his scholarship and his clear, precise style of writing. Although his archaeological and historical research was varied and wide in range, he is perhaps best known for his report on the Shek Pik site which he excavated in 1937. With his usual precision, he called the site 'protohistoric' rather than 'prehistoric', emphasising thereby, what we now know, the importance of early historic contacts in this area.

Although he left Hong Kong many years ago, he retained a very keen interest in this area to the end. His memory and knowledge of the place were astonishing. We had kept up a regular correspondence, and only six months ago he wrote to me at length on the possibility of Palaeolithic artefacts in Hong Kong, describing in detail an old find on Tsing Yi Island which he thought was of a Hoabinhian culture type. His death takes away one of the last of that brilliant small group of pre-war archaeologists to whom our debt can never sufficiently express."

PAUL TSUI KA CHEUNG

Paul Tsui Ka Cheung was born at 81 High Street, Sai Ying Pun, Hong Kong Island on 5 November 1916. He was the fifth child of Peter Tsui Yan Sau and Chin Kang Tai, who had been married in 1907. He was baptised a Catholic at St Anthony's Church on 26 November with godfather George Bristow.

Peter Tsui came to Hong Kong as a boy with his father Tsui Wang Mo, a Hakka from Tsim Hang Village, Ng Wah County in Kwangtung (Guangdong) Province. Wang Mo operated a beach kiosk at Cable Bay in Hong Kong (now Cyberport).

Peter Tsui attended school at the Catholic De La Salle Brothers St Joseph's College. Kang Tai was a staunch Protestant, educated and raised by the missionaries of the Basle Mission.

Peter later became a teacher, first at Ying Wah School, then at St Joseph's College. In 1918 he resigned from his teaching post to establish his own private school. At this time he sent his wife and five children back to Tsim Hang Village.

Wah Yan College commenced operation at 60 Hollywood Road on 16 December 1919. The school was successful and grew steadily. A Kowloon branch at 70 Portland Street was opened in 1922. In 1932 Peter resigned as headmaster and handed Wah Yan School to the Jesuits.

In 1930, Peter invested in the Pak On Rubber Estate at Jesselton, British North Borneo (now Kota Kinabalu, Sabah, Malaysia).

During the first four years of his life Paul did not see his father, only becoming aware of him when his mother, brothers and sisters moved back to Hong Kong from Tsim Hang Village in 1921. Then followed many changes of home: (a) Noel Bothelo's house in Chancery Lane, at the back of Victoria Prison on Old Bailey Street, (b) the rear portion of the first floor of 33 Mosque Junction, (c) the ground floor flat at 2 Robinson Road, (c) a servants' quarters outhouse at the same premises (a school building vacated by St Joseph's College), (d) a flat on High Street, and (e) a ground floor flat in Elegant Street (off Caine Road).

In 1924 Peter bought from his brother-in-law, the Rev Pang Lok Saam, part of an orchard and two paddy fields in Shung Him Tong Village, Fanling. On this site he built a house "Shek Lo" and the family moved there in February 1925.

Paul's education started at a special pre-primary class at Pui Ching School for Girls in Caine Road operated by the Precious Blood Sisters under the overall supervision of the Canossian Sisters. There he learnt to read and write, to say prayers and how to behave in church. After the move in 1925 he attended Tsung Him Primary School in Fanling. From 1 December 1928, he transferred to Wah Yan School on Hong Kong Island, travelling to and from Fanling by train. To ease the travel, from August 1931 to the end of 1933 he attended the Kowloon Branch of Wah Yan College. Amongst other subjects he learnt Putonghua and Chinese classics.

In 1930 Peter brought home a young woman Chiu Ying and their son and asked Kang Tai to accept her as his concubine. Paul's mother, in shock, anger, fury and despair, could not accept the situation and the concubine and son moved elsewhere to a house opposite the walled village of Ma Wat. Kang Tai subsequently lost two children from meningitis and pneumonia as a result of the trauma. Thereafter, the two families essentially lived separately and Kang Tai had no more children beyond the twelve she had borne so far. Peter was away for the silver anniversary in 1932 of his wedding with Kang Tai, which caused her much grief. However, she raised one of the sons of Chiu Ying who was left with her to connect the two families.

The families came together again during the war in 1941 when they all escaped to Tsim Hang Village in Ng Wah County, Kwangtung Province. They stayed there until the end of the war when Peter and his concubine moved with their family to the rubber estate in North Borneo. They lived between there and Hong Kong for the rest of their lives, begetting seven more children making a total of fifteen. Chiu Ying moved back to Hong Kong after the death of Kang Tai and lived at Shek Lo.

In 1935 Paul sat for the School Leaving Certificate (Public) Examination, the Hong Kong Government Junior Clerical Service Recruitment Examination and the Chinese Maritime Customs Recruitment Examination in Canton (Guangzhou), but failed to pass any of them. He therefore returned to Wah Yan College to repeat Year One.

Halfway through the year, Paul was invited by the bishop to a teaching job at Yang Ching School in Rabaul, New Britain. He sailed there from Hong Kong on TSMV *Neptuna*, a journey taking three weeks. He taught catechism, Chinese, physical education, outdoor activities at the school at a salary of A$12 per month for fifteen months, leaving following a major volcano eruption.

On return to Hong Kong in 1937, he passed the entrance examination for Lingnan University in Canton and enrolled in an economics and political science course at a refugee campus established by the university at Fanling. At the same time, he sat and passed the Hong Kong University Matriculation Examination. This allowed him to switch to the University of Hong Kong (HKU) in September 1938. Lacking sufficient funds, he took a pupil teacher job at Wah Yan College, offered on the condition that he attended the evening teachers' training course at the Government Technical Institute. Happily, the rubber estate was prospering and soon thereafter Paul's father agreed to pay for Paul's education at HKU.

At HKU Paul started studying letters and philosophy, but after Year One he switched to Chinese literature. Initially he resided at St John's Hall, but later moved to May Hall which was quieter.

In December 1941, the Japanese invaded Hong Kong. Paul was awarded a wartime degree by HKU. He travelled to Fanling to find that the family home there had been trashed but fortunately stored grain and rice had been left. Some while later he moved out of Hong Kong to rejoin the family at Li Long, originally a seminary for the training of pastors and preachers for the Basle Mission.

Anxious to contribute to the security of the motherland, Paul travelled on to Kukong (now Shaoguan), the wartime capital of

Kwantung Province, at the confluence of the East River and the West River north of Canton. His father had given Paul CNC$300 as "lifeline capital". In Kukong, he was interviewed by Dr Lindsay Ride (an Australian who had been a Rhodes scholar, professor of physiology at HKU, commandant of the Field Ambulance Corps of the Hong Kong Volunteer Defence Force and was later vice-chancellor at HKU and knighted) together with Richard Lee Ming Chak of the Lee Hysan family. Ride invited Paul to join an enterprise to assist prisoners of war escaping from Hong Kong. They were awaiting approval from Chungking (Chongqing) for the enterprise. Paul's salary would be CNC$400, plus a food and living allowance of CNC$400 per month. While awaiting the approval, Paul occupied a bed space at the Field Hospital of the Kwangtung Branch of the International Red Cross and assisted Ride amongst others by walking to and from the town to collect and deliver telegrams from/to overseas inquiring about the fate of persons in Hong Kong.

In October 1942 Duggie Clague (later Sir Douglas Clague, taipan of the John D. Hutchison Group) arrived in Kukong and joined Lindsay Ride's group.

Paul frequently had breakfast and lunch at one of the two 300-table tea-houses in Kukong. These were the community meeting venues with little attention paid to rank. The food was good and plentiful. By then Paul was a heavy smoker of more than two packets of twenty cigarettes per day.

In June 1942 Paul left Kukong for Wai Chow (now Huizhou) which at the time was flooded.

Duggie Clague, officer commanding designate for the Advanced Headquarters of the British Army Aid Group (BAAG, now the name of Ride's enterprise), arrived in Wai Chow with thirty other members the following month. Paul was appointed secretary and official interpreter for the BAAG. The duties of BAAG were: (a) arrange escape routes from Hong Kong and South China; (b) to collect intelligence and to communicate it to the military attaché at the British Embassy in Chungking (a Japanese HQ plane crashed in the area and a set of all the proposed secret airfields in China and the region was salvaged from the wreck together telegraphic code books and movement orders); (c) to issue daily news bulletins reporting Allied successes; (d) liaison with the Chinese Nationalist Army, the civilian authorities and Chinese Communist guerrillas. BAAG used space in the Seventh Day Adventist Mission. The Quaker Medical Team later reinforced the BAAG Medical and Aid Group that provided services to the community, air raid victims, prisons and refugee camps as well as to Group members.

In 1943 Paul met Rose Lin, a refugee from Hong Kong who was the daughter of retired major general Lin Yin Hung from Wong Mo Ling near Daya Bay. They married at St Joseph's Church in Wai Chow on 12 January 1944. Duggie Clague sent a white veil and pink cashmere scarf as wedding presents. Ronnie Holmes (later Hong Kong colonial secretary) gave his only decent suit. The wedding banquet was interrupted by a Japanese air raid. Later, when the Japanese advanced, Rose was evacuated to Ho Yuan (Heyuan) and later Hing Ning (Xingning) with other family members of the BAAG where she stayed with Paul's sister, Agnes, and later gave birth to a daughter on 22 September 1945. Paul's brother Mark subsequently married Rose's sister, Agnes.

In November 1944, Paul was awarded the MBE for his contribution to the BAAG. On 21 July 1945, he was commissioned as second lieutenant in the Intelligence Corps of the British Army with the temporary rank of captain.

On 15 August 1945, the Japanese army surrendered. Paul returned to Hong Kong with the BAAG on 9 September. He remained a captain (Staff Officer Grade III) in the British military administration of Hong Kong and for a year served as assistant district officer (New Territories, civil affairs, hereafter NTCA). He was joined by his wife, Rose, and daughter, Margaret, and they rented a house at 4 Ashley Road, Tsim Sha Tsui, from the family of Sir Robert Hotung. They occupied this house for a few years and three sons were born there. His duties as district officer (NTCA) included recruiting and organizing village labour to clear landslides and setting up a network of rice retailers to assure a reliable supply of rice.

A turning point then occurred in Paul's career when he was invited to apply to become a cadet officer in the colonial service. Until then only British subjects of European descent had qualified for this career. Paul was flown to Ceylon to be interviewed for the position and then was provisionally appointed as cadet officer subject to a medical examination and successfully completing the First Devonshire Course at Oxford University in England.

He flew to the United Kingdom in October 1946 and spent fifteen happy months at the Queen's College in Oxford. The course included agricultural studies, legal studies, historical and economic studies, geographical studies and anthropological studies. He had attachments to the School of Oriental and African Studies, the London School of Economics and the Borough of North Finchley, all in London. He used his spare time well, visiting Kew Gardens, the London Zoo, the

Wax Museum, Westminster Abbey, St Paul's Cathedral, the Cambridge Theatre, Covent Garden, the Albert Hall, Canterbury Cathedral, the Lake District, North Wales and Edinburgh.

Paul successfully passed the course and returned to Hong Kong by sea on the SS *Empire Brent*, arriving on 27 January 1948. Now a substantiated cadet officer III, he was assigned to the Imports and Exports Department of the government where he served for two years, during which period the name was changed to the Commerce and Industry Department. The duties of the department included issue of import and export licences and collecting duties on tobacco, alcoholic drinks, perfume, etc. The issue of certificates of origin involved factory visits which acquainted Paul with the conditions of manufacturing in the post-war period in Hong Kong.

His next posting was district officer, Southern District, an area encompassing the islands to the west and south of Hong Kong Island as well as the Sai Kung and Clear Water Bay (then known as Hang Hau) Peninsulas. This was the time when he wrote the notes that are reproduced in this publication. Around the same time, he studied in his spare time and passed the law examination that was the prerequisite for his promotion to cadet officer II.

Thereafter, Paul's career steadily progressed. He had a year as district officer, Yuen Long, attended a Second Devonshire Course at Oxford for nine months, and enjoyed the extraordinary variety that a career in the administrative service of the Hong Kong government offered: the Sanitation Department, the founding of the Public Enquiry Service, the Resettlement Department, the Working Party on Slum Clearance, the Working Party on Local Administration, the Aberdeen Technical College Executive Committee, the Board of Licencing Justices, the Chinese Permanent Cemeteries Board of Management, the Chinese Temples Committee, the Chinese Recreation Ground and the Public Square Yaumati Management Committee, the Li Po Chun Charitable Trust, the Narcotics Advisory Committee, Sir Robert Black Trust Fund, the Tung Wah Hospital Advisory Board, the Po Leung Kuk Board of Directors, the Community Relief Trust Fund, the Hong Kong Housing Authority and Housing Board.

Then followed two years in what at first was called the Chinese Affairs Department but changed its name to the Home Affairs Department, where he was deputy secretary and acted as secretary during the substantive officers' leave. He played an active role in rallying the community to support the Hong Kong government during the disturbances associated with the Cultural Revolution in the Mainland in 1967–68. For this he was labelled a "Yellow Skinned Running Dog" and "Assassination Target No. 1" by the local leftists and was awarded the OBE by Her Majesty, the Queen of England. During this period he was appointed an ex-officio member of the supreme governing bodies of Hong Kong, the Executive, Legislative and Urban Councils.

In 1970 for a year he was commissioner for resettlement (public housing). He finished his career with two years as commissioner for labour and mines during which he was awarded the CBE. He retired at the highest (Grade A) administrative officer, one of only eleven, at a salary of HK$12,500 per month in 1973.

He retired to Canada between 1989 and 1993. He died in Hong Kong in 1994. The present chief executive of Hong Kong, Donald Tsang, as well as other senior local administrative officers, were the pall-bearers at his funeral. By then Paul and Rose had eleven children.

AUSTIN COATES

Austin Coates (1922–97) was a British civil servant, writer and traveller. He was the son of noted English composer Eric Coates, who, amongst many other pieces, composed the theme march for the popular film, *The Dam Busters*.

Austin was born in London in 1922, educated at Stowe and in Paris.

Between 1942 and 1947, he was in Royal Air Force Intelligence, serving in India, Burma, Singapore and Indonesia.

In 1949 he joined the Hong Kong government as assistant colonial secretary. Later, he became a district officer and magistrate.

Between 1953 and 1955, he gave a series of lectures on Macao history under the auspices of the Instituto Portugues de Hong Kong.

From 1957 to 1959 he was Chinese affairs officer, magistrate and secretary to the governor of Sarawak; from 1959 to 1962, he was first secretary in the British High Commission in Kuala Lumpur and Penang, Malaya.

He left government service in 1962, already author of a number of books.

In 1966 he settled in Hong Kong and continued travelling and writing extensively.

In 1967 he was commissioned by the British government to write the Corona Library book on the Western Pacific.

He was elected a fellow of the Royal Asiatic Society, London.

Coates was the guest of many prominent Asians, among them the Tagore family, the Indian painter Jamini Roy, and Mahatma Gandhi. After his visit with Gandhi, he decided that understanding between East and West was one of the most important goals in the world.

His book, *City of Broken Promises*, was made into an extremely successful musical for the Hong Kong Arts Festival in 1978. The show was also staged in San Diego in 1979, starring Teresa Carpio.

Books by Austin Coates

Invitation to an Eastern Feast (London: Hutchinson, 1953).

Personal and Oriental (London: Hutchinson, 1957): The book describes his experience as a traveller in the East (Japan, Hong Kong, the Philippines, Burma, India, Pakistan and Istanbul).

The Road (London: Hutchinson, 1959): A novel of building the road to the new Shek Pik Reservoir on Lantau Island, Hong Kong.

Prelude to Hong Kong (London: Routledge and Kegan Paul, 1966); second edition as *Macao and the British, 1637–1842* (Hong Kong: Oxford University Press, 1988): Opening with a vivid description of the first English voyage to China in 1637, this book traces the ensuing course of British relations with China, through Canton and the Portuguese settlement of Macao, and ends amid the din and commotion of the Opium War and the foundation of Hong Kong as a British colony in 1842.

Basutoland (London: Her Majesty's Stationery Office, 1966).

City of Broken Promises (Hong Kong: Oxford University Press, 1967, 314 pp.): A novel based on the life of Martha Merop, a Chinese orphan in Macao who rose to great success in business and her liaison with Thomas Kuyck van Mierop, a principal of the British East India Company.

Myself a Mandarin (London: Frederick Muller, 1968): It describes the author's experience as a special magistrate in the New Territories.

Rizal, Philippine Nationalist and Martyr (Hong Kong: Oxford University Press, 1968).

Western Pacific Islands (London: Her Majesty's Stationery Office, 1970).

China, India and the Ruins of Washington (New York: John Day, 1972): It demonstrates the supremacy of the Chinese and Indian civilization in contrast to the Western civilization.

Islands of the South (London: Heinemann, 1974).

Numerology (London: Frederick Muller, 1974).

A Mountain of Light: The Story of the Hongkong Electric Company (Hong Kong: Heinemann, 1977).

A Macao Narrative (Hong Kong: Heinemann, 1978, 146 pp.).

Whampoa: Ships on the Shore (Hong Kong: South China Morning Post, 1980): A book about the founding of the Hongkong and Whampoa Dock Company and the transformation of Hong Kong from a sleepy village to the seventh biggest port of the world.

China Races (Hong Kong: Oxford University Press, 1984): A history of racing on the China Coast.

The Commerce in Rubber: The First 250 Years (Singapore: Oxford University Press, 1987).

Quick Tidings of Hong Kong (Hong Kong: Oxford University Press, 1990): A history of telecommunications in Hong Kong.

JAMES HAYES

James Hayes was born in 1930, and was educated at Waid Academy, Anstruther, Fife (1942–44) and Tiffin Boys' School, Kingston-on-Thames, Surrey (1944–49).

He has bachelor's and master's degrees in history from the University of London, and in 1975 was awarded a PhD by the university for his thesis on the history and institutions of the Hong Kong region.

He did national service in 1952–54, serving with the Duke of Wellington's Regiment in Korea in 1953. In Hong Kong, he also served with the Royal Hong Kong Regiment (the Volunteers) between 1956 and 1966.

He became a cadet officer class II in the Hong Kong civil service in 1956. His first posting was assistant secretary in the Colonial Secretariat with policy responsibility for the Royal Observatory, the Fire Brigade, and the Agriculture, Fisheries and Forestry Department. This had also been Austin Coates' first post, and Hayes recalls seeing a file on agriculture on which Coates had scribbled gaily in the margin, "Bring up next muck-spreading time", a direction which must have mystified the registry clerks more than somewhat!

He then spent a year's language study at the University of Hong Kong. He was district officer, South (1957–60) and Islands (1961–62). He served in the Resettlement Department (1962–65), acted for six months in 1966 as chief assistant secretary for Chinese affairs; was assistant colonial secretary (general) (1966–67); and was again in the Secretariat for Chinese Affairs in 1967–71, where he was one of the architects of the city district officer scheme.

After serving in the Commerce and Industry and Urban Services Departments in 1972–74, he was town manager and district officer, Tsuen Wan (1975–82), then deputy commissioner of labour (1982–84), and, finally, from 1985 to 1987, regional secretary, New Territories.

James Hayes is also a scholar and author. He was awarded an honorary doctorate of letters by the University of Hong Kong in 1992, and was made an honorary fellow of the Hong Kong University of Science and Technology in 2008. He is an honorary research fellow at the Centre of Asian Studies, the University of Hong Kong. He served as member, vice-president, and president of the Royal Asiatic Society between 1966 and 1989, and edited fourteen issues of the journal, 1967–80. He is an honorary fellow of the Hong Kong branch.

Books and selected publications by James Hayes

The Hong Kong Region 1850–1911: Institutions and Leadership in Town and Countryside (Hamden: Archon Books, 1977).

"Building a Community in a New Town: A Management Relationship with the New Population", in Leung Chi-keung, J. W. Cushman, and Wang Gungwu (eds.), *Hong Kong, Dilemmas of Growth* (Hong Kong: Centre of Asian Studies, the University of Hong Kong, 1980), pp. 309–340.

The Rural Communities of Hong Kong, Studies and Themes (Hong Kong: Oxford University Press, 1983).

"Collecting business papers of Chinese enterprises in Hong Kong", in Alan Birch, Y. C. Jao, and Elizabeth Sinn (eds.), *Research Materials for Hong Kong Studies* (Hong Kong: Centre of Asian Studies, the University of Hong Kong, 1984), pp. 47–55.

"Rural leadership in the Hong Kong region: Village autonomy in a traditional setting", in Goran Aijmer, *Leadership on the China Coast* (London: Curzon Press, 1984), pp. 32–52.

"Specialists and written materials in the village world", in David Johnson, Andrew J. Nathan and Evelyn Rawski (eds.), *Popular Culture in Late Imperial China* (Berkeley: University of California Press, 1985), pp. 75–111.

"Education and management in rural south China in the late Ch'ing", in *Proceedings of the Sixth International Symposium on Asian Studies, 1984* (Hong Kong: Asian Research Service, GPO Box 2232, 1986), Vol. 1, China, pp. 575–592.

"Stakenet and fishing canoe: Hong Kong and adjacent islands in the 19th and early 20th century; the sea and the shore in social, economic and political organization", in *Proceedings of the Eighth International Symposium on Asian Studies, 1986* (Hong Kong: Asian Research Service, GPO Box 2232, 1986), Vol. 1, China, pp. 573–598, and its sequel (with Jack Tin), "Some aspects of traditional life in Hong Kong: The village fisheries", in *Proceedings of the Ninth International Symposium on Asian Studies, 1987*, Vol. 1, China, pp. 53–63.

"Chinese customary law in the New Territories of Hong Kong", in *Proceedings of the Tenth International Symposium on Asian Studies, 1988*, Vol. 1 China (Hong Kong: Asian Research Service, GPO Box 2232, Hong Kong, 1989), pp. 455–476.

"East and West in Hong Kong: Vignettes from history and personal experience", in Elizabeth Sinn (ed.), *Between East and West: Aspects of Social and Political Development in Hong Kong* (Hong Kong: Centre of Asian Studies, the University of Hong Kong, 1990), pp. 7–24.

"Chinese customary law in the New Territories of Hong Kong, Part II", in *Proceedings of the Eleventh International Symposium on Asian Studies, 1990*, Vol. 1 China (Hong Kong: Asian Research Service, GPO Box 2232, Hong Kong, 1990).

"Government and village: Reactions to modern development in long-settled communities in the New Territories of Hong Kong", in Hugh D. R. Baker and Stephan Feuchtwang (eds.), *An Old State in New Settings: Studies in the Social Anthropology of China in Memory of Maurice Freedman* (Oxford: JASO, 1991), pp. 107–136.

"Chinese customary law in the New Territories of Hong Kong: The background to the operation of the New Territories Ordinance, 1899–1987", *Asian Profile*, Vol. 19, No. 2 (April 1991), pp. 97–136.

"Ancestral graves and the popular culture of China: Some examples from Hong Kong's New Territories", *International Association of Orientalist Librarians*, Bulletin 39 (1992), pp. 10–21.

Tsuen Wan: Growth of a New Town and Its People (Hong Kong: Oxford University Press, 1993). This Chinese edition was published in 1999 by the three rural committees of Tsuen Wan, Tsing Ye and Ma Wan, together with the Yuen Yuen Institute, Lo Wai, Tsuen Wan.

"The traditional background: Hong Kong villages in the 1950s", in P. H. Hase and Elizabeth Sinn (eds.), *Beyond the Metropolis: Villages in Hong Kong* (Hong Kong: Joint Publishing Company Limited, 1995), pp. 19–25.

Friends and Teachers: Hong Kong and Its People 1953–87 (Hong Kong: Hong Kong University Press, 1996).

"Hong Kong's own boat people, vignettes from life and history", in *The Hong Kong Anthropologist*, the annual Journal of the Hong Kong Anthropological Society, No. 11 (1998), pp. 2–12.

South China Village Culture (Hong Kong: Oxford University Press, 2001). This is a volume in the Oxford series, "Images of Asia".

"Colonial administration in British Hong Kong and Chinese customary law", in Elizabeth Sinn (ed.), *Hong Kong, British Crown Colony, Revisited* (Hong Kong: Centre of Asian Studies, the University of Hong Kong, 2001), pp. 63–104.

The Great Difference: Hong Kong's New Territories and Its People 1898–2004 (Hong Kong: Hong Kong University Press, 2006).

"Fertile and fortunate: Shanghai before the treaty port era", *Journal of the Hong Kong Branch, Royal Asiatic Society*, Vol. 48 (2008), pp. 175–203.

Index